T0212635

Lecture Notes in Computer Science   9598

Commenced Publication in 1973
Founding and Former Series Editors:
Gerhard Goos, Juris Hartmanis, and Jan van Leeuwen

More information about this series at http://www.springer.com/series/7407

Giovanni Squillero
Paolo Burelli et al. (Eds.)

Applications of Evolutionary Computation

19th European Conference, EvoApplications 2016
Porto, Portugal, March 30 – April 1, 2016
Proceedings, Part II

 Springer

Editors

see next page

ISSN 0302-9743   ISSN 1611-3349 (electronic)
Lecture Notes in Computer Science
ISBN 978-3-319-31152-4   ISBN 978-3-319-31153-1 (eBook)
DOI 10.1007/978-3-319-31153-1

Library of Congress Control Number: 2016933215

LNCS Sublibrary: SL1 – Theoretical Computer Science and General Issues

Printed on acid-free paper

This Springer imprint is published by Springer Nature
The registered company is Springer International Publishing AG Switzerland

Volume Editors

Giovanni Squillero
Politecnico di Torino, Italy
giovanni.squillero@polito.it

Paolo Burelli
Aalborg University, Copenhagen,
    Denmark
pabu@create.aau.dk

Jaume Bacardit
Newcastle University, UK
jaume.bacardit@newcastle.ac.uk

Anthony Brabazon
University College Dublin, Ireland
anthony.brabazon@ucd.ie

Stefano Cagnoni
University of Parma, Italy
cagnoni@ce.unipr.it

Carlos Cotta
Universidad de Málaga, Spain
ccottap@lcc.uma.es

Ivanoe De Falco
ICAR/CNR, Italy
ivanoe.defalco@na.icar.cnr.it

Antonio Della Cioppa
University of Salerno, Italy
adellacioppa@unisa.it

Federico Divina
Universidad Pablo de Olavide, Seville,
    Spain
fdivina@upo.es

A.E. Eiben
VU University Amsterdam,
    The Netherlands
a.e.eiben@vu.nl

Anna I. Esparcia-Alcàzar
Universitat Politècnica de València,
    Spain
aesparcia@pros.upv.es

Francisco Fernàndez de Vega
University of Extremadura, Spain
fcofdez@unex.es

Kyrre Glette
University of Oslo, Norway
kyrrehg@ifi.uio.no

Evert Haasdijk
VU University Amsterdam,
    The Netherlands
e.haasdijk@vu.nl

J. Ignacio Hidalgo
Universidad Complutense de Madrid, Spain
hidalgo@ucm.es

Ting Hu
Memorial University, St. John's, NL, Canada
ting.hu@mun.ca

Michael Kampouridis
University of Kent, UK
m.kampouridis@kent.ac.uk

Paul Kaufmann
University of Paderborn, Germany
paul.kaufmann@gmail.com

Michalis Mavrovouniotis
De Montfort University, UK
mmavrovouniotis@dmu.ac.uk

Antonio M. Mora García
Universidad de Granada, Spain
amorag@geneura.ugr.es

Trung Thanh Nguyen
Liverpool John Moores University, UK
T.T.Nguyen@ljmu.ac.uk

Robert Schaefer
AGH University of Science
    and Technology, Poland
schaefer@agh.edu.pl

Kevin Sim
Edinburgh Napier University, UK
k.sim@napier.ac.uk

Ernesto Tarantino
ICAR/CNR, Italy
ernesto.tarantino@na.icar.cnr.it

Neil Urquhart
Edinburgh Napier University, UK
n.urquhart@napier.ac.uk

Mengjie Zhang
Victoria University of Wellington,
    New Zealand
mengjie.zhang@ecs.vuw.ac.nz

Preface

This two-volume set contains the proceedings of EvoApplications 2016, the 19th European Conference on the Applications of Evolutionary Computation, that was held between March 30 and April 1, 2016, in Porto, Portugal, as a part of EVO*.

Over two decades, EVO* has become Europe's premier event in the field of evolutionary computing. Under the EVO* umbrella, EvoAPPS aims to show the modern applications of this research, ranging from proof of concepts to industrial case studies; EuroGP focuses on genetic programming; EvoCOP targets evolutionary computation in combinatorial optimization; and EvoMUSART is dedicated to evolved and bio-inspired music, sound, art, and design. The proceedings for all of these events are available in the LNCS series.

If EVO* coalesces four different conferences, EvoAPPS exhibits an even higher granularity: It started in 1998 as a collection of small independent workshops and it slowly brought together such vibrant and heterogeneous communities into a single organic event. At the same time, the scientific contributions started to show content more suited to a conference, and workshops evolved into tracks. The change is not over yet: As the world of evolutionary computation is in a constant turmoil, EvoAPPS is mutating to adapt. The scope broadened to include all nature-inspired and bio-inspired computational techniques and computational intelligence in general. New tracks appear every year, while others are merged or suspended.

The conference provides a unique opportunity for students, researchers, and professionals to meet and discuss the applicative and practical aspects of evolutionary computation, and to link academia to industry in a variety of domains.

The 2016 edition comprised 13 tracks focusing on different application domains: EvoBAFIN (business analytics and finance); EvoBIO (computational biology); Evo-COMNET (communication networks and other parallel and distributed systems); EvoCOMPLEX (complex systems); EvoENERGY (energy-related optimisation); EvoGAMES (games and multi-agent systems); EvoIASP (image analysis, signal processing, and pattern recognition); EvoINDUSTRY (real-world industrial and commercial environments); EvoNUM (continuous parameter optimization); EvoPAR (parallel architectures and distributed infrastructures); EvoRISK (risk management, security, and defence); EvoROBOT (evolutionary robotics); and EvoSTOC (stochastic and dynamic environments).

This year, we received 115 high-quality submissions, most of them well suited to fit in more than one track. We selected 58 papers for full oral presentation, while 17 works were given limited space and were shown as posters. All such contributions, regardless of the presentation format, appear as full papers in these two volumes (LNCS 9597 and 9598).

Many people contributed to this edition: We express our gratitude to the authors for submitting their works, and to the members of the Program Committees for devoting such huge effort to review papers pressed by our tight schedule.

The papers were submitted, reviewed, and selected through the MyReview conference management system, and we are grateful to Marc Schoenauer (Inria, Saclay-Île-de-France, France) for providing, hosting, and managing the platform.

We thank the local organizers, Penousal Machado and Ernesto Costa (University of Coimbra, Portugal), as well as the Câmara Municipal do Porto and Turismo do Porto for the local assistance.

We thank Pablo García Sánchez (Universidad de Granada, Spain) for maintaining the EVO* website and handling publicity.

We thank the invited speakers, Richard Forsyth and Kenneth Sörensen, for their inspiring presentations.

We thank the Institute for Informatics and Digital Innovation at Edinburgh Napier University, UK, for the coordination and financial administration.

And we express our gratitude to Jennifer Willies for her dedicated and continued involvement in EVO*. Since 1998, she has been essential for building our unique atmosphere.

February 2016

Giovanni Squillero	Paolo Burelli
Jaume Bacardit	Anthony Brabazon
Stefano Cagnoni	Carlos Cotta
Ivanoe De Falco	Antonio Della Cioppa
Federico Divina	A.E. Eiben
Anna I. Esparcia-Alcázar	Francisco Fernández de Vega
Kyrre Glette	Evert Haasdijk
J. Ignacio Hidalgo	Ting Hu
Michael Kampouridis	Paul Kaufmann
Michalis Mavrovouniotis	Antonio M. Mora Garcia
Trung Thanh Nguyen	Robert Schaefer
Kevin Sim	Ernesto Tarantino
Neil Urquhart	Mengjie Zhang

Organization

Organizing Committee

EvoApplications Coordinator

Giovanni Squillero   Politecnico di Torino, Italy

EvoApplications Publication Chair

Paolo Burelli   Aalborg University, Copenhagen, Denmark

Local Chairs

Penousal Machado   University of Coimbra, Portugal
Ernesto Costa   University of Coimbra, Portugal

Publicity Chair

Pablo García-Sánchez   University of Granada, Spain

Webmaster

Pablo García Sánchez   University of Granada, Spain

EvoBAFIN Chairs

Anthony Brabazon   University College Dublin, Ireland
Michael Kampouridis   University of Kent, UK

EvoBIO Chairs

Jaume Bacardit   Newcastle University, UK
Federico Divina   Universidad Pablo de Olavide, Seville, Spain
Ting Hu   Memorial University, St. John's, NL, Canada

EvoCOMNET Chairs

Ivanoe De Falco   ICAR/CNR, Italy
Antonio Della Cioppa   University of Salerno, Italy
Ernesto Tarantino   ICAR/CNR, Italy

EvoCOMPLEX Chairs

Carlos Cotta   Universidad de Málaga, Spain
Robert Schaefer   AGH University of Science and Technology, Poland

EvoENERGY Chairs

Paul Kaufmann   University of Paderborn, Germany
Kyrre Glette   University of Oslo, Norway

EvoGAMES Chairs

Paolo Burrelli   Aalborg University of Copenhagen, Denmark
Antonio M. Mora García   Universidad de Granada, Spain

EvoIASP Chairs

Stefano Cagnoni   University of Parma, Italy
Mengjie Zhang   Victoria University of Wellington, New Zealand

EvoINDUSTRY Chairs

Kevin Sim   Edinburgh Napier University, UK
Neil Urquhart   Edinburgh Napier University, UK

EvoNUM Chair

Anna I. Esparcia-   Universitat Politècnica de València, Spain
  Alcázar

EvoPAR Chairs

Francisco Fernández   University of Extremadura, Spain
  de Vega
J. Ignacio Hidalgo   Universidad Complutense de Madrid, Spain

EvoRISK Chair

Anna I. Esparcia-   Universitat Politècnica de València, Spain
  Alcázar

EvoROBOT Chairs

Evert Haasdijk   VU University Amsterdam, The Netherlands
A.E. Eiben   VU University Amsterdam, The Netherlands

EvoSTOC Chairs

Michalis Mavrovouniotis	De Montfort University, UK
Trung Thanh Nguyen	Liverpool John Moores University, UK

Program Committees

Robert K. Abercrombie	Oak Ridge National Laboratory, USA [EvoRISK]
Rami Abielmona	University of Ottawa, Canada [EvoRISK]
Eva Alfaro	Instituto Tecnológico de Informàtica, Spain [EvoBAFIN]
Jhon E. Amaya	UNET, Venezuela [EvoCOMPLEX]
Michele Amoretti	University of Parma, Italy [EvoIASP]
Anca Andreica	Universitatea Babeş-Bolyai, Romania [EvoCOMPLEX]
Ignacio Arnaldo	MIT, USA [EvoPAR]
Maria Arsuaga Rios	CERN [EvoINDUSTRY]
Farrukh Aslam Khan	National University of Computer and Emerging Sciences, Pakistan [EvoCOMNET]
Jason Atkin	University of Nottingham, UK [EvoINDUSTRY]
Joshua Auerbach	Ecole Polytechnique Fédérale de Lausanne, Switzerland [EvoROBOT]
Lucia Ballerini	University of Dundee, UK [EvoIASP]
Tiago Baptista	Universidade de Coimbra, Portugal [EvoCOMPLEX]
Bahriye Basturk Akay	Erciyes University, Turkey [EvoINDUSTRY]
Hans-Georg Beyer	Vorarlberg University of Applied Sciences, Austria [EvoNUM]
Leonardo Bocchi	University of Florence, Italy [EvoIASP]
Anthony Brabazon	University College Dublin, Ireland [EvoBAFIN]
Juergen Branke	University of Warwick, UK [EvoSTOC]
Nicolas Bredeche	Institut des Systémes Intelligents et de Robotique, France [EvoROBOT]
Paolo Burelli	Aalborg University, Denmark [EvoGAMES]
David Camacho	Universidad Autónoma de Madrid, Spain [EvoGAMES]
Jose Carlos Ribeiro	Politechnique Institute of Leiria, Portugal [EvoPAR]
Nabendu Chaki	University of Calcutta, India [EvoRISK]
Ying-ping Chen	National Chiao Tung University, Taiwan [EvoNUM]
Kay Chen Tan	National University of Singapore [EvoRISK]
Hui Cheng	Liverpool John Moores University, UK [EvoSTOC]
Anders Christensen	University Institute of Lisbon, ISCTE-IUL, Portugal [EvoROBOT]
Mario Cococcioni	NATO Undersea Research Centre, Italy [EvoRISK]
Jose Manuel Colmenar	URJC, Spain [EvoPAR]
Ernesto Costa	University of Coimbra, Portugal [EvoSTOC]
Antonio Córdoba	Universidad de Sevilla, Spain [EvoCOMPLEX]
Fabio D'Andreagiovanni	Zuse Institute Berlin, Germany [EvoCOMNET]
Sergio Damas	European Center for Soft Computing, Spain [EvoIASP]

Fabio Daolio	Shinshu University, Japan [EvoIASP]
Christian Darabos	University of Pennsylvania, USA [EvoBIO]
Ivanoe De Falco	ICAR - CNR, Italy [EvoIASP]
Antonio Della Cioppa	University of Salerno, Italy [EvoIASP]
Laura Dipietro	Cambridge, USA [EvoIASP]
Josep Domingo-Ferrer	Rovira i Virgili University, Spain [EvoRISK]
Stephane Doncieux	Institut des Systémes Intelligents et de Robotique, France [EvoROBOT]
Marco Dorigo	Université Libre de Bruxelles, Belgium [EvoROBOT]
Jitesh Dundas	Edencore Technologies, Indian Institute of Technology, India [EvoBIO]
Marc Ebner	Ernst Moritz Arndt Universität Greifswald, Germany [EvoIASP, EvoNUM]
Andries P. Engelbrecht	University of Pretoria, South Africa [EvoSTOC]
A. Sima Etaner-Uyar	Istanbul Technical University, Turkey [EvoSTOC]
Thomas Farrenkopf	University of Applied Sciences, Mittelhessen, Germany [EvoINDUSTRY]
Carlos Fernandes	ISR-Lisbon, Portugal [EvoCOMPLEX]
Stenio Fernandes	Federal University of Pernambuco, UFPE, Brazil [EvoRISK]
Florentino Fernández	Universidad de Vigo, Spain [EvoBIO]
Antonio J. Fernandez Leiva	University of Málaga, Spain [EvoGAMES]
Antonio Fernández-Ares	Universidad de Granada, Spain [EvoGAMES]
Gianluigi Folino	ICAR-CNR, Italy [EvoPAR]
Francesco Fontanella	University of Cassino, Italy [EvoIASP]
Alex Freitas	University of Kent, UK [EvoBIO]
José E. Gallardo	Universidad de Málaga, Spain [EvoCOMPLEX]
Pablo García Sànchez	University of Granada, Spain [EvoGAMES]
Antonios Gasteratos	Democritus University of Thrace, Greece [EvoCOMNET]
Carlos Gesherson	UNAM, Mexico [EvoCOMPLEX]
Mario Giacobini	Università di Torino, Italy [EvoBIO]
Raffaele Giancarlo	Università degli Studi di Palermo, Italy [EvoBIO]
Rosalba Giugno	University of Catania, Italy [EvoBIO]
Antonio Gonzalez Pardo	Basque Center for Applied Mathematics, Spain [EvoGAMES]
Casey Greene	Dartmouth College, USA [EvoBIO]
Michael Guckert	University of Applied Sciences, Mittelhessen, Germany [EvoINDUSTRY]
Johan Hagelbäck	Blekinge Tekniska Högskola, Sweden [EvoGAMES]
John Hallam	University of Southern Denmark, Denmark [EvoGAMES]
Heiko Hamann	University of Paderborn, Germany [EvoROBOT]
Jin-Kao Hao	University of Angers, France [EvoBIO, EvoCOMNET]
Jacqueline Heinerman	VU University Amsterdam, The Netherlands [EvoROBOT]
Malcom Heywood	Dalhousie University, Canada [EvoBAFIN]

Ronald Hochreiter	WU Vienna University of Economics and Business, Austria [EvoBAFIN]
Rolf Hoffmann	Technical University of Darmstadt, Germany [EvoCOMNET]
Joost Huizinga	University of Wyoming, USA [EvoROBOT]
Oscar Ibàñez	University of Granada, Spain [EvoIASP]
José Ignacio Hidalgo	Universidad Complutense de Madrid, Spain [EvoIASP]
Rodica Ioana Lung	Babes-Bolyai University, Germany [EvoGAMES]
Juan L. Jiménez Laredo	ILNAS/ANEC Normalisation, Luxembourg [EvoCOMPLEX, EvoPAR]
Michael Kampouridis	University of Kent, UK [EvoBAFIN]
Iwona Karcz-Dulęba	Politechnika Wrocławska, Poland [EvoCOMPLEX]
Ahmed Kattan	EvoSys.biz, Saudi Arabia [EvoBAFIN]
Shayan Kavakeb	Liverpool John Moores University, UK [EvoSTOC]
Edward Keedwell	University of Exeter, UK [EvoBIO]
Graham Kendall	University of Nottingham, UK [EvoCOMNET, EvoINDUSTRY]
Mario Koeppen	Kyushu Institute of Technology, Japan [EvoIASP]
Oliver Kramer	University of Oldenburg, Germany [EvoENERGY]
Wacław Kuś	Politechnika Śląska, Poland [EvoCOMPLEX]
William Langdon	University College London, UK [EvoNUM, EvoPAR]
Kenji Leibnitz	National Institute of Information and Communications Technology, Japan [EvoCOMNET]
Changhe Li	China University of Geosciences, China [EvoSTOC]
Antonios Liapis	University of Malta, Malta [EvoGAMES]
Federico Liberatore	Universidad Rey Juan Carlos, Spain [EvoGAMES]
Piotr Lipinski	University of Wroclaw, Poland [EvoBAFIN]
Francisco Luís Gutiérrez Vela	University of Granada, Spain [EvoGAMES]
Francisco Luna	Universidad de Málaga, Spain [EvoPAR]
Gabriel Luque	Universidad de Málaga, Spain [EvoCOMPLEX]
Evelyne Lutton	INRIA, France [EvoIASP]
Chenjie Ma	Fraunhofer Institute for Wind Energy and Energy System Technology, Germany [EvoENERGY]
Tobias Mahlmann	Lund University, Sweden [EvoGAMES]
Domenico Maisto	ICAR-CNR, Italy [EvoCOMNET]
Elena Marchiori	Radboud Universiteit van Nijmegen, The Netherlands [EvoBIO]
Davide Marocco	University of Naples, Italy [EvoCOMNET]
Ingo Mauser	FZI Karlsruhe, Germany [EvoENERGY]
Michalis Mavrovouniotis	De Montfort University, UK [EvoSTOC]
Michael Mayo	University of Waikato, New Zealand [EvoBAFIN]
Jorn Mehnen	Cranfield University, UK [EvoSTOC]
Juan J. Merelo	Universidad de Granada, Spain [EvoCOMPLEX, EvoNUM]

Pablo Mesejo Santiago	INRIA, France [EvoIASP]
Salma Mesmoudi	Institut des Systémes Complexes, France [EvoNUM]
Krzysztof Michalak	Wroclaw University of Economics, Poland [EvoBAFIN]
Martin Middendorf	University of Leipzig, Germany [EvoENERGY]
Jose Miguel Holguín	S2 Grupo, Spain [EvoRISK]
Maizura Mokhtar	Edinburgh Napier University, UK [EvoENERGY]
Jean-Marc Montanier	Barcelona Supercomputing Center, Spain [EvoROBOT]
Roberto Montemanni	IDSIA, Switzerland [EvoCOMNET]
Javier Montero	Universidad Complutense de Madrid, Spain [EvoRISK]
Frank W. Moore	University of Alaska Anchorage, USA [EvoRISK]
Antonio M. Mora García	University of Granada, Spain [EvoGAMES]
Maite Moreno	S2 Grupo, Spain [EvoRISK]
Vincent Moulton	University of East Anglia, UK [EvoBIO]
Jean-Baptiste Mouret	INRIA Larsen Team, France [EvoROBOT]
Nysret Musliu	Vienna University of Technology, Austria [EvoINDUSTRY]
Antonio Nebro	Universidad de Málaga, Spain [EvoCOMPLEX]
Ferrante Neri	De Montfort University, UK [EvoIASP, EvoNUM, EvoSTOC]
Frank Neumann	University of Adelaide, Australia [EvoENERGY]
Geoff Nitschke	University of Cape Town, South Africa [EvoROBOT]
Stefano Nolfi	Institute of Cognitive Sciences and Technologies, Italy [EvoROBOT]
Michael O'Neill	University College Dublin, Ireland [EvoBAFIN]
Una-May O'really	MIT, USA [EvoPAR]
Conall O'Sullivan	University College Dublin, Ireland [EvoBAFIN]
Kai Olav Ellefsen	University of Wyoming, USA [EvoROBOT]
Carlotta Orsenigo	Politecnico di Milano, Italy [EvoBIO]
Ender Ozcan	University of Nottingham, UK [EvoINDUSTRY]
Patricia Paderewski Rodríguez	University of Granada, Spain [EvoGAMES]
Peter Palensky	Technical University of Delft, The Netherlands [EvoENERGY]
Anna Paszyńska	Uniwersytet Jagielloński, Poland [EvoCOMPLEX]
David Pelta	University of Granada, Spain [EvoSTOC]
Sanja Petrovic	University of Nottingham, UK [EvoINDUSTRY]
Nelishia Pillay	University of KwaZulu-Natal, South Africa [EvoINDUSTRY]
Clara Pizzuti	ICAR CNR, Italy [EvoBIO]
Riccardo Poli	University of Essex, UK [EvoIASP]
Petr Pošík	Czech Technical University in Prague, Czech Republic [EvoNUM]
Mike Preuss	TU Dortmund, Germany [EvoGAMES, EvoNUM]
Abraham Prieto	University of La Coruña, Spain [EvoROBOT]
Jianlong Qi	Ancestry.com Inc., USA [EvoBio]
Michael Raymer	Wright State University, USA [EvoBIO]

Hendrik Richter	Leipzig University of Applied Sciences, Germany [EvoSTOC]
Diederik Roijers	University of Amsterdam, The Netherlands [EvoROBOT]
Simona Rombo	Università degli Studi di Palermo, Italy [EvoBIO]
Claudio Rossi	Universidad Politecnica De Madrid, Spain [EvoROBOT]
Guenter Rudolph	University of Dortmund, Germany [EvoNUM]
Jose Santos Reyes	Universidad de A Coruña, Spain [EvoBIO]
Sanem Sariel	Istanbul Technical University, Turkey [EvoINDUSTRY, EvoROBOT]
Ivo Fabian Sbalzarini	Max Planck Institute of Molecular Cell Biology and Genetics, Germany [EvoNUM]
Robert Schaefer	University of Science and Technology, Poland [EvoCOMNET]
Thomas Schmickl	University of Graz, Austria [EvoROBOT]
Marc Schoenauer	INRIA, France [EvoNUM]
Sevil Sen	Hacettepe University, Turkey [EvoCOMNET]
Noor Shaker	Aalborg University, Denmark [EvoGAMES]
Chien-Chung Shen	University of Delaware, USA [EvoCOMNET]
Bernhard Sick	University of Kassel, Germany [EvoENERGY]
Sara Silva	INESC-ID Lisbon, Portugal [EvoIASP]
Anabela Simões	Institute Polytechnic of Coimbra, Portugal [EvoSTOC]
Moshe Sipper	Ben-Gurion University, Israel [EvoGAMES]
Georgios Sirakoulis	Democritus University of Thrace, Greece [EvoCOMNET]
Stephen Smith	University of York, UK [EvoIASP]
Maciej Smołka	Akademia Górniczo-Hutnicza, Poland [EvoCOMPLEX]
Andy Song	RMIT, Australia [EvoIASP]
Stefano Squartini	Università Politecnica delle Marche, Italy [EvoENERGY]
Giovanni Squillero	Politecnico di Torino, Italy [EvoGAMES, EvoIASP]
Andreas Steyven	Edinburgh Napier University, UK [EvoINDUSTRY]
Kasper Stoy	IT University of Copenhagen, Denmark [EvoROBOT]
Guillermo Suárez-Tangil	Royal Holloway University of London, UK [EvoRISK]
Shamik Sural	Indian Institute of Technology, Kharagpur, India [EvoRISK]
Ernesto Tarantino	ICAR/CNR, Italy [EvoCOMNET]
Andrea Tettamanzi	University of Nice Sophia Antipolis/I3S, France [EvoBAFIN]
Olivier Teytaud	INRIA, France [EvoNUM]
Trung Thanh Nguyen	Liverpool John Moores University, UK [EvoSTOC]
Ruppa Thulasiram	University of Manitoba, Cananda [EvoBAFIN]
Jon Timmis	University of York, UK [EvoROBOT]
Renato Tinós	Universidade de São Paulo, Brazil [EvoSTOC]
Julian Togelius	New York University, USA [EvoGAMES]
Marco Tomassini	Lausanne University, Switzerland [EvoPAR, EvoCOMPLEX]

Alberto Tonda   Politecnico di Torino, Italy [EvoCOMPLEX, EvoGAMES]
Pawel Topa   AGH University of Science and Technology, Poland
                         [EvoCOMNET]
Vicenç Torra   University of Skövde, Sweden [EvoRISK]
Krzysztof Trojanowski   Polish Academy of Sciences, Poland [EvoSTOC]
Andy Tyrrell   University of York, UK [EvoENERGY]
Roberto Ugolotti   Henesis srl, Italy [EvoIASP]
Ryan Urbanowicz   University of Pennsylvania, USA [EvoBIO]
Tommaso Urli   Csiro Data61, Australia [EvoGAMES]
Andrea Valsecchi   European Center of Soft Computing, Spain [EvoIASP]
Leonardo Vanneschi   Universidade Nova de Lisboa, Portugal [EvoBIO,
                         EvoIASP]
Sebastien Varrette   Université du Luxemburg, Luxemburg [EvoPAR]
Nadarajen Veerapen   University of Stirling, UK [EvoINDUSTRY]
Roby Velez   University of Wyoming, USA [EvoROBOT]
Antonio Villalón   S2 Grupo, Spain [EvoRISK]
Marco Villani   University of Modena and Reggio Emilia, Italy
                         [EvoCOMNET]
Markus Wagner   University of Adelaide, Australia [EvoENERGY]
Jaroslaw Was   AGH University of Science and Technology, Poland
                         [EvoCOMNET]
Tony White   Carleton University, Canada [EvoCOMNET]
Alan Winfield   University of the West of England, UK [EvoROBOT]
Bing Xue   Victoria University of Wellington, New Zealand
                         [EvoIASP, EvoBIO]
Shengxiang Yang   De Monfort University, UK [EvoINDUSTRY, EvoSTOC]
Georgios N. Yannakakis   University of Malta, Malta [EvoGAMES]
Xin Yao   University of Birmingham, UK [EvoSTOC]
Mengjie Zhang   Victoria University of Wellington, New Zealand [EvoBIO]
Nur Zincir-Heywood   Dalhousie University, Canada [EvoCOMNET, EvoRISK]

Sponsoring Organizations

Institute for Informatics and Digital Innovation at Edinburgh Napier University, UK
World Federation on Soft Computing (technical sponsor of the EvoCOMNET track)
Câmara Municipal do Porto, Portugal
Turismo do Porto, Portugal
University of Coimbra, Portugal

Contents – Part II

Contents – Part I

EvoCOMPLEX

EvoENERGY

EvoGAMES

EvoIASP

EvoINDUSTRY

EvoNUM

Local Fitness Meta-Models with Nearest Neighbor Regression

Oliver Kramer[(✉)]

Computational Intelligence Group, Department of Computing Science,
University of Oldenburg, Oldenburg, Germany
oliver.kramer@uni-oldenburg.de

Abstract. In blackbox function optimization, the results of fitness function evaluations can be used to train a regression model. This meta-model can be used to replace function evaluations and thus reduce the number of fitness function evaluations in evolution strategies (ES). In this paper, we show that a reduction of the number of fitness function evaluations of a (1+1)-ES is possible with a combination of a nearest neighbor regression model, a local archive of fitness function evaluations, and a comparatively simple meta-model management. We analyze the reduction of fitness function evaluations on set of benchmark functions.

Keywords: (1+1)-ES · Meta-models · Nearest neighbor regression

1 Introduction

In expensive optimization problems, the reduction of the number of fitness function calls has an important part to play. Evolutionary operators produce candidate solutions $\mathbf{x}_i$ in the solution space that are evaluated on fitness function f. If a regression method $\hat{f}$ is trained with the pattern label pairs $(\mathbf{x}_i, f(\mathbf{x}_i))$, $i = 1, \ldots, N$, the model $\hat{f}$ can be used to interpolate and extrapolate the fitness of a novel solution $\mathbf{x}'$ that has been generated by the evolutionary operators. Model $\hat{f}$ is also known as meta-model or surrogate in this context. Almost all kinds of regression methods can be employed for this purpose. An important question is how to manage the meta-model. Past fitness function evaluations are stored in an archive. The questions come up, which patterns are used to train the meta-model, how often the meta-model should be tuned, and when the fitness function is used or when the meta-model is employed. Some methods take into account the fitness evaluation of the meta-model. Others use the meta-model for pre-screening, i.e., each solution is first evaluated on the meta-model. The most successful ones are evaluated on the real fitness function and then selected for being parents in the following generation.

In this paper, we analyze nearest neighbor regression when optimizing with a (1+1)-ES and Rechenberg's step size control technique. Nearest neighbor regression is based on the idea that the label of an unknown pattern $\mathbf{x}'$ should get the label of the k closest pattern in the training set. This method is also known

© Springer International Publishing Switzerland 2016
G. Squillero and P. Burelli (Eds.): EvoApplications 2016, Part II, LNCS 9598, pp. 3–10, 2016.
DOI: 10.1007/978-3-319-31153-1_1

as instance-based and non-parametric approach. It does not induce a functional model like linear regression. The objective of this paper is to show that a comparatively simple hybridization can result in a very effective optimization strategy. The paper is structured as follows. Nearest neighbor regression is introduced in Sect. 2. The integration of the k-nearest neighbors (kNN) meta-model is presented in Sect. 3. Related work is discussed in Sect. 4. The approach is experimentally analyzed in Sect. 5. Conclusions are drawn in Sect. 6.

2   Nearest Neighbors

Nearest neighbor regression, also known as kNN regression, is based on the idea that the closest patterns to a target pattern $\mathbf{x}'$, for which we seek the label, deliver useful label information. Based on this idea, kNN assigns the class label of the majority of the k-nearest patterns in data space. For this sake, we have to be able to define a similarity measure in data space. In $\mathbb{R}^d$, it is reasonable to employ the Minkowski metric (p-norm)

$$\|\mathbf{x}' - \mathbf{x}_j\|^d = \left(\sum_{i=1}^{d} |(x_i)' - (x_i)_j|^p \right)^{1/p}, \tag{1}$$

with parameter $p \in \mathbb{N}$. The distance measure corresponds to the Euclidean distance for $p = 2$ and the Manhattan distance for $p = 1$. In other data spaces, adequate distance functions have to be chosen, e.g., the Hamming distance in $\mathbb{B}^d$. For regression tasks, kNN can also be applied. For this sake, the task is to learn a function $\hat{f} : \mathbb{R}^d \to \mathbb{R}$ known as regression function. For an unknown pattern $\mathbf{x}'$, kNN regression computes the mean of the function values of its k-nearest neighbors

$$f(\mathbf{x}') = \frac{1}{K} \sum_{i \in \mathcal{N}_K(\mathbf{x}')} y_i \tag{2}$$

with set $\mathcal{N}_K(\mathbf{x}')$ containing the indices of the k-nearest neighbors of pattern $\mathbf{x}'$ in the training data set $\{(\mathbf{x}_i, f(\mathbf{x}_i))\}_{i=1}^N$. Standardization of patterns is usually applied as they can come in different units.

The choice of k defines the locality of kNN. For $k = 1$, little neighborhoods arise in regions, where patterns from different classes are scattered. For larger neighborhood sizes, e.g., $k = 20$, patterns with labels in the minority are ignored. Neighborhood size k is usually chosen with the help of cross-validation.

3   Algorithm

In this section, we introduce the meta-model-based ES. The main ingredients of meta-model approaches are an archive, which stores past fitness function evaluations, a meta-model maintenance mechanism, e.g., for parameter tuning and regularization, and the meta-model integration mechanism that defines how it is applied to save fitness function evaluations. One possibility is to test each candidate solution with a certain probability on the meta-model and use the predicted value instead of the real fitness function evaluations in the course of the evolutionary optimization process. We employ a different meta-model management that is tailored to the (1+1)-ES.

Algorithm 1 shows the pseudo-code of the (1+1)-ES with meta-model (MM-(1+1)-ES) and Rechenberg's adaptive step size control [15]. If the solution $\mathbf{x}'$ has been evaluated on f, both will be combined to a pattern and as pattern-label pair $(\mathbf{x}', f(\mathbf{x}'))$ included to the meta-model training set, i.e., an archive A. To improve the computational complexity with an archive of constant size, the oldest solutions are removed from the training set, if a defined size is exceeded. After this step, model $\hat{f}$ can be re-trained, which means that a novel neighborhood size k may be chosen with cross-validation.

The meta-model integration we employ is based on the idea that solutions are only evaluated, if they are promising. Let $\mathbf{x}$ be the solution of the last generation and let $\mathbf{x}'$ be the novel solution generated with the mutation operator. If the fitness prediction $\hat{f}(\mathbf{x}')$ of the meta-model indicates that $\mathbf{x}'$ employs a better fitness than the a-th last solution $\mathbf{x}_{-a}$ of the archive A, the solution is evaluated on the real fitness function f. The a-th last solution defines a fitness threshold that assumes that $\hat{f}(\mathbf{x}')$ may underestimate the fitness of $\mathbf{x}'$. The evaluations of candidate solutions that are worse than the threshold are saved and may lead to a decrease of the fitness function evaluations. A tuning of the model, i.e., the neighborhood size k of kNN, may be reasonable in certain optimization settings. Last, the question for the proper regression model has to be answered. In our blackbox optimization scenario, we assume that we do not know anything about the curvature of the fitness function. Of course, it makes sense to employ a polynomial model in case of spherical fitness function conditions. But in general, we cannot assume to have such information.

Algorithm 1. MM-(1+1)-ES

 1: initialize $\mathbf{x}$
 2: **repeat**
 3:     adapt σ with Rechenberg
 4:     $\mathbf{z} \sim \sigma \cdot \mathcal{N}(\mathbf{0}, \mathbf{I})$
 5:     $\mathbf{x}' = \mathbf{x} + \mathbf{z}$
 6:     **if** $\hat{f}(\mathbf{x}') \leq f(\mathbf{x}_{-a})$ **then**
 7:         evaluate $f(\mathbf{x}')$
 8:         add $(\mathbf{x}', f(\mathbf{x}'))$ to archive A
 9:         tune model (e.g., choose best k for kNN)
10:         **if** $f(\mathbf{x}') \leq f(\mathbf{x})$ **then**
11:             replace $\mathbf{x}$ with $\mathbf{x}'$
12:         **end if**
13:     **end if**
14: **until** termination condition

4   Related Work

Meta-models, also known as surrogates, are prominent approaches to reduce the number of fitness function evaluations in evolutionary computation. Most work in meta-modeling concentrates on fitness function surrogates, few also on reducing constraint functions evaluations. In this line of research early work concentrated on neural networks [6] and on Kriging [1]. Kriging belongs to the class of Gaussian process regression models and is an interpolation method. It uses covariance information and is based on piecewise-polynomial splines. Neural networks and Kriging meta-models have been compared in [18]. An example for the recent employment of Kriging models is the differential evolution approach by Elsayed *et al.* [5].

Various kinds of ways can be used to process the data for saving fitness function evaluations. Cruz-Vega *et al.* [3] employ granular computing to cluster points and adapt the parameters with a neuro-fuzzy network. Verbeeck *et al.* [17] propose a tree-based meta-model and concentrate on multi-objective optimization. Martínez and Coello [13] also focus on multi-objective optimization while employing a support vector regression meta-model. Loshchilov *et al.* [10] combine a one-class support vector machine with a regression approach as meta-model in multi-objective optimization. Ensembles of support vector methods are also used for in the approach by Rosales-Pérez [16] in multi-objective optimization settings. Ensembles combine multiple classifiers to reduce the fitness prediction error.

Kruisselbrink *et al.* [8] employ the Kriging model in CMA-ES based optimization. The approach puts an emphasis on the generation of archive points of improving the meta-model. Local meta-models for the CMA-ES are learned in the approach by Bouzarkouna *et al.* [2], who train a full quadratic local model for each sub-function in each generation. Also Liao *et al.* [9] propose a locally weighted meta-model, which only evaluates the most promising candidate solutions. The local approach is similar to the nearest neighbor methods we use in the experimental part, as kNN is a local method.

There exists a line of research that concentrates on surrogate-assisted optimization for the CMA-ES. For example, the approach by Loshchilov *et al.* [11] adjusts the life length of the current surrogate model before learning a new surrogate as well as its hyper-parameters. A variant with larger population sizes [12] leads to a more intensive exploitation of the meta-model.

Preuss *et al.* [14] propose to use a computationally cheap meta-model of the fitness function and tune the parameters of the evolutionary optimization approach on this surrogate. Kramer *et al.* [7] combine two nearest neighbor meta-models, one for the fitness function, and one for the constraint function with an adaptive penalty function in a constrained continuous optimization scenario.

Most of the work sketched here positively reported savings in fitness function evaluations, although machine learning models and meta-model managing strategies vary significantly.

5   Experimental Analysis

In this section, we experimentally analyze the meta-model based ES. Besides the convergence behavior, we analyze the influence of the neighborhood size k and the archive design.

5.1   Comparison Between Approaches

Table 1 shows the experimental results of 25 runs of the (1+1)-ES and the MM-(1+1)-ES on the Sphere function and on Rosenbrock for kNN as meta-model with $k = 1$. Both algorithms get a budget of 5000 function evaluations. The results show the mean values and the corresponding standard deviations. On the Sphere function, we employ the setting $a = 10$. We can observe that the ES with meta-model leads to a significant reduction of fitness function evaluations. For $d = 2$, the standard deviation even falls below a value that is measurable due to a limited machine accuracy. The results are statistically significant, which is confirmed by the Wilcoxon test. On Rosenbrock, we set $a = 50$. The MM-(1+1)-ES significantly outperforms the (1+1)-ES for $d = 2$, but only in few runs on Rosenbrock leading to no statistical superiority for $d = 10$.

Table 1. Experimental comparison of (1+1)-ES and the MM-(1+1)-ES on the Sphere function and on Rosenbrock.

problem	N	(1+1)-ES		MM-(1+1)-ES		Wilx.
		mean	dev	mean	dev	p-value
Sphere	2	2.067e-173	0.0	2.003e-287	0.0	0.0076
	10	1.039e-53	1.800e-53	1.511e-62	2.618e-62	0.0076
Rosenbrock	2	0.260	0.447	8.091e-06	7.809e-06	0.0076
	10	0.519	0.301	2.143	2.783	0.313

Figure 1 compares the evolutionary runs of the (1+1)-ES and the MM-(1+1)-ES on the Sphere function for (a) $d = 2$ and (b) $d = 10$. As of the very beginning of the optimization process, even the worst evolutionary runs with meta-model are better than the best evolutionary runs of the (1+1)-ES without meta-model. This effect is even more significant for $d = 2$.

5.2   Analysis of Archive Size

Now, we analyze the archive size N as it is the basis of the training set and has an important influence on the regression model quality. In Fig. 2, we compare the two archive sizes $N = 20$ and $N = 500$ on the Sphere function, again with the same settings, i.e., 5000 fitness function evaluations in each run and for $d = 2$ and $d = 10$. The runs show that a too small archive lets the search become less

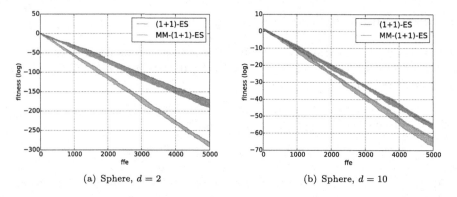

(a) Sphere, $d = 2$ (b) Sphere, $d = 10$

Fig. 1. Comparison of (1+1)-ES and MM-(1+1)-ES on the Sphere function with (a) $d = 2$ and (b) $d = 10$.

stable in both cases. Some runs may get stuck because of inappropriate fitness function estimates. Further analyses have shown that an archive size of $N = 500$ is an appropriate choice.

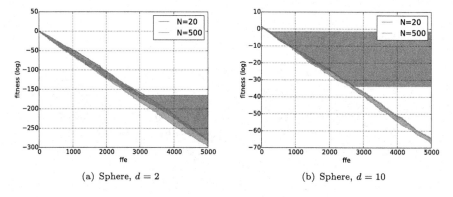

(a) Sphere, $d = 2$ (b) Sphere, $d = 10$

Fig. 2. Comparison of meta-model size $N = 20$ and $N = 500$ on the Sphere function with (a) $d = 2$ and (b) $d = 10$.

Our analysis of the neighborhood size k have shown that the choice $k = 1$ yields the best results in all cases. Larger choices slow down the optimization or let the optimization process stagnate, similar to the stagnation we observe for small archive sizes. Hence, we understand the nearest neighbor regression meta-models with $k = 1$ as local meta-models, which also belong to the most successful ones in literature, see Sect. 4.

6   Conclusions

Meta-modeling in evolutionary optimization is frequent and well-known technique that allows saving expensive fitness function evaluations. In this paper, we applied nearest neighbor regression in a (1+1)-ES with Gaussian mutation and Rechenberg's adaptive step size control. Fitness function evaluations are saved, if the predicted fitness is worse than the a-th best element in the archive. In particular on the Sphere function, the optimization runs of the (1+1)-ES with meta-model outperform the native ES. The results have been confirmed with the Wilcoxon test. It turns out that the small neighborhood size $k = 1$ is the best choice. Further, the archive should be large enough to offer diversity that can be exploited for fitness predictions during the evolutionary optimization process. However, one has to keep in mind that inexact fitness predictions based on bad surrogates may disturb the evolutionary process and result in deteriorations instead of improvements. Various other ways to employ a regression model as meta-model are possible. Pre-selection is only one method.

In the future, we will concentrate on a detailed comparison between various regression models and meta-model management strategies, similar to the rigorous comparison of classifiers on a large benchmark data set by Delgado et al. [4].

A   Benchmark Functions

In this work, we employ the following benchmark problems:

- Sphere $f(\mathbf{x}) = \sum_{i=1}^{d}(x_i)^2$
- Rosenbrock $f(\mathbf{x}) = \sum_{i=1}^{d-1}\left(100(x_i^2 - x_{i+1})^2 + (x_i - 1)^2\right)$

References

1. Armstrong, M.: Basic Linear Geostatistics. Springer, Heidelberg (1998)
2. Bouzarkouna, Z., Auger, A., Ding, D.Y.: Local-meta-model CMA-ES for partially separable functions. In: Genetic and Evolutionary Computation Conference (GECCO), pp. 869–876 (2011)
3. Cruz-Vega, I., Garcia-Limon, M., Escalante, H.J.: Adaptive-surrogate based on a neuro-fuzzy network and granular computing. In: Genetic and Evolutionary Computation Conference (GECCO), pp. 761–768 (2014)
4. Delgado, M.F., Cernadas, E., Barro, S., Amorim, D.G.: Do we need hundreds of classifiers to solve real world classification problems? J. Mach. Learn. Res. **15**(1), 3133–3181 (2014)
5. Elsayed, S.M., Ray, T., Sarker, R.A.: A surrogate-assisted differential evolution algorithm with dynamic parameters selection for solving expensive optimization problems. In: Proceedings of IEEE Congress on Evolutionary Computation (CEC), pp. 1062–1068 (2014)
6. Jin, Y., Olhofer, M., Sendhoff, B.: On evolutionary optimization with approximate fitness functions. In: Genetic and Evolutionary Computation Conference (GECCO), pp. 786–793 (2000)
7. Kramer, O., Schlachter, U., Spreckels, V.: An adaptive penalty function with meta-modeling for constrained problems. In: Proceedings of IEEE Congress on Evolutionary Computation (CEC), pp. 1350–1354 (2013)

8. Kruisselbrink, J.W., Emmerich, M.T.M., Deutz, A.H., Bäck, T.: A robust optimization approach using kriging metamodels for robustness approximation in the CMA-ES. In: Proceedings of IEEE Congress on Evolutionary Computation (CEC), pp. 1–8 (2010)

9. Liao, Q., Zhou, A., Zhang, G.: A locally weighted metamodel for pre-selection in evolutionary optimization. In: Proceedings of IEEE Congress on Evolutionary Computation (CEC), pp. 2483–2490 (2014)

10. Loshchilov, I., Schoenauer, M., Sebag, M.: A mono surrogate for multiobjective optimization. In: Genetic and Evolutionary Computation Conference (GECCO), pp. 471–478 (2010)

11. Loshchilov, I., Schoenauer, M., Sebag, M.: Self-adaptive surrogate-assisted covariance matrix adaptation evolution strategy. In: Genetic and Evolutionary Computation Conference (GECCO), pp. 321–328 (2012)

12. Loshchilov, I., Schoenauer, M., Sebag, M.: Intensive surrogate model exploitation in self-adaptive surrogate-assisted cma-es (saacm-es). In: Genetic and Evolutionary Computation Conference (GECCO), pp. 439–446 (2013)

13. Martínez, S.Z., Coello, C.A.C.: A multi-objective meta-model assisted memetic algorithm with non gradient-based local search. In: Genetic and Evolutionary Computation Conference (GECCO), pp. 537–538 (2010)

14. Preuss, M., Rudolph, G., Wessing, S.: Tuning optimization algorithms for real-world problems by means of surrogate modeling. In: Genetic and Evolutionary Computation Conference (GECCO), pp. 401–408 (2010)

15. Rechenberg, I.: Evolutionsstrategie - Optimierung technischer Systeme nach Prinzipien der biologischen Evolution. Frommann-Holzboog, Stuttgart (1973)

16. Rosales-Pérez, A., Coello, C.A.C., Gonzalez, J.A., García, C.A.R., Escalante, H.J.: A hybrid surrogate-based approach for evolutionary multi-objective optimization. In: Proceedings of IEEE Congress on Evolutionary Computation (CEC), pp. 2548–2555 (2013)

17. Verbeeck, D., Maes, F., Grave, K.D., Blockeel, H.: Multi-objective optimization with surrogate trees. In: Genetic and Evolutionary Computation Conference (GECCO), pp. 679–686 (2013)

18. Willmes, L., Bäck, T., Jin, Y., Sendhoff, B.: Comparing neural networks and kriging for fitness approximation in evolutionary optimization. In: Proceedings of IEEE Congress on Evolutionary Computation (CEC), pp. 663–670 (2003)

Validating the Grid Diversity Operator: An Infusion Technique for Diversity Maintenance in Population-Based Optimisation Algorithms

Ahmed Salah[1]([✉]), Emma Hart[2], and Kevin Sim[2]

[1] Faculty of Science, Mansoura University, Mansoura, Egypt
a_salah@mans.edu.eg
[2] School of Computing, Edinburgh Napier University, Edinburgh, UK
{e.hart,k.sim}@napier.ac.uk

Abstract. We describe a novel diversity method named Grid Diversity Operator (GDO) that can be incorporated into population-based optimization algorithms that support the use of *infusion* techniques to inject new material into a population. By replacing the random infusion mechanism used in many optimisation algorithms, the GDO guides the containing algorithm towards creating new individuals in sparsely visited areas of the search space. Experimental tests were performed on a set of 39 multimodal benchmark problems from the literature using GDO in conjunction with a popular immune-inspired algorithm (opt-ainet) and a sawtooth genetic algorithm. The results show that the GDO operator leads to better quality solutions in all of the benchmark problems as a result of maintaining higher diversity, and makes more efficient usage of the allowed number of objective function evaluations. Specifically, we show that the performance gain from using GDO increases as the dimensionality of the problem instances increases. An exploration of the parameter settings for the two main parameters of the new operator enabled the performance of the operator to be tuned empirically.

Keywords: Grid · Diversity · Optimization · Evolutionary algorithms · Artificial immune systems

1 Introduction

Managing the diversity of a population has been recognized as one of the most influential factors within an Evolutionary Algorithm (EA) right from their inception. From an exploration and exploitation perspective, an increase in diversity correlates with the exploration phase of an optimization algorithm whilst a decrease correlates with the exploitation phase. Maintaining a diverse population through the use of exploration operators is key to achieving a balance between the two phases [1].

© Springer International Publishing Switzerland 2016
G. Squillero and P. Burelli (Eds.): EvoApplications 2016, Part II, LNCS 9598, pp. 11–26, 2016.
DOI: 10.1007/978-3-319-31153-1_2

The term *diversity* refers to differences among individuals which can be at either the genotype or phenotype level. Typically, diversity is classified into three categories: diversity maintenance, diversity control, and diversity learning. Diversity *maintenance* encourages exploring multiple promising pathways through the search space simultaneously in hopes of reaching higher fitness peaks. On the other hand, diversity *control* methods use population diversity, individual fitness, and/or fitness improvements as feedback to steer the evolutionary process towards exploration or exploitation. The main difference between diversity control and diversity learning methods is that in the former case, the short-term history (e.g., current population) is often used during diversity computation, whilst in the latter case, long-term history is used in combination with machine learning techniques to find (un)explored search areas. More recent research focuses on diversity *learning*, using cultural learning or self-organizing maps for example in order to discover promising locations within the search space.

In this work we extend research within the field of diversity *maintenance*. We describe a novel genotypic diversity learning method named Grid-Diversity-Operator (GDO) that makes use of the long-term history of all populations in order to suggest a biased distribution for new individuals. GDO is not specific to any particular algorithm, but can be used with any optimization algorithm that supports the use of *infusion* techniques — that is, insertion of new individuals after a certain number of generations or special initialization techniques. The new operator is tested within two population-based optimization algorithms that support infusion techniques, replacing the infusion step that causes random generation of new individuals. Experiments are conducted on a set of 13 benchmark multimodal optimization problems in different search space dimensions. We proposed an outline of this operator in a recent short paper [2]. Here we provide a full description of the operator and the motivation behind it. In addition, we undertake the first thorough evaluation of the operator using a well-known set of 39 multimodal benchmark problems, ranging in dimensionality from 2 to 30 (in contrast, the earlier work considered a small set of 2-dimensional functions only as proof-of-concept). Finally, we incorporate the GDO operator into two different optimisation algorithms to provide evidence that GDO is not tied to any particular algorithm, but can be used as a generic operator. A comprehensive statistical analysis is conducted to provide a more accurate explanation of GDO's performance.

This paper is structured as follows. Section 2 will cover related work in the literature regarding diversity preserving techniques. The Grid Diversity Operator (GDO) will be introduced in detail in Sect. 3. Section 4 will briefly introduce the algorithms that will be used in the experimental analysis. The experimental protocol followed here with the benchmark problems adopted and the testing procedure is introduced in Sect. 5. The obtained results and discussion along with the statistical analysis will be presented in Sect. 6, while the final comments will be given in Sect. 7.

2   Related Work

It is widely accepted that maintaining high levels of diversity within a population greatly contributes to algorithm performance [3]. Recent and detailed survey of the area is given in by Črepinšek *et al.* [4] in which a taxonomy of approaches for managing exploration and exploitation within a search algorithm is defined. Three approaches are identified: diversity *maintenance*, diversity *learning* and diversity *control*.

Of most relevance to our work is the branch of this taxonomy that concerns methods for *diversity maintenance*, which the authors divide into two classes: non-niching and niching. Within the former class, four approaches are defined — population-based, selection-based, crossover/mutation-based and hybrid. Our interest lies within population-based methods, which are sub-classified into methods that vary population size, eliminate duplicates, infusion techniques and external archives. We provide a short overview of relevant work in the area of infusion-techniques, given that the GDO operator proposed in this paper is inspired by this — the reader is referred to the survey article for detailed examples of each of the other techniques.

Infusion techniques typically involve the insertion of new individuals into the population after a certain number of generations or managing initialisation of the population (e.g. [5]). Early approach to this were described by Grefenstette in [6] in which random immigrants are inserted into the population every generation, and Koza [7] whose 'decimation' algorithm replaced a random fraction of the population each generation. More recently, Koumousis and Katsaras [8] proposed a SawTooth GA that utilised a periodic re-initialization step at which a number of new, random individuals are added to the population. This algorithm introduced an operator that varied the population size over time — the population size is linearly decreased over a number of generations, then suddenly increased to its original value through the addition of random immigrants, hence the sawtooth name. This was shown to achieve both better population diversity and overall performance on a set of continuous optimization benchmarks compared to a standard GA.

In addition, our proposed approach has some similarities to previous diversity *learning* methods. This class of methods is less well-studied than the maintenance methods — the idea is that the long-term history of the population can be used in combination with a machine learning techniques to learn which areas of the space have not been explored. For example, Leung et al. [9] propose a history-driven evolutionary algorithm (HdEA), in which a binary space-partitioning tree (BSP) is used to record evaluated individuals and their associated fitness throughout the evolution process. A guided anisotropic search governs the search direction based on the history of the BSP tree which results in an individual being either exploitatively mutated towards the nearest historical optimum or randomly mutated if the individual is already a local optimum. Other learning methods [10] use self-organising maps (SOMs) to learn about unexplored areas of the space using information from the whole evolutionary history; this information is used to determine *novelty* and thus encourage exploration.

Another class of algorithms that also exploit variable population size with random immigration are the *aiNET* series of algorithms, first proposed in [11]. Steps intrinsic to the core algorithm maintain diversity by suppressing individuals that are genotypically similar (thus varying population size) while additionally adding a small number of new, randomly generated individuals to the population each iteration. Andrews and Timmis [12] further increased the diversity of aiNET with respect to function optimization through adding an immune-inspired mutation operator. De França *et al.* [13] evaluated the diversity mechanisms of opt-ainet compared to fitness-sharing methods using a set of continuous optimization benchmarks, finding that the diversity mechanisms within opt-ainet both obtained better solutions but also maintained more diverse solutions.

Most *infusion techniques* described in the literature rely on *random* initialisation methods; in contrast, we propose an operator that biases the generation of new individuals towards areas of the search space that are under-explored by the current population. The proposed GDO operator shares some similarities with previous learning approaches in that is exploits the history of the complete evolutionary process, however unlike [9] we focus in its use to guide exploration rather than exploitation. The operator can be used with any optimization algorithm that uses an infusion-technique. In this paper, we test its validity by replacing the random generation step of both Opt-aiNET and SawTooth algorithms which we will briefly introduce to give better understanding in Sects. 4.1 and 4.2.

3   Grid Diversity Operator (GDO)

The Grid Diversity Operator (GDO) is a novel approach for achieving exploration and exploitation balance through maintaining diversity. It can be defined as a hybrid, non-niching, population-based, genotype diversity maintaining and learning technique. Simply put, GDO is a special infusion technique for initializing new individuals that are inserted into a population after a certain number of generations. Instead of randomly initializing individuals over the whole domain, GDO tries to initialize them in unexplored locations. A memory archive is used to store information collected throughout the run regarding the distribution of the individuals, and is used to infer rarely visited locations.

The basic idea behind GDO is to split the feasible space (the domain) into smaller sub-spaces using the grid size parameter $G_{sz} \in \Re^n$, which defines the number of intervals per dimension, where n is the number of dimensions for the problem. This process will form a 2D grid for 2-dimensional problem, 3D grid for a 3-dimension problem, and so on as demonstrated in Fig. 1. The GDO will then try to distribute new individuals into the grid slots that have received fewer visits over time, thus increasing the explorative power of the algorithm.

First, a memory archive is initialized as an empty dictionary that has n component *keys*, where each of them matches a single value. The *keys* refer to the indices of a slot within the grid, while the *value* represents the number of individuals that have previously been placed in this slot. The memory archive is designed in this manner for efficiency: as most of initialized memory will be

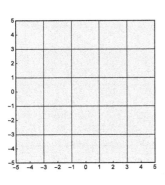

(a) Grid for 2-dimensional problem

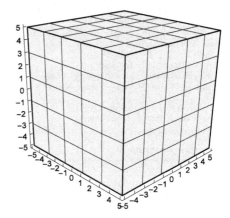

(b) Grid for 3-dimensional problem

Fig. 1. Examples of grid for 2-dimensional and 3-dimensional problems

sparse, by using the dictionary only visited locations are stored, and therefore the use of memory is efficient regardless of problem dimensionality.

An example of the memory archive is shown in Table 1 that represent a sample of a memory archive with two entries for a 10-dimensional problem.

Table 1. Example of memory archive for 10-dimensional problem

Slot	Dictionary key (dimensions indexes)										N (value)
	d_1	d_2	d_3	d_4	d_5	d_6	d_7	d_8	d_9	d_{10}	
1	5	2	1	4	3	1	5	2	8	1	7
2	7	3	4	1	3	6	3	2	5	3	4

For the first entry, the slot located at key $< 5, 2, 1, 4, 3, 1, 5, 2, 8, 1 >$ has been assigned a value of 7, indicating that 7 solutions have appeared in this slot during the run so far. The slot located at key $< 7, 3, 4, 1, 3, 6, 3, 2, 5, 3 >$ has been assigned a value of 4 indicating 4 solutions visited this slot. For each iteration in the containing algorithm, every new individual is processed to identify its slot to update the memory archive. For each individual being processed, if there is an entry in the archive with a key that matches the identified slot, the value corresponding to this entry is increased by one. Otherwise, a new entry is added to the archive with a value of 1. The process of updating the memory archive is demonstrated in Listing 1.

After processing all individuals, the updated archive is used to initialize new individuals. For each new individual required, we pick a slot $S_{(d_1, d_2, ..., d_n)}$ at random and calculate its *distribution probability* P according to Eq. 1.

$$P = e^{(-N)} \tag{1}$$

Listing 1. GDO Update Archive Method

```
Input: MemoryArchive, Individuals, Resolution
Output: MemoryArchive
  for Individualᵢ ∈ Individuals do
    Key ← FindSlot(Individualᵢ)
    if KeyExist(MemoryArchive,Key) then
      IncreaseValue(MemoryArchive,Key,1)
    else
      AddKey(MemoryArchive,Key)
      SetValue(MemoryArchive,Key,1)
    end if
  end for
  Return (MemoryArchive)
```

where N is the value matching the slot key in the archive or zero if the slot does not yet belong to the archive. Finally, the calculated probability P is compared to the probability threshold parameter of the algorithm, P_{th}, and if $P > P_{th}$ then a new individual is initialized randomly in this specific slot. If not, another slot is picked at random, and the steps are repeated until the individual is initialized successfully. The Grid Diversity Operator, GDO, is described in Listing 2.

Listing 2. Grid Diversity Operator

```
Input: MemoryArchive,N_new,G_sz,P_th
Output: S_new
  S_new ← ∅
  for i = 1 to N_new do
    distributed ← FALSE
    while ¬ distributed do
      Key ← PickSlotAtRandom(G_sz)
      if KeyExist(MemoryArchive,Key) then
        V ← GetValueOfKey(MemoryArchive,Key)
        P ← e^{-V}
      else
        P ← 1
      end if
      if P > P_th then
        Individual ← CreateNewIndividualInSlot(Key)
        InsertIndividual(S_new,Individual)
        distributed ← TRUE
      end if
    end while
  end for
  Return (S_new)
```

The process of updating the memory archive and distributing new individuals continues until the algorithm terminates, at which point the final population is expected to be more diverse than simply using a random initialisation procedure.

4   Selected Infusion-Supported Algorithms for Experimentation

To assess the performance of the proposed Grid Diversity Operator (GDO), we incorporate it within two algorithms from the literature that feature a distinct diversity maintenance mechanism. The first of the two is the artificial immune algorithm *opt-ainet* [11] and the other one is the *sawtooth* genetic algorithm [14].

4.1 Opt-aiNET Algorithm

Opt-aiNET, proposed in [11], is a well-known Artificial Immune System optimization algorithm. Opt-aiNET evolves a population that consists of a network of candidate solutions to the function being optimised known as antibodies. These undergo a process of evaluation against the objective function, clonal expansion, mutation, selection and interaction between themselves resulting in a population of dynamically changing size. Overtime, Opt-aiNET creates a memory set of antibodies that represent the best candidate solutions to the objective function. One of the main features of opt-ainet algorithm is that it has a defined, separated, diversity mechanism that injects a small number of new randomly created solutions into the population following each cycle of clonal expansion and mutation (the *AppendNewRandomCells()* shown in the bottom of Listing 3). This diversity-maintaining step makes it an ideal candidate for GDO injection.

Listing 3. Opt-aiNET Algorithm

Input: $Initial_{size}$, N_{clones}, $Supp_{th}$, Err_{th}, $Mutation_{par}$, Div_{ratio}
Output: Population
  Population $\leftarrow \emptyset$
  AppendNewRandomCells(Population,$Initial_{size}$)
  while $\neg$ StopCondition() **do**
    EvaluateCells(Population)
    Clones $\leftarrow$ CloneCells(Population,N_{clones}, Err_{th})
    MutatedClones $\leftarrow$ HypermutateClones(Clones,$Mutation_{par}$)
    EvaluateCells(MutatedClones)
    for $Cell_i \in$ Population **do**
      BestClone $\leftarrow$ GetBestClonePerCell(MutatedClones,i)
      if F(BestClone) < F(Population[i]) **then**
        Population[i] $\leftarrow$ BestClone
      end if
    end for
    SupressLowAffinityCells(Population, $Supp_{th}$)
    AppendNewRandomCells(Population,Div_{ratio})
  end while
  Return (Population)

Most of the parameters of opt-ainet were set as suggested by the authors of the algorithm [11]. We set the number of clones per cell $N_{clones} = 10$, suppression threshold $Supp_{th} = 0.2$, error threshold $Err_{th} = 0.001$, mutation parameter $\mu = 100$ and diversity ratio $D_{rate} = 40\%$. The only change was the value of initial population size which we set to 50 instead of the suggested value of 20. This change was important to help the algorithm with high dimensional problems and has been applied to all opt-ainet versions whether injected with GDO or not.

4.2   SawTooth Algorithm

The SawTooth Genetic Algorithm [14] is an algorithm that follows a sawtooth scheme defined by population size mean $\bar{n}$, an amplitude D and period of variation T in order to balance periods of exploration and exploitation. During every period, the population size decreases linearly. At the start of the next period, a population re-initialization occurs by appending randomly generated individuals to the population as shown in Listing 4. To achieve variable sized population, the population size is calculated every generation according to Eq. 2 below.

$$n(t) = \text{int}\left\{ \bar{n} + D - \frac{2D}{T-1}\left[t - T \cdot \text{int}(\frac{t-1}{T}) - 1\right]\right\} \tag{2}$$

where $\text{int}(\cdot)$ is the floor function. The authors suggested a range for the optimum values of the T and D parameters of the sawtooth based on experimentation. The optimum normalized amplitude $D/\bar{n}$ being from 0.9 to 0.96 and the optimum normalized period ranging from $T/\bar{n} = 0.5$ to 0.7. In this paper, we choose $\bar{n} = 80$, $T = 50$, $D = 75$ and these values complies with the ranges suggested by the algorithm authors. We also set the initial population size to 200, crossover rate to 0.7 and mutation rate to 0.95.

Listing 4. SawTooth Algorithm

```
Input: Population_size, CrossOver_rate, Mutation_rate, ñ, D, T
Output: Population
  t ← 0
  Population ← InitializePopulation(Population_size)
  EvaluatePopulation(Population)
  while ¬ StopCondition() do
    t ← t + 1
    n(t) ← CalculateNextPopulationSize(ñ,D,T,t)
    if mod(t,T) = 0 then
      NewIndividuals ← InitializePopulation(2 * D)
      InsertIndividuals(Population,NewIndividuals)
    end if
    NextPopulation ← ∅
    for i = 1 to n(t) by 2 do
      [P₁, P₂] ← SelectParents()
      [C₁, C₂] ← CrossOver(P₁, P₂, CrossOver_rate)
      Mutate(C₁, Mutation_rate), Evaluate(C₁)
      Mutate(C₂, Mutation_rate), Evaluate(C₂)
      [B₁, B₂] ← SelectBestTwoIndividuals([P₁, P₂, C₁, C₂])
      InsertIndividuals(NextPopulation, [B₁, B₂])
    end for
    Population ← NextPopulation
  end while
  Return (Population)
```

---

5   Experimental Protocol

The aim of the experiments is to assess if the performance of both algorithms introduced above improves when injected with GDO in order to determine whether GDO can help in achieving better quality solutions.

In addition, since GDO introduces two parameters – grid size G_{sz} and probability threshold P_{th}, we investigate settings for each parameter in a brief empirical investigation. For the grid size parameter, G_{sz}, the chosen values were (100, 500, 1000) while for the probability threshold P_{th} the values were (0.01, 0.10, 0.20). Thus, nine configurations of GDO are tested in an attempt to find the most successful configuration as shown in Table 2.

Table 2. Configuration code names and parameter values

	GDO-1	GDO-2	GDO-3	GDO-4	GDO-5	GDO-6	GDO-7	GDO-8	GDO-9	GDO-free
G_{sz}	100	100	100	500	500	500	1000	1000	1000	N/A
P_{th}	0.01	0.10	0.20	0.01	0.10	0.20	0.01	0.10	0.20	N/A

The experimental tests are performed using opt-ainet and sawtooth algorithms on a set of benchmark problems from the *CEC 2014* benchmark suite [15]. The selected problems (F_4 through F_{16}) are multimodal problems where many of which have huge numbers of local optima that make the assessment process challenging. The rest of the problems from the suite (three unimodal functions F_1–F_3 and 14 hybrid/composite functions F_{17}–F_{30}) are interesting as well but have different features and therefore were not selected.

Following the evaluation criteria defined in *CEC 2014* benchmark suite [15], both algorithms with and without GDO (including the nine GDO configurations) will be tested against the 13 problems in 2, 10 and 30 dimensional space. For every function, each algorithm/configuration is run 25 times with a maximum number of function evaluations equal to (Max_{FES}) of $10,000 * D$ where D is the number of dimensions. Therefore, Max_{FES} is 20,000 for two dimensions, 100,000 for 10 dimensions and 300,000 for 30 dimensions. The best quality solution is noted for each run, along with the number of function evaluations at which the best solution is found: if this is less than Max_{FES} this indicates stagnation of the algorithm. A detailed statistical analysis is conducted to assess the results.

6   Results and Discussion

Due to space limitations, it is not possible to provide detailed tables for the results of all experiments with 39 test cases (13 problems over three different dimensions) for both algorithms with 10 configurations each — a total of 780 results. In an attempt to summarize the data, we count the number of problem instances in which a GDO configuration outperforms the corresponding GDO-free algorithm. Table 3 displays this information for the nine GDO configurations used within the two algorithms over the three different dimensions for the benchmark problems. For instance, the first entry denotes that opt-ainet injected with GDO-1 outperformed GDO-free opt-ainet in 5 problems out of 13 instances with respect to solution quality (while in the remaining 8 problems either GDO-free was better or there were no significant difference). Two observations are clear from Table 3. Firstly, the number of instances in which GDO outperforms GDO-free increases as the dimensionality of the instances increases, i.e., its benefit increases with dimensionality. Secondly, the GDO-9 configuration performs best out of the 9 different parameterisations, when considered across all problem dimensions.

Table 3. Number of functions (out of 13) where a GDO configuration achieves better quality solutions than GDO-free

Config.	Dimension [Opt-aiNET]			Dimension [SawTooth]		
	2	10	30	2	10	30
GDO-1	5	8	12	7	7	11
GDO-2	8	8	11	9	8	12
GDO-3	6	8	11	9	9	8
GDO-4	5	6	10	6	10	11
GDO-5	8	7	10	5	10	9
GDO-6	9	8	10	8	9	9
GDO-7	8	5	12	7	9	10
GDO-8	9	6	11	7	5	10
GDO-9	10	9	12	9	9	10

Since the algorithm/configuration that optimises both objectives (minimises convergence speed and minimises summed fitness) is always considered more successful, additionally we provide graphs that summarise the results concerning both solution quality and convergence speed.

Figures (2-7) summarise the results where every graph contains the result of 9 GDO configurations and one without GDO injection (GDO-free) for a specific algorithm and problem dimensionality. The different configurations are distributed on the graph's horizontal axes. The bars on top denote the sum of the normalized fitness for the configuration over the 13 problem instances; similarly, the bars on the bottom refer to the sum of the normalized of number of evaluations used before algorithm termination. Error bars are provided as an indication of difference in means between configurations for both quality (top) and efficiency (bottom).

Beginning with the optainet algorithm results, we see in Fig. 2 that all GDO-injected configurations of opt-ainet for two-dimensional problems outperform the GDO-free version with respect to solution quality. However, all the GDO configurations use more evaluation cycles than the GDO-free version. For the ten-dimensional tests, Fig. 3 shows that 7 out of 9 of the GDO configurations of opt-ainet — all but GDO-7 and GDO-8 — surpasses the GDO-free version in terms of solution quality. In this case, GDO-4,GDO-5, and GDO-8 configurations used less number of function evaluations than the GDO-free version while GDO-1, GDO-3, GDO-6 were similar to GDO-free and the rest were the worst. Figure 4

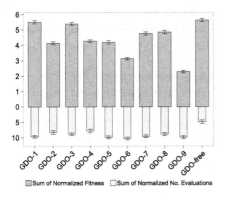

Fig. 2. Opt-aiNET (2 Dimensions)

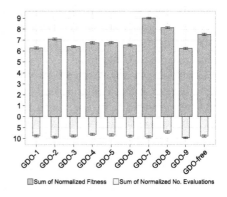

Fig. 3. Opt-aiNET (10 Dimensions)

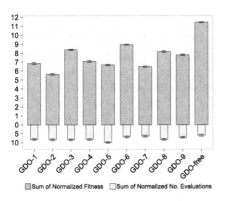

Fig. 4. Opt-aiNET (30 Dimensions)

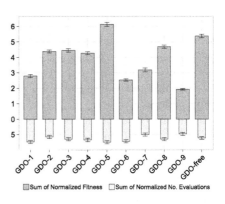

Fig. 5. Sawtooth (2 Dimensions)

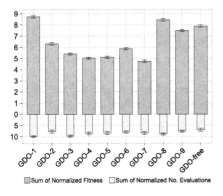

Fig. 6. Sawtooth (10 Dimensions)

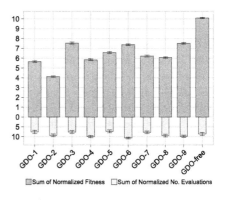

Fig. 7. Sawtooth (30 Dimensions)

shows the results for the 30-dimensional tests where we see that all instances of GDO configured opt-ainet algorithm were able to achieve better results than GDO-free opt-ainet, while all GDO configurations have utilised more function evaluations than GDO-free.

The sawtooth algorithm for two-dimensional problems results (Fig. 5) shows that only the GDO-5 configuration failed to achieve better quality solutions than GDO-free sawtooth. In this case, GDO-2, GDO-7 and GDO-9 configurations shown faster convergence while the rest were the worst with respect to number of evaluations used. The results of the ten-dimensional sawtooth tests are shown in Fig. 6 were seven GDO-injected sawtooth algorithm instances outperform the GDO-free version for quality. However, all GDO versions utilise more evaluation cycles in this case. Finally, for the 30-dimensional tests, the results in Fig. 7 shows that all GDO instances outperformed the GDO-free version with respect to solution quality while GDO configurations (GDO-1, GDO-3, GDO-5 and GDO-7) shown better utilisation to their allowance of function evaluations than GDO-free version of sawtooth.

It is interesting to note that some GDO configurations are able to achieve better results with fewer evaluations than GDO-free, e.g. in Figs 3, 5 and 7. However, these are exceptions: in general, the GDO configurations utilise more function evaluations than the GDO-free configurations for both opt-ainet and sawtooth algorithms due to the additional exploration ability facilitated by the GDO operator.

In summary, in terms of solution quality, the majority of GDO configurations provide an improvement in all the six experiments defined by (*algorithm/dimensionality*). Out of them we note GDO-9 which performed better than the rest of the configurations in three instances as shown in Figs 2, 3 and 5 followed by GDO-2 as a second best beating all configurations in two other instances as shown in Figs 4 and 7. The benefit increases with dimensionality, with GDO-9 winning 12/13 instances when used with opt-ainet, and 10/13 cases when used with the saw-tooth algorithm for 30 dimensions problems according to Table 3. To further demonstrate the performance of GDO-9 configuration, Tables 4 and 5 shows the median for all the runs over the 13 functions using the algorithms opt-ainet and sawtooth respectively. Comparing the results of GDO-free in both tables we can see that sawtooth capable of achieving better results than opt-ainet for all problems especially in higher dimensions where opt-ainet tends to diverge notably with functions 4 and 15. For function 15, and while GDO-9 helped sawtooth to make significant improvements for the same function for all dimensions, GDO-9 help for opt-ainet was much more important and the performance drastically improved over the poor results of GDO-free opt-ainet especially in 10 and 30 dimensions.

With respect to number of evaluations, it is clear that most GDO-injected algorithms require more function evaluations to converge than the GDO-free version of the algorithm. Recall that each algorithm is given a fixed number of function evaluations (Max_{FES}), but terminates before this in case of no improvement which in many cases reflect being stuck in a local optima. This is clearly

Table 4. Median of Opt-äiNET algorithm results: GDO-free against GDO-9 over the 3 dimensions for the 13 functions

Fn	Dimension [GDO-free]			Dimension [GDO-9]		
	2	10	30	2	10	30
4	**400.4180170**	**15076.92619**	**41522.53368**	**400.0067050**	**1382.623949**	**25880.42417**
5	510.1203580	520.2005640	520.6004260	504.8840650	520.0004160	520.2967840
6	601.0638150	617.8945810	650.5898950	600.3518890	611.1999250	643.6837470
7	713.9159790	1404.468688	2867.517032	700.1278140	806.8315380	1676.329282
8	802.0770040	1012.395469	1285.506462	800.3384240	894.4034050	1244.529391
9	901.2102780	1154.560341	1436.587623	901.8637690	1000.095998	1437.381565
10	1092.261931	4307.027543	10982.91621	1014.131418	2713.651994	9295.601940
11	1222.144605	4525.313885	13180.02786	1114.076847	3326.207438	9469.204918
12	1200.005679	1202.956275	1206.578518	1200.004737	1201.733634	1203.362254
13	1301.264216	1307.748503	1309.748985	1300.567841	1303.354419	1308.982133
14	1400.341883	1612.230827	1783.065834	1400.100778	1435.029170	1728.472052
15	**1500.167713**	**3556487.517**	**98467228.88**	**1500.048914**	**4426.678862**	**7485.869758**
16	1600.156432	1604.219537	1614.369001	1600.156378	1603.941531	1613.563775

Table 5. Median of SawTooth algorithm results: GDO-free against GDO-9 over the 3 dimensions for the 13 functions

Fn	Dimension [GDO-free]			Dimension [GDO-9]		
	2	10	30	2	10	30
4	**400.0142740**	**727.9746490**	**9170.752282**	**400.0065010**	**651.3681520**	**8160.968453**
5	507.9617110	520.0003550	520.0021610	514.0684370	520.0002660	520.0022920
6	600.1977440	611.0157390	641.2174190	600.1320940	610.4406700	639.3822960
7	700.9761200	759.2174120	1189.677979	700.2440890	769.6810190	1157.155085
8	800.0000000	838.8033260	944.6040920	800.9949590	827.8587750	946.8881300
9	901.9899180	941.7881410	1094.533675	900.9949590	950.7426880	1093.022542
10	1000.312173	1140.325716	4056.051603	1000.312173	1269.169192	3983.343406
11	1218.438335	2277.064899	5145.395051	1218.438335	2454.652439	5542.767461
12	1200.000579	1200.698797	1201.134650	1200.000604	1200.664882	1201.048815
13	1300.062153	1303.208397	1305.765346	1300.018911	1303.365590	1306.059232
14	1400.004992	1404.043086	1570.808730	1400.125393	1405.069313	1567.143899
15	**1500.090571**	**1952.314124**	**2576.155155**	**1500.019729**	**1948.094013**	**2391.734321**
16	1600.356445	1603.382337	1612.343996	1600.074448	1603.331906	1612.063572

observed in Figs 2 and 4 that show that the GDO-free algorithms stagnates early, therefore recording a low number of function evaluations, but has the worst solutions from the quality perspective. The larger number of evaluations used by GDO in the majority of experiments reflects the extra effort used by the operator to continue searching in novel parts of the solution space, therefore using the allowed budget of evaluations to avoid convergence to local optima. Thus in general GDO facilitates longer running times, leading to improved solution quality, though clearly there is computational cost to this in terms of utilised CPU time.

6.1   Statistical Analysis

Although the graphical summaries given above provide some insight and intuition regarding GDO performance, a proper statistical analysis is required to justify any claims. The statistical analysis in this section addresses the following questions:

- Does GDO injection helps improve solution quality when compared to the equivalent GDO-free algorithm?
- Which GDO configuration (from the proposed nine configurations) shows the best performance in terms of solution quality?

Since we have two algorithms, each with 10 possible configurations (9 GDO, 1 GDO-free) and each is tested against 13 problems in each of three different dimensions, there are 780 datasets. Each dataset contains the results of running a specific algorithm/configuration on a specific problem and specific dimension 25 times.

Before conducting any statistical analysis of the results, we utilise a Shapiro-Wilk normality test (with a level of significance of $\alpha = 0.05$) to determine with the datasets follow a normal distribution in order to identify appropriate statistical tests. The samples in 194 groups out of 780 were found to appear normally distributed while 586 groups were not and therefore non-parametric tests are used.

We used the non-parametric MannWhitney U test (with a level of significance of $\alpha = 0.05$) to test for difference in means and Table 6 shows the number of cases (out of 78) where a GDO configuration outperformed GDO-free and those where GDO-free dominated. The test shows that if there is a statistically difference in solution mean between GDO-injected and GDO-free algorithm, then the probability that the GDO-injected algorithm will lead to better quality solutions is twice that of the GDO-free one: the final row of Table 6 shows that GDO configurations are better in 158 cases, as opposed to 75 for the GDO-free. In addition, the results for the GDO-9 configuration across all experiments appear to be statistically better with respect to solution quality than any other GDO configuration when compared to GDO-free algorithms results. As shown in Table 6, the probability that GDO-9 will likely have better quality solutions is three times that of the GDO-free algorithm.

Table 6. MannWhitney U test results (aggregated)

Config	GDO-1	GDO-2	GDO-3	GDO-4	GDO-5	GDO-6	GDO-7	GDO-8	GDO-9	Total
Better	14	18	15	19	17	22	16	19	**18**	158
Worse	9	9	9	9	9	9	8	7	**6**	75

7   Conclusion

Diversity control methods for search algorithms fall into three classes: diversity maintenance, diversity learning and diversity control. We have described a new operator, GDO, that directs exploration into previously unvisited areas of the search space. While mainly falling into the class of diversity maintenance algorithms, it also draws inspiration from diversity learning methods in making use of historical information from the evolution process.

The GDO operator was shown to achieve effective exploration through testing on the benchmark problems when incorporated within opt-ainet and sawtooth algorithms. When compared to the GDO-free versions of the algorithms, it was shown that GDO can significantly help a supported algorithm to achieve better quality solutions in most cases. Importantly, Table 3 showed that the benefit of the GDO operator increases as the dimensionality of the problems increases. It was noticeable however that GDO-injected algorithms tend to use more function evaluation cycles than the GDO-free versions, i.e. does not stagnate prematurely. GDO forces the algorithm to explore more of the space thus using more evaluations while exploring but without preventing the algorithm's internal exploitation method from finding improved solutions. This added exploration ability helps the injected algorithm to both escape local optima and to achieve better solutions.

Within the current implementation, the grid size is fixed for all dimensions such that setting the grid size parameter $G_{sz} = 100$ in three-dimensional problem will correspond to a grid resolution of $100 \times 100 \times 100$. There is no obligation to set the grid resolution equally for all dimensions and in fact, it may be better to set different values depending on the dimensionality of the problems. In this paper, we empirically studied the values of grid size and probability threshold parameters of GDO by selecting three appropriate values for each. Although the nine proposed configurations of GDO (each with different parameters values) behaved differently, certain configurations were found able to offer the best possible performance across the tests, in particular the GDO-9 configuration where $G_{sz} = 1000$ and $P_{th} = 0.20$. The GDO-9 configuration was able to outperform GDO-free in all cases as shown in Figs 2–7 and all other GDO configurations as well in three cases as shown in Figs 2, 3 and 5. In addition, the statistical analysis conducted in this paper confirmed GDO-9's superiority. However, by incorporating feedback from search process it seems clear that this could be achieved autonomously, so that both parameters in fact could adapt over a single run to achieve better performance. We aim to address this issue in future work.

References

1. Soza, C., Becerra, R.L., Riff, M.C., Coello Coello, C.A.: Solving timetabling problems using a cultural algorithm. Appl. Soft Comput. **11**(1), 337–344 (2011)
2. Salah, A., Hart, E.: Grid diversity operator for some population-based optimization algorithms. In: Proceedings of the Companion Publication of the 2015 on Genetic and Evolutionary Computation Conference, pp. 1475–1476. ACM (2015)

3. Michalewicz, Z.: Genetic Algorithms + Data Structures = Evolution Programs. Springer, New York (1996)
4. Črepinšek, M., Liu, S., Mernik, M.: Exploration and exploitation in evolutionary algorithms. ACM Comput. Surv. **45**(3), 1–33 (2013)
5. Li, Z., Wang, X.: Chaotic differential evolution algorithm for solving constrained optimization problems. Inf. Technol. J. **10**, 2378–2384 (2011)
6. Grefenstette, J.J.: Genetic Algorithms for Changing Environments. Elsevier, Amsterdam (1992)
7. Koza, J.R.: Genetic Programming: On the Programming of Computers by Means of Natural Selection. MIT Press, Cambridge (1992)
8. Koumousis, V.K., Katsaras, C.P.: A saw-tooth genetic algorithm combining the effects of variable population size and reinitialization to enhance performance. IEEE Trans. Evol. Comput. **10**(1), 19–28 (2006)
9. Leung, S.W., Yuen, S.Y., Chow, C.K.: Parameter control by the entire search history: case study of history-driven evolutionary algorithm. In: 2010 IEEE Congress on Evolutionary Computation (CEC), pp. 1–8, July 2010
10. Amor, H.B., Rettinger, A.: Intelligent exploration for genetic algorithms: using self-organizing maps in evolutionary computation. In: Proceedings of the 7th Annual Conference on Genetic and Evolutionary Computation, GECCO 2005, pp. 1531–1538. ACM, New York (2005)
11. de Castro, L.N., Timmis, J.: An artificial immune network for multimodal function optimization. In: Proceedings of the 2002 Congress on Evolutionary Computation, 2002. CEC 2002, vol. 1, pp. 699–704 (2002)
12. Andrews, P.S., Timmis, J.: On diversity and artificial immune systems: incorporating a diversity operator into aiNet. In: Apolloni, B., Marinaro, M., Nicosia, G., Tagliaferri, R. (eds.) WIRN 2005 and NAIS 2005. LNCS, vol. 3931, pp. 293–306. Springer, Heidelberg (2006)
13. De França, F.O., Coelho, G.P., Zuben, V., F.J.: On the diversity mechanisms of opt-aiNet: a comparative study with fitness sharing. In: IEEE World Congress on Computational Intelligence, WCCI 2010–2010 IEEE Congress on Evolutionary Computation, CEC 2010 (2010)
14. Koumousis, V., Katsaras, C.: A Saw-tooth genetic algorithm combining the effects of variable population size and reinitialization to enhance performance. IEEE Trans. Evol. Comput. **10**(1), 19–28 (2006)
15. Liang, J.J., Qu, B.Y., Suganthan, P.N.: Problem definitions and evaluation criteria for the CEC 2014 special session and competition on single objective real-parameter numerical optimization. Technical report, Computational Intelligence Laboratory, Zhengzhou University, Zhengzhou, China, December 2013

Benchmarking Languages for Evolutionary Algorithms

J.J. Merelo[1]([✉]), Pedro Castillo[1], Israel Blancas[1], Gustavo Romero[1],
Pablo García-Sanchez[1], Antonio Fernández-Ares[1], Víctor Rivas[2],
and Mario García-Valdez[3]

[1] Department of Computer Architecture and Technlogy and CITIC,
University of Granada, Granada, Spain
jmerelo@geneura.ugr.es
[2] Department of Informatics, University of Jaén, Jaén, Spain
[3] Instituto Politécnico de Tijuana, Tijuana, Mexico
http://citic.ugr.es

Abstract. Although performance is important, several other issues should be taken into account when choosing a particular language for implementing an evolutionary algorithm, such as the fact that the speed of different languages when carrying out an operation will depend on several factors, including the size of the operands, the version of the language and underlying factors such as the operating system. However, it is usual to rely on compiled languages, namely Java or C/C++, for carrying out any implementation without considering other languages or rejecting them outright on the basis of performance. Since there are a myriad of languages nowadays, it is interesting however to measure their speed when performing operations that are usual in evolutionary algorithms. That is why in this paper we have chosen three evolutionary algorithm operations: bitflip mutation, crossover and the fitness function OneMax evaluation, and measured the speed for several popular, and some not so popular, languages. Our measures confirm that, in fact, Java, C and C++ not only are the fastest, but also have a behaviour that is independent of the size of the chromosome. However, we have found other compiled language such as Go or interpreted languages such as Python to be fast enough for most purposes. Besides, these experiments show which of these measures are, in fact, the best for choosing an implementation language based on its performance.

1 Introduction

It is usual in soft computing to try and use languages such as C++ or Java with the rationale that they are the fastest tools available for the kind of algorithms and problems that are solved by them. However, while there are benchmarks available for the languages at large and they have been tested in a variety of environments, it remains to be seen how fast are these languages at the tasks that are done usually by evolutionary algorithms.

© Springer International Publishing Switzerland 2016
G. Squillero and P. Burelli (Eds.): EvoApplications 2016, Part II, LNCS 9598, pp. 27–41, 2016.
DOI: 10.1007/978-3-319-31153-1_3

Let us emphasize that running speed is not everything. In many cases, ease of integration with existing or legacy tools, coding speed or availability of parallel or distributed frameworks are much more important than how fast a single CPU program runs. There are also restrictions inherent to the application at hand, such as embedded or web based systems where certain languages must be used. However, in some cases, mainly when the size of the problem or the running time of the algorithm call for the maximum running speed available, it is interesting at least to know which languages can be used to obtain the best performance.

In this case it is quite clear that efficiency matters, as said in [1]. And, as a matter of fact, in general and restricting the concept of *speed* to *speed of the compiled/interpreted application* it might be the case that some languages are faster to others, as evidenced by benchmarks such as [2,3]. Taken in general or even restricting it to some particular set of problems such as floating point computation, some compiled languages tend to be faster than interpreted languages.

But, in the same spirit of the *There is no free lunch* theorem [4] we can affirm there is a *no fast lunch* theorem for the implementation of evolutionary optimization, in the sense that, while there are particular languages that might be the fastest for particular problem sizes and specially fitness functions, these two constraints cannot be disregarded, and, specially, for non-trivial problem sizes and limiting ourselves to the realm of evolutionary algorithm operators, some interpreted and unpopular languages such as `awk` can be the fastest option available for evolving particular data structures such as regular expressions [5]. Besides, benchmarks should include a wide variety of sizes, because the time needed to perform particular operations does not always depends linearly on size, as performance is also affected by technical details for instance the implementation of loops and memory management.

For the purposes of this paper, we will try to be more extensive on the number of languages tested than on the number of operations. In general, evolutionary algorithms use fitness functions that can have any size and shape, so it is not easy to cover them all and further characterization might be needed. We are going to focus on combinatorial optimization problems, which are usually represented using bit strings and the two most characteristics operators: mutation and crossover. Besides, a fitness function usually employed as a baseline for evolutionary algorithms will be used: OneMax.

In general, programs do not spend the majority of the time applying them; this place in the ranking rather goes to selection operators and other higher-level, population-handling ones, as well as usually the fitness function. However, these functions are well covered by usual benchmarks so you can usually rely on any of them for choosing a language to implement your evolutionary algorithm. This makes the scope or interest of this paper certainly restricted to the set of problems in which classical bit-wise operations are performed and where fitness does not take most of the time: Testing new operators or implementing parallel or other kind of algorithms on functions such as HIFF, OneMax or Royal Road.

Finally, our intention is not so much to choose the winner of the competition of fastest language for evolutionary algorithms as much as to check the variety of speeds available and to know what order of magnitude these differences are. This work can be used to aid in making the decision of a language for implementing an evolutionary algorithm, or at least to justify the choice of non-mainstream languages such as Python, Lua, JavaScript or Go, which, in fact, do reach interesting performance marks.

Coming up next, we will present the state of the art of the analysis of evolutionary algorithm implementations. Next we will present in Sect. 3 the tests we have used for this paper and its rationale along with the languages we have chosen for carrying them out. Finally, in Sect. 4 we will present the results of measuring the performance of eleven languages running some widely used evolutionary algorithm operators and functions mutation, crossover and OneMax. Finally, we will draw the conclusions and present future lines of work.

2   State of the Art

In fact, the examination of the running time of an evolutionary algorithm has received some attention from early on. Implementation matters [6,7], which implies that paying attention to the particular way an algorithm is implemented might result in speed improvements that outclass those achieved by using the *a priori* fastest language available. In fact, careful coding and choosing the right tools [8,9] in interpreted languages can make them as fast, or even faster, than compiled languages such as Java.

However, most papers devoted to the implementation of evolutionary algorithms in languages other than C or Java try to prove that, for the particular type of problems used in scientific computing in general, the running speed is not as important as coding speed or even learning speed, since most scientific programs are, in fact, run a few times while a lot of time is spent on coding them. That is why expressive languages such as Perl, JavaScript or Python are, in many cases, superior to these fast-to-run languages.

Even so and when speed matters, the benchmarks performed in those papers were restricted to particular problem sizes and to very specific languages. They also test a single language for the whole evolutionary algorithm; however, it might happen that, since different operations are involved, the ranking varies depending on the operation and on the size. This can be of special interest in environments such as the Kappa architecture [10] or web-services based frameworks [11,12] where different parts of an application might be written in different languages. An advantage of these loosely connected systems is that they might change the language used for a particular operation if they encounter performance issues, as opposed to monolithic architectures written in a single language.

This is, as a matter of fact, the state of the art such as we have found them. That is why we think it is important to take real measures so that decisions on implementation are based on facts strictly related to evolutionary algorithms, instead of relying on common lore.

Next we will explain the operations used for the benchmark and how they have been tested.

3   Experimental Setup

First, a particular problem was chosen for testing different languages and also data representations: performing bit-flip mutation on a binary string. In fact, this is not usually the part of the program an evolutionary algorithm spends the most time in [7]. In general, that is the fitness function, and then reproduction-related functions: chromosome ranking, for instance. However, mutation is an operation that is performed quite frequently and sometimes once for every newly generated individual; it is also quintessential to the algorithm itself and one of the pillars of the canonical genetic algorithm, so it allows the comparison of the different languages in the proper context. Crossover is also part of that canonical algorithm, and MaxOnes or Count-Ones or OneMax is a fitness function frequently used as baseline for comparison of evolutionary algorithms.

In this section we will outline first the specifics of the implementation and the rationale behind them in subsection 3.1, to proceed to outline the different data structures that have been used here in subsection 3.2 to finally present the different languages that have been tested and the peculiarities of its implementation 3.3.

3.1   Functions and Operators Included in the Benchmark

Essentially, mutation is performed by

1. Generating a random integer from 0 to the length of the chromosome.
2. Choosing the bit in that position and flipping it
3. Building a chromosome with the value of that bit changed.

These operations deal mostly with random number generation and then list, string or vector processing. In general, copying and creating strings could depend on the length, but its implementation might vary from one language to another.

The next operation, *two-point crossover* is performed as follows:

- Generating two random integers with a range from 0 to the length of the chromosomes.
- Building two new chromosomes including the original from position 0 to the first point, interchanged bits from the first point to the second, and the original ones from the second position to the end of the strings.

A priori this operation seems quite similar to the first one. However, it involves copying of strings, an operation that will scale in a different way than simply running over a string and modifying one bit.

Finally, *OneMax* follows this procedure

- Generate a random string.
- Run over the string.
- If the bit is set to one, add one to a counter.
- Return the counter.

Despite its apparent simplicity, counting the number of ones in a string is an extremely complicated operation, which is in fact used by human resources teams to examine the prowess of candidates in the creation of algorithms and in the knowledge of a language. The straightforward way of carrying it out is using a loop that looks, one by one, at the bits in the string and adds 1 to the counter. However, in most cases that might not be the fastest way. At any rate, this fitness function is quite similar to others that decode the bits of a binary chromosome and, even if it is quite clearly not the only existent fitness function, it is one that is widely used in evolutionary algorithms and whose speed can be applied to other similar functions.

Being as it is a loop, we should expect that the time needed would grow linearly with the chromosome size. We will check whether this is true or not for the different languages below.

3.2   Available Data Structures

Chromosomes can be represented in several different ways: an array or vector of Boolean values, or any other scalar value that can be assimilated to it, or as a bitstring using generally "1" for true values or "0" for false values. Different data structures will have an impact on the result, since the operations that are applied to them are, in many cases, completely different and thus the underlying implementation is more or less efficient. Besides, languages use different native data structures to represent this information. In general, it can be divided into three different fields:

- *Strings*: representing a set bit by 1 and unset by 0, it is a data structure present in all languages and simple to use in most.
- *Vector of Boolean values*: not all languages have a specific primitive type for the Boolean false and true values; for those who have, sometimes they have specific implementations that make this data structure the most efficient.
- *Bitsets*: bits are bits, and you can simply use bits packed into bytes for representing chromosomes, with 32 bits packed in a single 4 byte data structure. Memory-wise the most efficient, without low-level access operators it can indeed be the slowest, and in any case not too efficient for decoding to function parameters.

The memory and speed efficiency of these data structures is different, and it is advisable for anyone implementing an evolutionary algorithm to check all possible data structures before committing, out of inertia, to the easiest one. Once again, implementation matters [6,7], and differences in evolutionary algorithm performance for different data structures can be, indeed, quite big.

Not all data structures, however, are easily available for every language, or easy to deal with. We will check in the next subsection which ones have been used in every language.

3.3   Languages Tested

Eleven languages have been chosen for performing the benchmark. The primary reason for choosing these languages was the availability of open source implementations for the authors, but also we have tried to be inclusive in by considering languages that represent different philosophies in language design and also languages traditionally used in the implementation of evolutionary algorithms together with others that are not so popular. These languages are presented in the next two subsections.

Table 1. Languages used and file written to carry out the benchmark. No special flags were used for the interpreter or compiler

Language	Version	URL	Data structures
Scala	2.11.7	https://git.io/benchscl	String, Bit Vector
Lua	5.2.3	https://github.com/JJ/LunEO	String
Perl	v5.20.0	https://git.io/bperl	String, Bit Vector
JavaScript	node.js 5.0.0	https://git.io/bnode	String
Python	2.7.3	https://git.io/vBSYb	String
Go	go1.2.1	https://github.com/JJ/goEO	Bit Vector
Julia	0.2.1	https://github.com/iblancasa/JuliEO	Bit Vector
C	4.8.2	http://git.io/v8kvU	char string
C++	4.8.4	http://git.io/v8T57	String
Java	1.8.0_66	http://git.io/v8TdR	Bitset
PHP	5.5.9	http://git.io/v8k9g	String

Compiled Languages. Compiled languages are represented by Scala, Java, Go, C and C++. Scala and Java both use Java bytecode and run in the Java Virtual Machine. Go, C and C++ compile to a binary that runs natively on the target operating system. Also JavaScript is compiled to native machine code when it runs in the V8 JavaScript Engine used by both Node.js and Google Chrome web browser. This list of languages is rather comprehensive, including the most popular compiled languages in scientific computing as well as two languages that are emerging in popularity: Go and Scala.

Scala [13] is a strongly-typed functional language that compiles to a Java Virtual Machine bytecode. Scala is in many cases faster than Java [3] due to its more efficient implementation of type handling. In this paper, two different representations were used in Scala: `String` and `Vector[Boolean]`. They both have the same underlying type, `IndexedSeq` and in fact the overloading of operators allows us to use the same syntax independently of the type. The benchmark is available under a GPL license, at the URL shown in Table 1. As far as we know, there are no evolutionary algorithm frameworks published in Scala; however, its increasing popularity within the programmer, and, over all, data science community makes it quite likely to find one in the near future.

Java is probably the most popular language within the evolutionary algorithm community, with several well established free software frameworks available, such as JCLEC [14] or ECJ [15]. It is a multi-platform language, with compiled *bytecode* running in any operating system with a native Java virtual machine (JVM).

C is still one of the most popular languages, despite being free of the paradigms that are so popular nowadays: object orientation or concurrency. It is not so popular within the EA community, who usually opt for its object-oriented version, C++, that has representatives such as ParadisEO [16] or MALLBA [17].

Go [18] was a language initially introduced by Google, but that has dug into the niches that were dominated formerly by C. It is also a compiled language designed for system programming, and is concurrent.

These four languages will be tested against a set of interpreted languages described below.

Interpreted Languages. Interpreted languages are represented by Lua, PHP, Julia, Perl, Python and JavaScript. These include three popular languages, Perl, Python and JavaScript, Python being also the most popular in the scientific computing environment, and then three other languages seldom seen in the EC arena, but popular in their particular niches.

Lua [19] is a popular embedded language designed for easy implementation, including a minimalist grammar so that its interpreter can be reduced to a few hundreds of Ks. Indeed, it can be lately found in places such as Game Engines. There are no known frameworks for EAs written in Lua, although an evolutionary algorithm that runs in a Canon camera has been found in YouTube.

Perl is an interpreted language that has been used extensively for evolutionary algorithms [8,20,21] with satisfactory results, although its main emphasis is not on speed, but on freedom for the programmer to code using any style and paradigm. The module and tools ecosystem built around it make it extremely interesting for any kind of programming. Our implementation of Perl uses different open-source libraries and in some cases also different versions of the language. Perl produces new even minor versions every spring, with the current version being 5.22. Odd minor versions such as 5.23 are considered development versions, with features that are, later on, implemented on the stable versions of the language.

JavaScript is an interpreted language that follows an ECMA standard and with many different implementations, including the different ones embedded in all internet browsers. Node.js is one such implementation of JavaScript with an asynchronous input/output loop and designed to run standalone in servers or the command line. Node.js uses a the V8 JIT compiler to create native machine code when in reads the script, and has been used lately as part the NodEO library [22] and the volunteer computing framework NodIO [23], as well as other libraries focused on the browser [24].

PHP is a language that is most commonly known for the creation of web sites. In fact, it is an interpreted and general-purpose language with emphasis on string processing. There is no known implementation of evolutionary algorithms in PHP.

Python is probably the most popular scripting language, head to head with JavaScript. It is directly supported by major corporations such as Google, and its different implementations (including one for the Java Virtual Machine, Jython) make it a very strong contender in the evolutionary algorithm arena. In fact, several evolutionary algorithm frameworks, such as DEAP [25] use it.

Julia [26] is the last language tested. It is a modern, interpreted and dynamic language used mainly for technical computing. It cannot be called exactly popular, but we have included it for completeness, mainly.

The interpreter versions and repositories for these languages are shown in the Table 1.

3.4   Implementation Notes

When available, open source implementations of the operators and OneMax were used. In all cases except in Scala, implementation took less than one hour and was inspired by the initial implementation made in Perl. Adequate data and control structures were used for running the application, which applies mutation to a single generated chromosome a hundred thousand times. The length of the mutated string starts at 16 and is doubled until reaching 2^{15}, that is, 32768. This upper length was chosen to have an ample range, but also so small as to be able to run the benchmarks within one hour. Results are shown next. In some cases and when the whole test took less than one hour, length was taken up to 2^{16}.

In most cases, and specially in the ones where no implementation was readily available, we wrote small programs with very little overhead that called the functions directly. That means that using classes, function-call chains, and other artifacts, will add an overhead to the benchmark; besides, this implies that the implementation is not exactly the same for all languages. However, this inequality reflects what would be available for anyone implementing an evolutionary algorithm and, when we think it might have an influence on the final result, we will note it.

Every program used also provides native capabilities for measuring time, using system calls to check the time before and after operations were performed. This might result in a slightly different behavior, but it is the best system available, so it is what we used.

All programs produced the same output, a comma separated value of language, operand length and time. Results were eventually collated and are available in the same repository than this paper, together with the code in several languages that was produced also specially for this.

4   Results and Analysis

We will examine first the performance for the bitflip operation, that is graphed in Fig. 1.

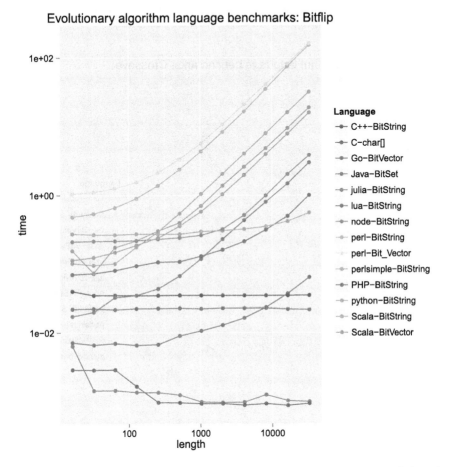

Fig. 1. Plot of the time needed to perform $100\,\mathrm{K}$ mutations in strings with lengths increasing by a factor of two from 16 to 2^{15}. Please note that x and y both have a logarithmic scale.

We can look at this figure in several ways. First, let us look at the *flat* lines, which represent those languages whose speed is independent of the length. These are C and Java, a fact that can be explained by the fact that the PHP implementation might, in fact, be very close to that of C. The rest of the languages are affected by operand length one way or another, but it is interesting to note that Java and Go are, in fact, faster when the length grows. The rest generally takes longer to process longer strings, although some languages such as Perl (in a *simple* implementation that uses the *Algorithm::Evolutionary::Simple* library), Lua or Node.js have a *flat* segment that goes up to considerable lengths.

Please note that, although the fastest in this case are always compiled languages, PHP is faster than C++ and some languages such as Python faster than Scala. Perl is the slowest, but the *simple* implementation beats Scala for non-trivial lengths of the string. This leads to the blurring the clear-cut distinction between *fast* compiled languages and *slow* scripting languages.

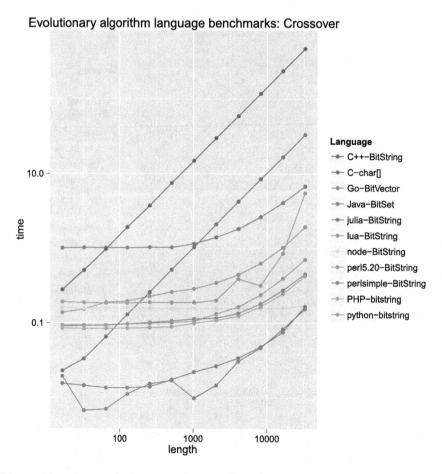

Fig. 2. Plot of time needed to perform 100 K crossover operations in strings with lengths increasing by a factor of two from 16 to 2^{15}. Please note that x and y both have a logarithmic scale.

This scenario is even more blurred if we look at the measures taken for the crossover operator in Fig. 2. The fastest overall are Go and, once again, Java, but the slowest is another compiled language, C++, at least for sizes bigger than 64. The scaling of this performance is similar for C, which becomes the second

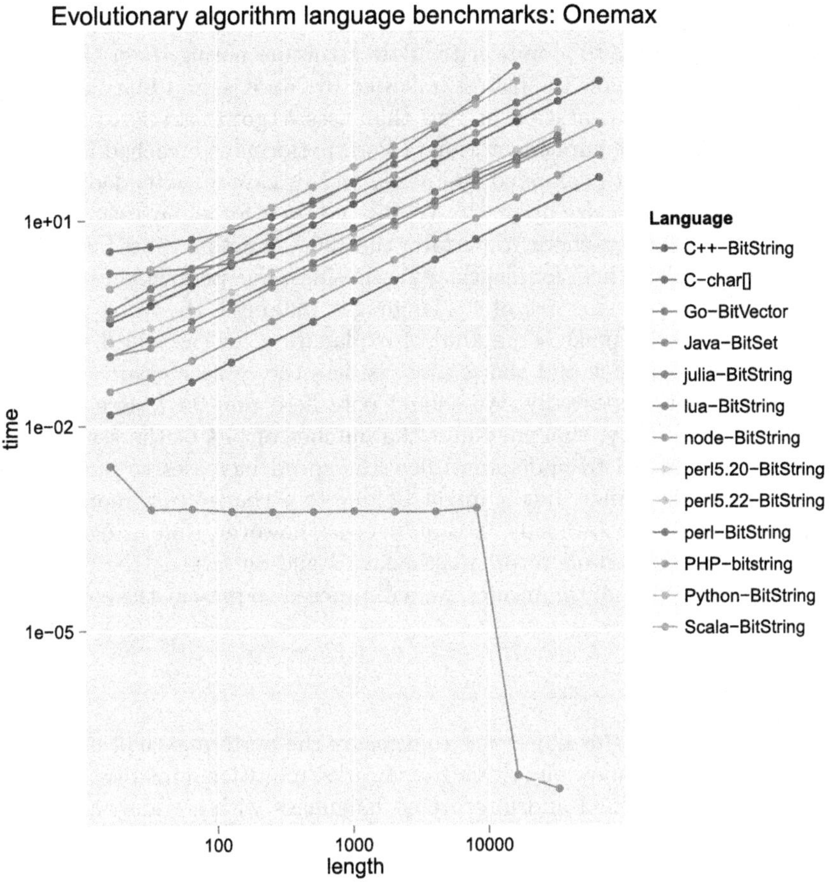

Evolutionary algorithm language benchmarks: Onemax

Fig. 3. Plot of time needed to perform $100\,\mathrm{K}$ OneMax evaluations in strings with lengths increasing by a factor of two from 16 to 2^{15}. Please note that x and y both have a logarithmic scale.

slowest for sizes bigger than 1024. The Node.js JavaScript interpreter performs quite consistently and independently of the size, but Perl and PHP are also quite fast for a significant portion of the size range.

Python, on the other hand, shows in this occasion a behavior that degrades for the high end of the size range, becoming slower than Lua or any Perl implementation.

In this case, the underlying operation is string or array copy. Languages such as Node.js, Perl or PHP whose main focus is in data processing are optimized for that kind of thing. System-oriented languages such as C or C++, on the other hand, are not so that they fare much worse in this kind of operation.

If we proceed to the last benchmark, the OneMax function, which is shown in Fig. 3 we see that the behaviour for all languages except Java is essentially the same, scaling linearly with the chromosome size due to the fact that the only way of counting ones is to loop over the data structure noting them and adding one to a counter. C is now the fastest, followed by Node.js, and Lua the slowest, followed by the implementation of Perl that uses `Algorithm::Evolutionary`. However, in the *Simple* implementation of Perl, performance reached is on a par with Scala and Go, both compiled languages. In this case we extended, for some languages, chromosome size up to 2^{16}. We did not do it for all languages, since it took them more than one hour to perform the 100 K function calls. Lua was not even able to complete them for length $= 2^{14}$. Besides, Lua performance increased with length faster than the rest of the languages, making it the worst of the set.

We should maybe make some kind of explanation for the speed shown for Java, which is consistent and shows more or less the same measures when we run the benchmark repeatedly. We should note here that Java uses a built-in primitive, `cardinality`, that measures the number of bits of the set that are, effectively, set. We fail to understand how the speed increases so much in the high end of the size range, but it might be due to a change to a more efficient implementation on the Java side. It is quite clear, however, that using this data structure and associated operators makes Java stand out among the rest.

After this overview of the results, we will proceed to present the conclusions.

5   Conclusions

Our main intention in this paper was to measure the performance of usual functions found in evolutionary algorithm frameworks: mutation, crossover and One-Max, for several compiled and interpreted languages with a wide range of use and popularity within the evolutionary computation and scientific community and programmers at large.

In general, performance results do not place interpreted and compiled languages in different categories. The performance of the implementation of a whole algorithm will depend on the times a particular function is called, and of course the time it takes to evaluate the fitness function, which is usually the bottleneck in any evolutionary algorithm. However, for these atomic functions interpreted and compiled languages go on a par, with no category emerging as a clear winner across all three functions tested. However, Java is almost always among the fastest across all three functions, together with C or Go. And if there is one loser, we could say it is Lua, whose performance for OneMax and big sizes is quite disappointing. However, Lua might be the only option for in-game or in-web server evolution, the same as Perl, which also acts as the slowest in some benchmarks and sizes, might be a good option to interface with databases or as a web server back-end.

Go has also emerged as a very interesting contender in the arena, with very fast performance for all functions and a graceful degradation with size, even if the implementation is quite novice and can probably be improved with more experience in that area. Scala, in that sense, is also a fast alternative to the more mainstream Java. On the other hand, we do not think that Julia will ever become popular, since its performance is never the worst, but not very good either.

Among interpreted languages, Python and the Node.js implementation of JavaScript offer the best results, although the difference to Perl is not too big in some benchmarks; besides, PHP, largely not considered a general-purpose language, is indeed the fastest in some of the tests.

All these tests lead us to the conclusion that there is no type of language that is superior to all others across all sizes and problems considered, although if you use Java you will be close to the best or reach the best performance; the performance of other compiled languages will vary wildly across languages and functions. There is no free lunch also in the implementation in evolutionary algorithms, and the fact that heterogeneous, asynchronous distributed architectures are now possible leads us to propose them as an alternative to the single-language frameworks that are the most usual nowadays.

Future lines of work might include a more extensive measurement of other operators such as tournament selection and other selection algorithms. A priori, they are essentially CPU integer operations and their behavior might be, in principle, very similar to the one shown in these operations. This remains to be proved, however, but it is left as future line of work.

It would also be interesting to mix and match different languages, choosing every one for its performance, in a hybrid architecture. Communication might have some overhead, but it might be offset by performance. Combining some compiled languages such as Go or C with others characterized by its speed in some string operations, like Perl or programming ease, might result in the best of both worlds: performance and rapid prototyping. Creating a whole multi-language framework along these lines is a challenge that might be interesting in the future.

Focusing only in the part of measuring algorithms and in the interest of reproductibility, we intend to create a Docker container with all the tests so that it can be downloaded and run in any machine, checking for differences in different underlying architectures.

Acknowledgements. This paper is part of the open science effort at the university of Granada. It has been written using `knitr`, and its source as well as the data used to create it can be downloaded from the GitHub repository https://github.com/geneura-papers/2015-ea-languages. It has been supported in part by GeNeura Team.

This work has been supported in part by projects TIN2014-56494-C4-3-P (Spanish Ministry of Economy and Competitiveness), SPIP2014-01437 (Dirección General de Tráfico), PRY142/14 (Fundación Pública Andaluza Centro de Estudios Andaluces en la IX Convocatoria de Proyectos de Investigación), and project V17-2015 of the Microprojects program 2015 from CEI BioTIC Granada.

References

1. Anderson, E., Tucek, J.: Efficiency matters!. ACM SIGOPS Oper. Syst. Rev. **44**(1), 40–45 (2010)
2. Prechelt, L.: An empirical comparison of seven programming languages. Computer **33**(10), 23–29 (2000)
3. Fulgham, B., Gouy, I.: The computer language benchmarks game (2010)
4. Wolpert, D.H., Macready, W.G.: No free lunch theorems for optimization. IEEE Trans. Evol. Comput. **1**(1), 67–82 (1997)
5. Langdon, W.B., Harrison, A.P.: Evolving regular expressions for genechip probe performance prediction. In: Rudolph, G., Jansen, T., Lucas, S., Poloni, C., Beume, N. (eds.) PPSN 2008. LNCS, vol. 5199, pp. 1061–1070. Springer, Heidelberg (2008)
6. Merelo, J.J., Romero, G., Arenas, M.G., Castillo, P.A., Mora, A.M., Laredo, J.L.J.: Implementation matters: programming best practices for evolutionary algorithms. In: Cabestany, J., Rojas, I., Joya, G. (eds.) IWANN 2011, Part II. LNCS, vol. 6692, pp. 333–340. Springer, Heidelberg (2011)
7. Grosan, C., Abraham, A.: Evolutionary algorithms. In: Grosan, C., Abraham, A. (eds.) Intelligent Systems. ISRL, vol. 17, pp. 345–386. Springer, Heidelberg (2011)
8. Merelo-Guervós, J.J., Castillo, P.A., Alba, E.: Algorithm: Evolutionary, a flexible Perl module for evolutionary computation. Soft Comput. **14**(10), 1091–1109 (2010)
9. Lee, W., Kim, H.Y.: Genetic algorithm implementation in Python. In: Fourth Annual ACIS International Conference on Computer and Information Science, 2005, pp. 8–11 (2005)
10. Erb, B., Kargl, F.: A conceptual model for event-sourced graph computing. In: Proceedings of the 9th ACM International Conference on Distributed Event-Based Systems. DEBS 2015, pp. 352–355. ACM, New York (2015)
11. García-Sánchez, P., González, J., Castillo, P.A., Merelo, J.J., Mora, A.M., Laredo, J.L.J., Arenas, M.G.: A distributed service oriented framework for metaheuristics using a public standard. In: González, J.R., Pelta, D.A., Cruz, C., Terrazas, G., Krasnogor, N. (eds.) NICSO 2010. SCI, vol. 284, pp. 211–222. Springer, Heidelberg (2010)
12. García-Sánchez, P., González, J., Castillo, P.A., García-Arenas, M., Merelo-Guervós, J.J.: Service oriented evolutionary algorithms. Soft. Comput. **17**(6), 1059–1075 (2013)
13. Odersky, M., Altherr, P., Cremet, V., Emir, B., Maneth, S., Micheloud, S., Mihaylov, N., Schinz, M., Stenman, E., Zenger, M.: An overview of the Scala programming language. Technical report, EPFL-Lausanne (2004)
14. Ventura, S., Romero, C., Zafra, A., Delgado, J.A., Hervás, C.: JCLEC: a java framework for evolutionary computation. Soft. Comput. **12**(4), 381–392 (2008)
15. Luke, S., Panait, L., Balan, G., Paus, S., Skolicki, Z., Bassett, J., Hubley, R., Chircop, A.: ECJ: a Java-based evolutionary computation research system. Downloadable versions and documentation can be found at the following (2006). http://cs.gmu.edu/eclab/projects/ecj
16. Liefooghe, A., Jourdan, L., Talbi, E.G.: A software framework based on a conceptual unified model for evolutionary multiobjective optimization: ParadisEO-MOEO. Eur. J. Oper. Res. **209**(2), 104–112 (2011)
17. Alba, E., Almeida, F., Blesa, M.J., Cabeza, J., Cotta, C., Díaz, M., Dorta, I., Gabarró, J., León, C., Luna, J.M., Moreno, L., Pablos, C., Petit, J., Rojas, A., Xhafa, F.: MALLBA: a library of skeletons for combinatorial optimisation. In: Monien, B., Feldmann, R.L. (eds.) Euro-Par 2002. LNCS, vol. 2400, pp. 927–932. Springer, Heidelberg (2002)

18. Pike, R.: The go programming language. Talk given at Google's Tech Talks (2009)
19. Ierusalimschy, R., De Figueiredo, L.H., Celes Filho, W.: Lua-an extensible extension language. Softw. Pract. Exper. **26**(6), 635–652 (1996)
20. Merelo, J.J., Castillo, P.A., Mora, A., Fernández-Ares, A., Esparcia-Alcázar, A.I., Cotta, C., Rico, N.: Studying and tackling noisy fitness in evolutionary design of game characters. In: Rosa, A., Merelo, J.J., Filipe, J., eds.: ECTA 2014 - Proceedings of the International Conference on Evolutionary Computation Theory and Applications, pp. 76–85 (2014)
21. Merelo-Guervós, J.J., Mora, A.M., Castillo, P.A., Cotta, C., García-Valdez, M.: A search for scalable evolutionary solutions to the game of MasterMind. In: IEEE Congress on Evolutionary Computation, pp. 2298–2305. IEEE (2013)
22. Merelo-Guervós, J.J., Castillo-Valdivieso, P.A., Mora-García, A., Esparcia-Alcázar, A., Rivas-Santos, V.M.: NodEO, a multi-paradigm distributed evolutionary algorithm platform in JavaScript. In: Genetic and Evolutionary Computation Conference, GECCO 2014, Vancouver, BC, Canada, July 12–16, 2014, pp. 1155–1162. ACM, Companion Material Proceedings (2014)
23. Merelo-Guervós, J.J., García-Sánchez, P.: Modeling browser-based distributed evolutionary computation systems. CoRR abs/1503.06424 (2015)
24. Rivas, V.M., Guervós, J.J.M., López, G.R., Arenas-García, M., Mora, A.M.: An object-oriented library in javascriptto build modular and flexiblecross-platform evolutionary algorithms. In: Esparcia-Alcázar, A.I., Mora, A.M. (eds.) EvoApplications 2014. LNCS, vol. 8602, pp. 853–862. Springer, Heidelberg (2014)
25. Fortin, F.A., Rainville, D., Gardner, M.A.G., Parizeau, M., Gagné, C., et al.: DEAP: Evolutionary algorithms made easy. J. Mach. Learn. Res. **13**(1), 2171–2175 (2012)
26. Bezanson, J., Karpinski, S., Shah, V.B., Edelman, A.: Julia: A fast dynamic language for technical computing (2012). arXiv preprint arXiv:1209.5145

On the Closest Averaged Hausdorff Archive
for a Circularly Convex Pareto Front

Günter Rudolph[1](✉), Oliver Schütze[2], and Heike Trautmann[3]

[1] Department of Computer Science, TU Dortmund University, Dortmund, Germany
guenter.rudolph@tu-dortmund.de
[2] Department of Computer Science, CINVESTAV, Mexico City, Mexico
schuetze@cs.cinvestav.mx
[3] Department of Information Systems, University of Münster, Münster, Germany
trautmann@uni-muenster.de

Abstract. The averaged Hausdorff distance has been proposed as an indicator for assessing the quality of finitely sized approximations of the Pareto front of a multiobjective problem. Since many set-based, iterative optimization algorithms store their currently best approximation in an internal archive these approximations are also termed archives. In case of two objectives and continuous variables it is known that the best approximations in terms of averaged Hausdorff distance are subsets of the Pareto front if it is concave. If it is linear or circularly concave the points of the best approximation are equally spaced.

Here, it is proven that the optimal averaged Hausdorff approximation and the Pareto front have an empty intersection if the Pareto front is circularly convex. But the points of the best approximation are equally spaced and they rapidly approach the Pareto front for increasing size of the approximation.

Keywords: Multi-objective optimization · Averaged hausdorff distance · Convex front · Optimal archives

1 Introduction

The goal in the *a posteriori* approach of multiobjective optimization is a finitely sized approximation of the Pareto front which itself is innumerable for continuous problems in general. The quality of the approximation is typically measured by scalar-valued quality indicators which are largely based on some notion of distance between approximation set and Pareto front. Popular indicators in the field of multiobjective evolutionary algorithms are the dominated hypervolume [1], generational distance [2], (inverted) generational distance [3], the ε-indicator [4], the R2-indicator [5] and others. Besides measuring the quality of an approximation, these indicators can also be used as a selection operator that drives a set-based optimization algorithm like an evolutionary algorithm towards the Pareto front (see e.g. [6–8]).

When used as quality indicator for assessing the approximation found by some set-based algorithm the indicator implicitly determines the characteristics of optimal approximations (like the distribution of solutions on the Pareto front). Since

© Springer International Publishing Switzerland 2016
G. Squillero and P. Burelli (Eds.): EvoApplications 2016, Part II, LNCS 9598, pp. 42–55, 2016.
DOI: 10.1007/978-3-319-31153-1_4

the Pareto front is typically known analytically in case of constructed benchmark problems, it is often possible to calculate the optimal approximation (or optimal archive of points) for the given indicator. As a consequence, it is then possible to assess the quality of the approximation during the optimization run by the difference between optimal approximation and current solution of the evolutionary algorithm or other set-based metaheuristics. For example, this has been achieved for the dominated hypervolume indicator in case of two objectives [9].

Here, we consider optimal approximations/archives for the recently proposed Δ_p-indicator which is based on the *averaged Hausdorff distance* between two sets [10]. In case of two objectives it is known that the optimal archives are subsets of the Pareto front if it is concave (which includes linear fronts) [11]. In addition, optimal archives and Δ_p-indicator values have been determined for linear and circularly concave Pareto fronts. The appealing feature of optimal Δ_p-archives is the uniform spacing between the archive points. Moreover, the Δ_p-indicator can be efficiently used for expert knowledge based multiobjective optimization using a specific archive technique approximating a predefined aspiration set [12].

But numerical experiments have revealed that the optimal Δ_p-archive for *convex* Pareto fronts is not a subset of the Pareto front which questioned the deployment of the Δ_p-indicator for quality assessments of multiobjective evolutionary algorithms in case of (at least piecewise) continuous Pareto fronts. Actually, we prove in case of a circularly convex Pareto front that the optimal Δ_p-archive and the Pareto front have an empty intersection. This seemingly disappointing result, however, does not discredit the Δ_p-indicator as a measuring device in benchmarks since we prove that the Euclidean distance of each archive point to the Pareto front decreases rapidly for increasing archive size. As a consequence, already for moderately sized archives the deviation from the Pareto front is irrelevant from a measuring point of view.

In Sect. 2 we present some mathematical results that exempt the proofs in subsequent sections from disturbing excursions. Section 3 introduces the averaged Hausdorff inframetric and its properties. The main results regarding optimal Hausdorff archives for circularly convex fronts can be found on Sect. 4. Finally, we draw conclusions in Sect. 5.

2   Mathematical Preliminaries

The results presented in this section will be helpful in Sect. 4. Neither of these results is new; the proofs are supplied only for making this work more self-contained.

A square matrix A of size $n \times n$ is termed positive semidefinite (p.s.d.) if $x'Ax \geq 0$ for all $x \in \mathbb{R}^n$ and positive definite (p.d.) if $x'Ax > 0$ for all $x \in \mathbb{R}^n \setminus \{0\}$. Here, x' denotes the transpose of vector x.

Lemma 1. *If A is a p.d. and B a p.s.d. $n \times n$-matrix then $A + B$ is p.d.*

Proof.
$$\forall x \in \mathbb{R}^n \setminus \{0\} : x'(A + B)x = \underbrace{x'Ax}_{>0} + \underbrace{x'Bx}_{\geq 0} > 0.$$

$\square$

A function $f : D \rightarrow \mathbb{R}$ with convex domain $D \subseteq \mathbb{R}^n$ is said to be convex, if $f(\gamma\, x + (1 - \gamma)\, y) \leq \gamma\, f(x) + (1 - \gamma)\, f(y)$ for all $x, y \in D$ and $\gamma \in [0, 1]$. If the inequality is strict the function is termed strictly convex.

Lemma 2. *Let $f : D \rightarrow \mathbb{R}$ be an additively decomposable function with $D = D_1 \times D_2$ and $f(x) = f_1(x_1) + f_2(x_2)$ where the sub-functions $f_1 : D_1 \rightarrow \mathbb{R}$ and $f_2 : D_2 \rightarrow \mathbb{R}$ are convex. Then:*

(a) $f : D \rightarrow \mathbb{R}$ is convex.
(b) $f : D \rightarrow \mathbb{R}$ is strictly convex if at least one sub-function is strictly convex.

Proof.

(a) Let $\gamma \in [0, 1]$ and $x, y \in D$. According to the definition of convexity we get

$$\begin{aligned}
f(\gamma\, x + (1 - \gamma)\, y) &= f_1(\gamma\, x_1 + (1 - \gamma)\, y_1) + f_2(\gamma\, x_2 + (1 - \gamma)\, y_2) \\
&\leq \gamma\, f_1(x_1) + (1 - \gamma)\, f_1(y_1) + \gamma\, f_2(x_2) + (1 - \gamma)\, f_2(y_2) \quad (1) \\
&= \gamma\, (f_1(x_1) + f_2(x_2)) + (1 - \gamma)\, (f_1(y_1) + f_2(y_2)) \\
&= \gamma\, f(x) + (1 - \gamma)\, f(y).
\end{aligned}$$

(b) Inequality (1) is strict if at least one sub-function is strictly convex.

$\square$

Optima of a differentiable function may be determined by inspection of its gradient and Hessian matrix. Sometimes a monotone transformation of the function can make the analysis easier.

Lemma 3. *Let $f : \mathbb{R}^n \rightarrow \mathbb{R}$ and $g : \mathbb{R} \rightarrow \mathbb{R}$ be continuously differentiable. Then:*

(a) $\{x \in \mathbb{R}^n : \nabla f(x) = 0\} \subseteq \{x \in \mathbb{R}^n : \nabla g(f(x)) = 0\}$.
(b) $\{x \in \mathbb{R}^n : \nabla f(x) = 0\} = \{x \in \mathbb{R}^n : \nabla g(f(x)) = 0\}$ if $g(\cdot)$ strictly monotone.
(c) If $g(\cdot)$ is twice differentiable and strictly monotone increasing on the range of $f(\cdot)$ then the location of the local minima of $g(f(x))$ are identical to those of $f(x)$.

Proof.

(a) Since $\nabla g(f(x)) = \nabla f(x) \cdot g'(f(x))$ the left hand side of the equation is zero if $\nabla f(x) = 0$ or the derivative of $g(\cdot)$ has a root in the range of $f(\cdot)$.
(b) Since $g(y)$ is strictly monotone we have either $g'(y) > 0$ or $g'(y) < 0$ but never $g'(y) = 0$ for all $y \in f(\mathbb{R}^n)$. Therefore $\nabla g(f(x))$ is zero if and only if $\nabla f(x) = 0$.
(c) Strictly increasing monotonicity implies $g'(y) > 0$ for all $y \in f(\mathbb{R}^n)$. Owing to part (b) of this Lemma it follows that the set of candidates for local optima are identical. The Hessian matrix of $g(f(x))$ is

$$\begin{aligned}
\nabla^2 g(f(x)) = \nabla(\nabla g(f(x))) &= \nabla(g'(f(x)) \cdot \nabla f(x)) = \\
g'(f(x)) \cdot \nabla^2 f(x) &+ \underbrace{g''(f(x)) \cdot \nabla f(x) \cdot \nabla f(x)^T}_{\text{zero matrix if } x = x^*} . \quad (2)
\end{aligned}$$

Insertion of a candidate solution x^* in the Hessian (2) leads to

$$\nabla^2 g(f(x^*)) = \underbrace{g'(f(x^*))}_{>0} \cdot \nabla^2 f(x^*)$$

revealing that the Hessian matrix of $f(x)$ is p.d. in x^* if and only if $\nabla^2 g(f(x^*))$ is p.d.   □

Finally, we recall the trigonometric identities

$$\left. \begin{array}{rcl} \sin(\alpha + \delta) - \sin(\alpha + \delta) &=& 2 \cos\alpha \sin\delta \\ \sin(\alpha + \delta) + \sin(\alpha + \delta) &=& 2 \sin\alpha \cos\delta \\ \cos(\alpha + \delta) - \cos(\alpha + \delta) &=& -2 \sin\alpha \sin\delta \\ \cos(\alpha + \delta) + \cos(\alpha + \delta) &=& 2 \cos\alpha \cos\delta \end{array} \right\} \quad (3)$$

which follow immediately from entries 4.3.34 to 4.3.37 in [13] whereas

$$\sin(\arctan x) = \frac{x}{\sqrt{1 + x^2}} \quad \text{and} \quad \cos(\arctan x) = \frac{1}{\sqrt{1 + x^2}} \quad (4)$$

can be extracted from entry 4.3.45 in [13].

3   Averaged Hausdorff Distance

The Hausdorff distance is a well known distance measure between two sets. Actually, it can be shown that it is a metric on sets.

Definition 1. *The value* $d_H(A, B) := \max(d(A, B), d(B, A))$ *is termed the Hausdorff distance between two sets* $A, B \subset \mathbb{R}^k$, *where*

$$d(B, A) := \sup\{d(u, A) : u \in B\} \quad and \quad d(u, A) := \inf\{\|u - v\| : v \in A\}$$

for a vector norm $\|\cdot\|$.

However, the value of the Hausdorff distance is strongly affected by single outliers which may lead to counter-intuitive assessments of closeness between sets [10]. Therefore, the indicators GD_p and IGD_p have been introduced in [10] to construct a new distance measure between finite sets that shares some properties with the Hausdorff distance but which is less prone to outliers.

Definition 2. *The value* $\Delta_p(A, B) = \max(\mathsf{GD}_p(A, B), \mathsf{IGD}_p(A, B))$ *with*

$$\mathsf{GD}_p(A, B) = \left(\frac{1}{|A|} \sum_{a \in A} d(a, B)^p \right)^{1/p} \quad and \quad \mathsf{IGD}_p(A, B) = \left(\frac{1}{|B|} \sum_{b \in B} d(b, A)^p \right)^{1/p}$$

for $p > 0$ *is termed the* averaged Hausdorff distance *between sets* A *and* B *as given in* Definition 1.

We note that $\Delta_p(\cdot, \cdot)$ is not a metric but fulfills all axioms of a distance.

Definition 3. (see [14, p. 1])
A function $d : X \times X \to \mathbb{R}$ is termed a distance *on set X if*

  1. $d(x, y) \geq 0$ *(nonnegativity)*
  2. $d(x, y) = d(y, x)$ *(symmetry)*
  3. $d(x, x) = 0$ *(reflexivity)*

for all $x, y \in X$.   □

The averaged Hausdorff distance is not a metric (but a *nearmetric*) only since the triangle property is not fully valid.

Definition 4. (see [14, p. 7])
A function $d : X \times X \to \mathbb{R}$ is termed a C-nearmetric *on set X if*

  1. *it is a distance on X*
  2. $d(x, y) > 0$ *for $x \neq y$*
  3. $d(x, y) \leq C\left(d(x, z) + d(z, y)\right)$

for all $x, y, z \in X$ and some $C \geq 1$.   □

Actually, it can be proven [10] that Δ_p is a $N^{1/p}$-nearmetric where N is the maximum cardinality of the sets.

The purpose of the averaged Hausdorff distance is to assess the quality of a finitely sized approximation Y of the Pareto front F^* by observing the value $\Delta_p(Y, F^*)$. In the context of the definitions in [10] the Pareto front F^* is assumed to be discretized appropriately.

Here, we consider only bi-objective problems and assume that the Pareto front is continuous and expressible in parametric form. Thus, let $F^* = \{\varphi(\omega) : \omega \in [0, \pi/2]\} \subset \mathbb{R}^2$ be the Pareto front and $Y \subset \mathbb{R}^2$ with $|Y| = m < \infty$ the approximation.

Two examples are the circularly concave Pareto front defined by $\varphi(\omega) = (\sin\omega, \cos\omega)'$ and the circularly convex Pareto front defined by $\varphi(\omega) = (1 - \cos\omega, 1 - \sin\omega)'$. An illustration of these Pareto fronts is given in Fig. 1.

Since here we deal with continuous Pareto fronts F^*, we have to adapt Δ_p to this context: let $Y = \{y_1, \ldots, y_m\} \subset f(X)$ where $y_1, \ldots, y_m$ are arranged in lexicographic order. We obtain

$$\mathsf{GD}_p(Y, F^*) = \left[\frac{1}{|Y|} \sum_{y \in Y} d(y, F^*)^p\right]^{\frac{1}{p}} \tag{5}$$

and

$$\mathsf{IGD}_p(Y, F^*) = \left[\frac{1}{L} \int_a^b d(\varphi(s), Y)^p \, ds\right]^{\frac{1}{p}} \tag{6}$$

where L is the length of the curve described by F^*. In case of the two examples we have of course $L = \pi/2$.

In [11] the optimal Δ_p-archive has been determined for the circularly concave front. Here, we consider the determination of the optimal Δ_p-archive in case of the circularly convex front.

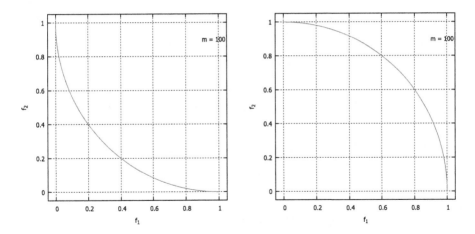

Fig. 1. Left: circularly convex Pareto front. Right: circularly concave Pareto front.

4   Construction of Optimal Archive

Since the averaged Hausdorff distance Δ_p between finite archive and innumerable Pareto front is the maximum of GD_p and IGD_p between the two sets we can find the optimal archive as follows: we know that solutions with minimum GD_p are archives with all members on the Pareto front which results in $GD_p = 0$ so that the corresponding Δ_p value is just the IGD_p value for the optimal GD_p archive. Next, find solutions with minimum IGD_p. If such a solution has a larger IGD_p value than the corresponding GD_p value then this solution is the minimal Δ_p archive. If this inequality is not valid then this approach fails.

Presume that this approach does work. First, we need a method to calculate the GD_p value for an arbitrary given archive. This method is developed in Sect. 4.1. Second, we need a method to determine an archive with minimal IGD_p value. We demonstrate the method in case of a circularly convex Pareto front in Sect. 4.2. Finally, a comparison with the corresponding GD_p value in Sect. 4.3 reveals that this solution is the optimal Δ_p archive.

4.1   Averaged Generational Distance (GD_p)

In the remainder of this work we assume that the vector norm used in Definition 1 is the Euclidean norm and that $\varphi(\cdot)$ is continuously differentiable.

Theorem 1. *Let* $F^* = \{\varphi(\omega) : \omega \in [0, \pi]\} \subset \mathbb{R}^2$ *and* $Y \subset \mathbb{R}^2$ *with* $|Y| = m < \infty$. *Then*

$$GD_p(Y, F^*) = GD_p(\{y^{(1)}, \ldots, y^{(m)}\}, F^*) = \left[\frac{1}{m} \sum_{i=1}^{m} d(\varphi(\omega_i^*), y^{(i)})^p \right]^{1/p}$$

where ω_i^* *is an appropriate solution of*

$$\frac{\partial}{\partial \omega_i} d(\varphi(\omega_i), y^{(i)}) \overset{!}{=} 0$$

for $i = 1, \ldots, m$ regardless of $p \geq 1$.

Proof. Suppose that $Y \cap F^* = \emptyset$ since elements of Y that are on F^* do not contribute to the value of GD_p. Let

$$g(\omega) = g(\omega_1, \ldots, \omega_m) = \frac{1}{m} \sum_{i=1}^{m} d(\varphi(\omega_i), y^{(i)})^p.$$

Partial derivation leads to

$$\frac{\partial}{\partial \omega_i} g(\omega)^{1/p} = \left[\frac{\partial}{\partial \omega_i} g(\omega) \right] \cdot \underbrace{\frac{1}{p} \cdot g(\omega)}_{> 0} \stackrel{!}{=} 0 \qquad (7)$$

for $i = 1, \ldots, m$. Therefore, it suffices to look at

$$\frac{\partial}{\partial \omega_i} g(\omega) = \frac{1}{m} \cdot \frac{\partial}{\partial \omega_i} d(\varphi(\omega_i), y^{(i)})^p =$$
$$\frac{1}{m} \left[\frac{\partial}{\partial \omega_i} d(\varphi(\omega_i), y^{(i)}) \right] \cdot \underbrace{p \cdot d(\varphi(\omega_i), y^{(i)})^{p-1}}_{> 0} \stackrel{!}{=} 0 \qquad (8)$$

for $i = 1, \ldots, m$. Thus, it is sufficient to solve

$$\frac{\partial}{\partial \omega_i} d(\varphi(\omega_i), y^{(i)}) \stackrel{!}{=} 0$$

independently for each $i = 1, \ldots, m$ and regardless of $p \geq 1$ to find the candidates for optimal angles ω_i^*. It remains to show that these candidates lead to a minimum, i.e., that the Hessian matrix of GD_p is positive definite (p.d.). The second partial derivatives of GD_p are obtained by partial derivation of (7) yielding

$$\frac{\partial}{\partial \omega_j} \left[\frac{\partial}{\partial \omega_i} g(\omega) \right] \cdot \frac{1}{p} \cdot g(\omega) = \frac{1}{p} \left(\frac{\partial^2 g(\omega)}{\partial \omega_i \, \partial \omega_j} g(\omega) + \frac{\partial g(w)}{\partial \omega_i} \cdot \frac{\partial g(w)}{\partial \omega_j} \right)$$

which can be expressed in matrix form as

$$p \cdot \nabla^2 \mathsf{GD}_p(\omega) = \underbrace{g(\omega) \cdot \nabla^2 g(\omega)}_{> 0} + \underbrace{\nabla g(\omega) \, \nabla g(\omega)^T}_{\text{p.s.d.}}.$$

The Hessian matrix of GD_p is p.d. if $\nabla^2 g(\omega)$ is p.d. (Lemma 1). Partial derivation of the first partial derivatives of $g(\omega)$ given in (8) w.r.t. ω_j yields

$$\frac{\partial^2 g(\omega)}{\partial \omega_i \partial \omega_j} = \frac{p}{m} \cdot \underbrace{\frac{\partial^2 d(\varphi(\omega_i), y^{(i)})}{\partial \omega_i \partial \omega_j}}_{= 0 \text{ if } i \neq j} \cdot d(\varphi(\omega_i), y^{(i)})^{p-1}$$

$$+ \; \frac{p}{m} \cdot \frac{\partial d(\varphi(\omega_i), y^{(i)})}{\partial \omega_i} \cdot \underbrace{\frac{\partial d(\varphi(\omega_i), y^{(i)})^{p-1}}{\partial \omega_j}}_{= 0 \text{ if } i \neq j}.$$

Thus, $\nabla^2 g(\omega)$ is a diagonal matrix $\text{diag}(d_1, \ldots, d_m)$ that is p.d. if every diagonal entry d_i is positive. Since

$$d_i = \frac{p}{m} \cdot \frac{\partial^2 d(\varphi(\omega_i), y^{(i)})}{\partial \omega_i^2} \cdot \underbrace{d(\varphi(\omega_i), y^{(i)})^{p-1}}_{>0}$$

$$+ \frac{p}{m} \cdot \underbrace{\left(\frac{\partial d(\varphi(\omega_i), y^{(i)})}{\partial \omega_i} \right)^2}_{\geq 0} \cdot \underbrace{(p-1)}_{\geq 0} \cdot \underbrace{d(\varphi(\omega_i), y^{(i)})^{p-2}}_{>0}$$

the sufficient condition reduces to

$$\frac{\partial^2 d(\varphi(\omega_i), y^{(i)})}{\partial \omega_i^2} \overset{!}{>} 0,$$

which is always fulfilled since the Euclidean norm is strictly convex.   □

Suppose the Pareto front is given by $F^* = \{\varphi(\omega) : \omega \in [0, \pi/2]\}$ with $\varphi(\omega) = (1 - \cos \omega, 1 - \sin \omega)^T$ and let $Y \subset [0, 1)^2$ with $|Y| = m < \infty$. Owing to Theorem 1 it suffices to solve

$$\frac{\partial d(\varphi(\omega), y)}{\partial \omega} \overset{!}{=} 0$$

for a single pair (ω, y). Lemma 3(c) asserts that the solution of the squared problem delivers the desired ω^*. Thus,

$$\frac{\partial d(\varphi(\omega), y)^2}{\partial \omega} = \frac{\partial}{\partial \omega}[(1 - \cos \omega - y_1)^2 + (1 - \sin \omega - y_2)^2]$$

$$= 2(1 - \cos \omega - y_1)(- \sin \omega) + 2(1 - \sin \omega - y_2) \cos \omega \overset{!}{=} 0$$

which is equivalent to

$$\frac{1 - y_2}{1 - y_1} \overset{!}{=} \frac{\sin \omega}{\cos \omega} = \tan \omega \quad \Leftrightarrow \quad \arctan\left(\frac{1 - y_2}{1 - y_1} \right) \overset{!}{=} \omega^*.$$

Insertion of ω^* and usage of the trigonometric identities (4) leads to

$$d(\varphi(\omega^*(y)), y) = \left[(1 - y_1)^2 + (1 - y_2)^2 - 2\sqrt{(1 - y_1)^2 + (1 - y_2)^2} + 1 \right]^{\frac{1}{2}} \quad (9)$$

which can be used as building block to calculate GD_p for an arbitrary given set $\{y^{(1)}, \ldots, y^{(m)}\} \subset [0, 1)^2$.

4.2   Averaged Inverted Generational Distance (IGD_p)

We solve the squared problem. At first the integrals are eliminated before we apply partial differentiation and proceed in the standard way to identify the optimal points of set Y analytically. To this end let $0 = \alpha_0 \leq \alpha_1 \leq \ldots \leq \alpha_m = \pi/2$.

$$\frac{\pi}{2} \text{IGD}_2(\{y^{(1)}, \ldots, y^{(m)}\}, F^*)^2 = \sum_{i=1}^{m} \int_{\alpha_{i-1}}^{\alpha_i} d(\varphi(\omega), y^{(i)})^2 \, d\omega$$

$$= \sum_{i=1}^{m} \int_{\alpha_{i-1}}^{\alpha_i} \left[(1 - \cos\omega - y_1^{(i)})^2 + (1 - \sin\omega - y_2^{(i)})^2 \right] d\omega$$

$$= \sum_{i=1}^{m} \int_{\alpha_{i-1}}^{\alpha_i} \left[(1 - y_1^{(i)})^2 + (1 - y_2^{(i)})^2 + 1 - 2(1 - y_1^{(i)})\cos\omega - (1 - y_2^{(i)})\sin\omega \right] d\omega$$

$$= \sum_{i=1}^{m} \left[(z_1^{(i)})^2 + (z_2^{(i)})^2 + 1 \right] (\alpha_i - \alpha_{i-1})$$

$$- 2 \sum_{i=1}^{m} z_1^{(i)} (\sin\alpha_i - \sin\alpha_{i-1})$$

$$+ 2 \sum_{i=1}^{m} z_2^{(i)} (\cos\alpha_i - \cos\alpha_{i-1}) \tag{10}$$

where we temporarily made the variable replacement $z_k^{(i)} = 1 - y_k^{(i)}$ for $k = 1, 2$. Partial differentiation of (10) w.r.t. $z_k^{(i)}$ leads to the first set of necessary conditions with $i = 1, \ldots, m$:

$$\frac{\partial V}{\partial z_1^{(i)}} = 2 z_1^{(i)} (\alpha_i - \alpha_{i-1}) - 2(\sin\alpha_i - \sin\alpha_{i-1}) \overset{!}{=} 0$$

$$\frac{\partial V}{\partial z_2^{(i)}} = 2 z_2^{(i)} (\alpha_i - \alpha_{i-1}) + 2(\cos\alpha_i - \cos\alpha_{i-1}) \overset{!}{=} 0$$

which are equivalent to

$$z_1^{(i)} \overset{!}{=} \frac{\sin\alpha_i - \sin\alpha_{i-1}}{\alpha_i - \alpha_{i-1}} \quad \text{and} \quad z_2^{(i)} \overset{!}{=} -\frac{\cos\alpha_i - \cos\alpha_{i-1}}{\alpha_i - \alpha_{i-1}}. \tag{11}$$

Thus, conditions (11) tell us how to choose set Y for any given partitioning of F^* with switch points $\varphi(\alpha_1), \ldots, \varphi(\alpha_{m-1})$. If $m = 1$ there is no switch point since $\alpha_0 = 0$ and $\alpha_1 = \pi/2$. Therefore

$$y_1^{(1)} = 1 - z_1^{(1)} = 1 - \frac{2}{\pi} \quad \text{and} \quad y_2^{(1)} = 1 - z_2^{(1)} = 1 - \frac{2}{\pi}$$

with optimal

$$\mathsf{IGD}_2(\{y^{(1)}\}, F^*) = \sqrt{1 - \frac{8}{\pi^2}} \approx 0.43524$$

and associated

$$\mathsf{GD}_2(\{y^{(1)}\}, F^*) = 1 - \frac{\sqrt{8}}{\pi} \approx 0.09968$$

after insertion in (10) and (9), respectively. For $m > 1$ we also need to consider the optimality conditions for the angles. Partial differentiation of (10) w.r.t. α_i leads to the second set of necessary conditions

$$
\frac{\partial V}{\partial \alpha_i} = (z_1^{(i)})^2 + (z_2^{(i)})^2 - z_1^{(i)} \cos \alpha_i - z_2^{(i)} \sin \alpha_i
$$
$$
- \left[(z_1^{(i+1)})^2 + (z_2^{(i+1)})^2 - z_1^{(i+1)} \cos \alpha_i - z_2^{(i+1)} \sin \alpha_i \right] \overset{!}{=} 0 \quad (12)
$$

where $i = 1, \ldots, m - 1$. After insertion of (11) in (12) we obtain

$$
2 \cos \alpha_i \left[\frac{\cos \alpha_{i+1}}{(\alpha_{i+1} - \alpha_i)^2} - \frac{\cos \alpha_{i-1}}{(\alpha_i - \alpha_{i-1})^2} \right] + 2 \sin \alpha_i \left[\frac{\sin \alpha_{i+1}}{(\alpha_{i+1} - \alpha_i)^2} - \frac{\sin \alpha_{i-1}}{(\alpha_i - \alpha_{i-1})^2} \right]
$$
$$
+ \cos \alpha_i \left[\frac{\sin \alpha_{i+1}}{\alpha_{i+1} - \alpha_i} + \frac{\sin \alpha_{i-1}}{\alpha_i - \alpha_{i-1}} \right] - \sin \alpha_i \left[\frac{\cos \alpha_{i+1}}{\alpha_{i+1} - \alpha_i} + \frac{\cos \alpha_{i-1}}{\alpha_i - \alpha_{i-1}} \right] \overset{!}{=} 0. \quad (13)
$$

If $m = 2$ conditions (13) reduce to a single equation with a single variable:

$$
\frac{2\,(1 - \cos \alpha_1)}{\alpha_1^2} - \frac{2\,(1 - \sin \alpha_1)}{(\pi/2 - \alpha_1)^2} - \frac{2 \sin \alpha_1}{\alpha_1} + \frac{2 \cos \alpha_1}{\pi/2 - \alpha_1} \overset{!}{=} 0. \quad (14)
$$

Evidently, the choice of $\alpha_1^* = \pi/4$ solves the equation. Insertion of $\alpha_1^* = \pi/4$ in the second derivative (i.e., the derivative of (14) w.r.t. α_1) yields a positive value (> 0.7) revealing that this choice is at least a *local* minimum. The positions of the corresponding archive points

$$
y^{(1)} = \begin{pmatrix} 1 - \dfrac{\sqrt{8}}{\pi} \\[2mm] 1 - \dfrac{4 - \sqrt{8}}{\pi} \end{pmatrix} \quad \text{and} \quad y^{(2)} = \begin{pmatrix} 1 - \dfrac{4 - \sqrt{8}}{\pi} \\[2mm] 1 - \dfrac{\sqrt{8}}{\pi} \end{pmatrix} \quad (15)
$$

are obtained after insertion of α_1^* in (11). These values lead to minimal

$$
\mathsf{IGD}_2(\{y^{(1)}, y^{(2)}\}, F^*) = \sqrt{1 - \frac{16\,(2 - \sqrt{2})}{\pi^2}} \approx 0.22441
$$

with corresponding

$$
\mathsf{GD}_2(\{y^{(1)}, y^{(2)}\}, F^*) = 1 - \frac{4\sqrt{2 - \sqrt{2}}}{\pi} \approx 0.0255046.
$$

A closer look at the coordinates in (15) discloses a symmetry in the solution. This observation gives rise to the **conjecture** that also solutions in the general case with $m > 2$ should exhibit this kind of symmetry, namely

$$
y_1^{(i)} = y_2^{(m-i+1)}
$$

for $i = 1, \ldots, m$. If the conjecture is true then the difference between consecutive angles must be equal, i.e., $\alpha_i^* - \alpha_{i-1}^* = \delta > 0$ for $i = 1, \ldots, m$ or, equivalently,

$$\alpha_i^* = i \cdot \delta \quad \text{with} \quad \delta = \frac{\pi}{2\,m}$$

for $i = 0, 1, \ldots, m$. As a consequence, $\alpha_{i-1} = \alpha_i - \delta$ and $\alpha_{i+1} = \alpha_i + \delta$. Using this setting of angles and the trigonometric identities in (3) the necessary conditions for the angles (13) are fulfilled leading to optimal archive points

$$y^{(i)} = \left(1 - \frac{\sin(i \cdot \delta) - \sin((i-1) \cdot \delta)}{\delta}, \; 1 + \frac{\cos(i \cdot \delta) - \cos((i-1) \cdot \delta)}{\delta} \right)^{\mathsf{T}}$$

where $\delta = \pi/(2\,m)$. Figure 2 shows optimal archives of size $m = 1, 2, 3, 4, 5$ and 10. As can be seen, the archive points move closer to the Pareto front for increasing cardinality.

Table 1 provides an impression about the minimal IGD_2 and corresponding GD_2 values for increasing archive size $m \in \mathbb{N}$. The results give rise to the educated guess that the IGD_2 value decreases with order $1/m$ whereas the GD_2 value decreases even with order $1/m^2$. A closer inspection provides at least numerical evidence of the relation

$$\mathsf{GD}_2 = 1 - \sqrt{1 - \mathsf{IGD}_2^2}$$

if IGD_2 is the optimal value.

It remains to show that the conjecture of equidistant angles is true. Notice that equidistant angles fulfill the necessary conditions (13) and (11). If there

Table 1. Minimal IGD_2 values and corresponding GD_2 values for increasing archive size $m \in \mathbb{N}$.

m	IGD_2	GD_2
1	4.3524×10^{-1}	9.96837×10^{-2}
2	2.2441×10^{-1}	2.55046×10^{-2}
3	1.5046×10^{-1}	1.13841×10^{-2}
4	1.1307×10^{-2}	6.41315×10^{-3}
5	9.0541×10^{-2}	4.10726×10^{-3}
6	7.5489×10^{-2}	2.85334×10^{-3}
7	6.4724×10^{-2}	2.09681×10^{-3}
8	5.6649×10^{-2}	1.60561×10^{-3}
9	5.0358×10^{-2}	1.26876×10^{-3}
10	4.5326×10^{-2}	1.02777×10^{-3}
100	4.5345×10^{-3}	1.02808×10^{-5}
1,000	4.5345×10^{-4}	1.02808×10^{-7}
10,000	4.5345×10^{-5}	1.02808×10^{-9}

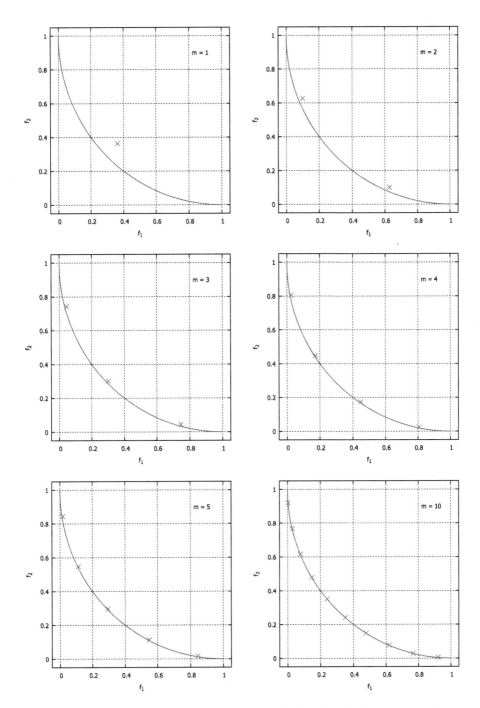

Fig. 2. Optimal archives of size $m = 1, 2, 3, 4, 5, 10$ for the circularly convex Pareto front.

is no other solution that fulfills the necessary conditions then there is no other candidate solution in the interior of $[0, \pi/2]^m$ with possibly better function value. Since (10) can be written as a sum of strictly convex sub-functions

$$f_i(z^{(i)}) = \delta_i \left((z_1^{(i)})^2 + (z_2^{(i)})^2 \right) + a_i \, z_1^{(i)} + b_i \, z_2^{(i)} + \delta_i$$

where $\delta_i := \alpha_i - \alpha_{i-1} > 0$ and $a_i, b_i \in \mathbb{R}$ for $i = 1, \ldots, m$, Lemma 2(b) asserts that (10) is strictly convex. As a consequence the local minimizer is unique and also the global solution.

4.3   Optimal Averaged Hausdorff Archive (Δ_p)

In the introduction to this Sect. 4 it was elaborated that the optimal IGD_p-archive is also the optimal Δ_p-archive if the associated GD_p value for this solution is smaller than the minimal IGD_p value. As seen in the preceding subsection, this relationship is true for archive sizes $m = 1, 2$ and it can be assured numerically exact for larger values of m by the formulas we have proven.

The expression (11) for the coordinates of the optimal archive points reveals that the archive elements will not be elements of the Pareto front regardless of the finite archive size m. Only if $m \to \infty$, which in turn implies $\delta \to 0$, the optimal archive points converge to $(1 - \cos\omega, 1 - \sin\omega)'$ for every $\omega \in (0, \pi/2)$.

5   Conclusions

The observation that the optimal Δ_p-approximation does not lie on the convex part(s) of the Pareto front has questioned the deployment of the Δ_p-indicator for the quality assessment of multiobjective evolutionary algorithms. The theoretical analysis presented here proves for circular convex fronts (1) that this observation is actually not a speculation but a fact and (2) that the deviation from the Pareto front decreases rapidly for increasing size of the approximation / population leading to insignificant inaccuracies already for population size ≥ 100. In summary, our results are not in opposition to the deployment of the Δ_p-indicator for benchmarking multiobjective evolutionary algorithms.

The results presented in this work can be extended in various ways. The 'numerically exact' proof that the IGD_p value for the optimal archive is larger than the associated GD_p value should be replaced by an analytic version. The method may be applied and extended, if necessary, to nonsymmetric convex test functions. Finally, we must admit that the extension of this approach from the 2-dimensional to 3-dimensional case is an open question.

Acknowledgment. Support from CONACYT project no. 207403 and DAAD project no. 57065955 is gratefully acknowledged. Additionally, Heike Trautmann acknowledges support by the European Center of Information Systems (ERCIS).

References

1. Zitzler, E., Thiele, L.: Multiobjective Optimization Using Evolutionary Algorithms - A Comparative Case Study. In: Eiben, A.E., Bäck, T., Schoenauer, M., Schwefel, H.-P. (eds.) PPSN 1998. LNCS, vol. 1498, pp. 292–301. Springer, Heidelberg (1998)
2. van Veldhuizen, D.A.: Multiobjective Evolutionary Algorithms: Classifications, Analyses, and New Innovations. PhD thesis, Department of Electrical and Computer Engineering. Graduate School of Engineering. Air Force Institute of Technology, Wright-Patterson AFB, Ohio, May 1999
3. Bosman, P., Thierens, D.: The balance between proximity and diversity in multiobjective evolutionary algorithms. IEEE Trans. Evol. Comput. **7**(2), 174–188 (2003)
4. Zitzler, E., Thiele, L., Laumanns, M., Fonseca, C., Grunert da Fonseca, V.: Performance assessment of multiobjective optimizers: an analysis and review. IEEE Trans. Evol. Comput. **7**(2), 117–132 (2003)
5. Hansen, M.P., Jaszkiewicz, A.: Evaluating the quality of approximations of the non-dominated set. IMM Technical Report IMM-REP-1998-7, Institute of Mathematical Modeling, Technical University of Denmark, Lyngby, March 1998
6. Beume, N., Naujoks, B., Emmerich, M.: SMS-EMOA: Multiobjective selection based on dominated hypervolume. Eur. J. Oper. Res. **181**(3), 1653–1669 (2007)
7. Gerstl, K., Rudolph, G., Schütze, O., Trautmann, H.: Finding evenly spaced fronts for multiobjective control via averaging Hausdorff-measure. In: Proceedings of 8th International Conference on Electrical Engineering, Computing Science and Automatic Control (CCE), pp. 1–6. IEEE Press (2011)
8. Brockhoff, D., Wagner, T., Trautmann, H.: R2 indicator based multiobjective search. Evol. Comput. **23**(3), 369–395 (2015)
9. Auger, A., Bader, J., Brockhoff, D., Zitzler, E.: Theory of the hypervolume indicator: Optimal μ-distributions and the choice of the reference point. In: Proceedings of the 10th ACM SIGEVO Workshop on Foundations of Genetic Algorithms FOGA 2009, pp. 87–102. ACM, New York (2009)
10. Schütze, O., Esquivel, X., Lara, A., Coello Coello, C.A.: Using the averaged Hausdorff distance as a performance measure in evolutionary multiobjective optimization. IEEE Trans. Evol. Comput. **16**(4), 504–522 (2012)
11. Rudolph, G., Schütze, O., Grimme, C., Domínguez-Medina, C., Trautmann, H.: Optimal averaged Hausdorff archives for bi-objective problems: theoretical and numerical results. Comput. Optim. Appl. (2015). doi:10.1007/s10589-015-9815-8, online 23 December 2015
12. Rudolph, G., Schütze, O., Grimme, C., Trautmann, H.: An aspiration set EMOA based on averaged Hausdorff distances. In: Pardalos, P.M., Resende, M.G.C., Vogiatzis, C., Walteros, J.L. (eds.) LION 8. LNCS, vol. 8246, pp. 153–156. Springer, Heidelberg (2014)
13. Abramowitz, M., Stegun, I. (eds.): Handbook of Mathematical Formulas. National Bureau of Standards, Washington (DC) (1972). 10th printing
14. Deza, M., Deza, E.: Encyclopedia of Distances, 3rd edn. Springer, Heidelberg (2014)

Evolving Smoothing Kernels for Global Optimization

Paul Manns[1] and Kay Hamacher[1,2,3(✉)]

[1] Department of Computer Science, TU Darmstadt, Darmstadt, Germany
[2] Department of Biology, TU Darmstadt, Darmstadt, Germany
[3] Department of Physics, TU Darmstadt, Darmstadt, Germany
hamacher@bio.tu-darmstadt.de

Abstract. The Diffusion-Equation Method (DEM) – sometimes synonymously called the Continuation Method – is a well-known natural computation approach in optimization. The DEM continuously transforms the objective function by a (Gaussian) kernel technique to reduce barriers separating local and global minima. Now, the DEM can successfully solve problems of small sizes. Here, we present a generalization of the DEM to use convex combinations of smoothing kernels in Fourier space. We use a genetic algorithm to incrementally optimize the (meta-)combinatorial problem of finding better performing kernels for later optimization of an objective function. For two test applications we derive and show their *transferability* to larger problems. Most strikingly, the original DEM failed on a number of the test instances to find the global optimum while our transferable kernels – obtained via evolutionary computations – were able to find the global optimum.

Keywords: Smoothing kernels · Diffusion-equation method · Fourier space

1 Introduction

Natural systems and the inspired natural computations are frequently converging to minima or maxima of some objective function, e.g. molecular structures in their natural state(s) [1,2]. The Diffusion-Equation-Method (DEM) is a global optimization approach that is inspired by the natural phenomena of diffusion – an approach also known as the (Gaussian) continuation method [3,4]. The DEM requires the analytic convolution of the objective function with the Green's function of the heat (or diffusion) equation – a Gaussian kernel. Despite some success stories, the general applicability was questioned due to a few counter-examples [1,5–7].

The general idea of our contribution is the *problem-specific* derivation of *adapted and modified DEM kernels* to improve both the convergence rate and the probability to obtain global minima. Note that we use a genetic algorithm to solve the (meta-)problem of optimized kernels for global optimization of some

© Springer International Publishing Switzerland 2016
G. Squillero and P. Burelli (Eds.): EvoApplications 2016, Part II, LNCS 9598, pp. 56–72, 2016.
DOI: 10.1007/978-3-319-31153-1_5

underlying objective function. The applicability and transferability of these kernels is based on performance evaluations for smaller and known test instances and their later evaluation for larger problem sizes.

Our contributions are: (a) we show how to improve DEM convergence for a given objective function by modifying the convolution kernel in Fourier space and thus its ability to serve as a "low-pass filter" for structures in the search space. To this end, a technique that evolves such an adapted kernel for a given objective function is necessary; (b) we show that genetic algorithms (GA) are an efficient way to evolve such problem-specific DEM-kernels; (c) we evolve superior DEM-kernels for the long-range potential of two-body interactions frequently encountered in molecular physics, namely the Lennard-Jones potential.

2   Related Work on the DEM

2.1   The "Classical" Diffusion Equation Method (DEM)

The idea of using the DEM as an global optimization approach was first proposed in [1]. The rationale is as follows: consider an objective function $f : \mathbb{R}^n \to \mathbb{R}$ which is at least twice continuously differentiable. The DEM approach models f as an initial value problem of a (time-dependent) diffusion process F_t. Then, f is the initial particle density $F_{t=0} = f$, see [1].

Typically, the diffusion equation models how the density distribution evolves in time. In real diffusive process, the density minima and the maxima become shallower and converge to an uniform distribution for $t \to \infty$ [5] while barriers separating optima are in general reduced in height. Now, in the DEM this process is simulated "backwards" in time to go from a less structured, smoothed function to the original objective function. In practice, one can choose a finite diffusion time $t = t_{End}$ that needs to be sufficiently large.

DEM implements an iterative procedure for minimization of the densities "in reverse order": we start at the "diffused" density $F_{t_{End}}$, obtain a (possible local) minimum via local minimization. Then the time t is reduced by a fixed time step h and the local optimization is executed again with the starting point set to the result of the previous step. This procedure is repeated while $t > 0$. Finally, when $t = 0$ is reached $F_{t=0} = f$ is (locally) minimized with the starting point that evolved during the diffusion process [1,5].

Algorithm 1 shows the computation in pseudocode. The reverse diffusion process for a one-dimensional 2π-periodic objective function is depicted in Fig. 1. In this algorithm localminimizer(f, x) minimizes the function f starting from x locally. It returns the location of the local minimum x_m. We use the conjugate-gradient method as implemented in the GSL-package [8] in the subsequent parts of our study.

2.2   DEM in Fourier Space

In the following, $\hat{g}$ denotes a function in Fourier space, its arguments are frequencies. In particular, $\hat{g}$ is used to denote the Fourier transform of a function g.

Algorithm 1. DEM

Require: Diffusion process F_t with $F_0 = f$
Require: Starting point x_{Start}
Require: Start time t_{End} (DEM starts at the end time of the diffusion process)
Require: Time stride h

$t := t_{End}$
while $t > 0$ **do**
    $x_{Start} := \texttt{localminimizer}(F_t, x_{Start})$
    $t := t - h$
end while
return $\texttt{localminimizer}(f, x_{Start})$

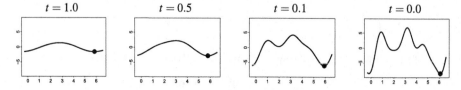

Fig. 1. DEM process for $f(x) = 3\cos(2x - 6.19) + 4\cos(2x + 4.87) + 4\cos(2x - 3.02) + 2\cos(2x - 4.68) + 3\cos(3x + 3.64) + 6\cos(5x - 3.87) + 4\cos(x + 3.55) + 5\cos(5x - 0.77)$. For each step, the current minimum is marked with a dot. Note, that in the DEM the time t runs backwards.

$\mathscr{F}$ denotes the Fourier transform (FT) operator and $\mathscr{F}^{-1}$ the inverse Fourier transform (IFT) operator.

The diffusion process is modeled by the heat equation – a partial differential equation of the form $\partial_t F(x,t) = \Delta_x F(x,t)$, where Δ_x is the Laplacian operator. It has a so-called fundamental solution. If one knows the fundamental solution Ψ of a partial differential equation (PDE) and an initial value f, then $\Psi * f$ with the convolution operator $*$ is a solution of the corresponding initial value problem [9].

Definition 1. *The fundamental solution of the heat equation is*

$$\Phi_t(x) = \begin{cases} \frac{1}{(2\sqrt{t\pi})^n} \exp\left(\frac{-\|x\|^2}{4t}\right) & t \geq 0 \\ 0 & t < 0 \end{cases} \tag{1}$$

As described above, the function $F = \Phi_t * f$ is the solution of the following initial value problem [1,7]:

$$F(x,0) = f(x)$$
$$\partial_t F(x,t) = \Delta_x F(x,t)$$

The convolution $\Phi_t * f$ is typically computed by transforming Φ_t and f into the Fourier space, executing a multiplication of the Fourier-transformed functions and perform an inverse Fourier Transform of the result. This is possible due to the Convolution Theorem.

Remark 1. The Convolution Theorem states $f * g = \mathscr{F}^{-1}(\hat{f} \cdot \hat{g})$ if the convolution and all intermediate transforms exist. Proofs and detailed information can be found in [10, chapter I.2] and [11, p. 19].

2.3   Limitations of the DEM

The DEM is applicable to a large class of objective functions. However, there is no guarantee that the global minimum of a function is obtained for every starting point when using the DEM. Hamacher [7] discussed objective functions for which the DEM returns the global minimum for starting points in only half of the domain of definition (e.g. the well-known Shubert function [12]) or for which the DEM never returned a global minimum[1].

2.4   Hybrid Metaheuristics and Genetic Algorithms

Metaheuristics are optimization techniques that provide good solutions without excessive computation demands (compared to deterministic techniques) but do not offer any guarantees about the optimality or quality of the solution [13]. We develop a hybrid metaheuristic which consists of three sequential steps:

1. Adaption of kernel functions for small- or mid-size optimization problems which were solved beforehand (training set).
2. Estimation of a proposed kernel functions for (more complex) objective functions.
3. Application of the modified DEM based on the best kernel to obtain the global minimum for a challenging, new problem.

Now, genetic algorithms are population-based metaheuristic approaches that can solve our problem stated above. This approach is especially useful if the problem setup does not provide any differentiability properties [13]. This is exactly the case for the combinatorial (meta-)optimization of new, modified DEM kernels where the goal is to improve upon the probability to obtain a global optimum for the underlying objective function. While a genetic algorithm mimics an evolutionary process with a set of candidate solutions, our fitness function is evaluated based on the performance of candidate kernel combinations to optimize via an adapted DEM procedure an underlying test function. We use mutation and crossover operators – as in standard GA protocols [14–16].

3   Improvements of the DEM

3.1   Starting Idea

The fundamental solution Φ_t of Eq. 1 is the normal distribution and has a Fourier transform $\hat{\Phi}_t(\omega) \sim \exp(-t\|\omega\|^2)$. This is still a Gaussian, but t changed its position from the denominator in the real-space Gaussian of Eq. 1 to the numerator

---

[1] For example, the DEM fails to converge to the global optimum of $g(x) = \cos(x - 3.4) + 6\cos(2x + 1) + 3\cos(3x + 2.5)$ for all starting points.

in the frequency ω based Fourier transform. The product of $\hat{\Phi}_t$ and $\hat{f}$ implements a low-pass filter that reduces the contribution of high frequencies. This motivated us to ask whether one can achieve better performance of a modified DEM for an objective function f if we use *band-pass filters* instead of a *low-pass filter* as in the classical DEM.

This would translate into: instead of multiplying $\hat{f}$ with $\hat{\Phi}_t$, $\hat{f}$ is multiplied with a frequency shifted version $\hat{\Phi}_t^{(1)}$ of $\hat{\Phi}_t$. Furthermore, one could possibly also apply a linear combination of band-pass-filters $\hat{\Phi}_t^{(i)}$ in Fourier space with weights α_i as in $\sum_{i=1}^n \alpha_i \hat{f} \hat{\Phi}_t^{(i)}$.

Section 3.1 shows that convex combinations of frequency shifted kernel functions $\hat{\Phi}_t^{(i)}$ conserve the necessary property of the original DEM

$$\lim_{t \to 0} \mathscr{F}^{-1} \left(\sum_{i=1}^n \alpha_i \hat{\Phi}_t^{(i)} \hat{f} \right) = f$$

for an important class of functions f. This means that the reverse time process converges to the initial problem as in the classical DEM. The adaption of a kernel function for f is then in itself an (meta-)optimization problem. As it is a combinatorial one, we leverage the power of genetic algorithms. Here, our population $\mathscr{P}$ of each step is a set of "chromosomes", which are convex combinations of kernel functions and their respective weights

$$\mathscr{P} = \left\{ (\hat{\Phi}^{(1,1)}, \alpha_{1,1}, \ldots \hat{\Phi}^{(1,n)}, \alpha_{1,n}); \ldots ; (\hat{\Phi}^{(|\mathscr{P}|,1)}, \alpha_{|\mathscr{P}|,1}, \ldots \hat{\Phi}^{(|\mathscr{P}|,n)}, \alpha_{|\mathscr{P}|,n}) \right\}$$
(2)

The fitness of any individual i of dimension n and representation $(\hat{\Phi}^{(i,1)}, \alpha_{i,1}, \ldots \hat{\Phi}^{(i,n)}, \alpha_{i,n})$ will be set to the fraction of starting points that converge to the (known) global minimum for some test instances (see below).

Definitions. We introduce the following definition to extend the Fourier-transformed convolution kernel for one-dimensional functions.

Definition 2. *A function $\hat{k}_{\alpha,\mu,t} : \mathbb{R} \to \mathbb{R}$ is called a **Generalized DEM Kernel** if it can be written as*

$$\hat{k}_{\alpha,\mu,t}(\omega) = \sum_{j=1}^n \alpha_j \exp(-t(\omega - \mu_j)^2)$$

with frequency shifts $\mu \in \mathbb{R}^n$, $\mu_k \neq \mu_l$ for $k \neq l$, $\sum_{j=1}^n \alpha_j = 1$, $\alpha_j \in (0,1]$, $t > 0$.

In a nutshell, a Generalized DEM Kernel is a convex combination of frequency-shifted Fourier transforms of the fundamental solution in the classical DEM.

Convergence of Generalized DEM Kernels for $t \to 0$. We show that $\mathscr{F}^{-1}(\hat{k}_{\alpha,\mu,t}\hat{f})$ converges to the objective function f for $t \to 0$. Furthermore – to simplify computations – it is sufficient to consider the real part of the (in general) complex term $\mathscr{F}^{-1}(\hat{k}_{\alpha,\mu,t}\hat{f})$. Both properties are a direct consequence of the following theorem.

Theorem 1. *Consider a **Generalized DEM Kernel** $\hat{k}_{\alpha,\mu,t}$ and an integrable, twice continuously differentiable objective function f. Then $k_{\alpha,\mu,t} * f$ is the solution of the initial value problem*

$$F(x,0) = f(x)$$

$$\partial_t F(x,t) = \partial_{xx} F(x,t) - \sum_{k=1}^{n} \alpha_k \left(i\, 2\mu_k \partial_x F(x,t) + \mu_k^2 F(x,t) \right)$$

Proof. Due to length restrictions we can only give a sketch of the proof here. The idea of the proof is based in operator semigroup theory. The main steps are as follows. Prove that $(T(t))_{t\geq 0}$ with $T(t)f := k_t * f$ for $t > 0$ and $T(0)f := f$ for $t = 0$ is a C_0-semigroup. The Hille-Yosida theorem [17] states that the generator A of the C_0-semigroup $(T(t))_{t\geq 0}$ provides a solution for the abstract Cauchy problem:

$$\partial_t F(t) = AF(t) \text{for } t \geq 0 \qquad \text{and} \qquad F(0) = f$$

The generator A can now be computed as $\partial_t T(t)f|_{t=0}$. Definitions and theorems can be found in [17,18]. The objective function is real-valued and the initial value of the given PDE. The modified DEM process $F(t)$ converges to f for $t \to 0$. f is real-valued, hence the imaginary part of $F(t)$ during the modified DEM process can be neglected.

3.2   DEM* – an Extended DEM

Based on the idea of problem-specific band-pass filters and the mathematical rigor from above, we can formulate the extended and improved DEM* algorithm in Algorithm 2. Now, then we have to develop a method of how to obtain suitable generalized DEM kernels $\hat{k}_{\alpha,\mu,t}$. In the following, we will show that this in itself poses a combinatorial optimization problem that can be solved by the previously discussed genetic algorithms.

Algorithm 2. DEM*

Require: Generalized DEM Kernel $\hat{k}_{\alpha,\mu,t}$
Require: Objective function f and its Fourier transform $\hat{f}$
Require: A starting point x_{Start}
Require: Start time t_{End}
Require: Time stride h
$t := t_{End}$
while $t > 0$ **do**
    $x_{Start} := \text{localminimizer}(\text{Re}\,\mathscr{F}^{-1}(\hat{k}_{\alpha,\mu,t}\hat{f}), x_{Start})$
    $t := t - h$
end while
    return $\text{localminimizer}(f, x_{Start})$

---

4   DEM* Kernel Optimization via Genetic Algorithms

The task of finding suitable and efficient Generalized DEM Kernels for an objective function f can be stated as follows:

$$\underset{\hat{k}}{\arg\max} \, |\{x \in \mathscr{I} \; : \; |\text{DEM}^*(x, \hat{f}, \hat{k}) - x_{\min}| \leq \varepsilon\}| \tag{3}$$

with a discretization $\mathscr{I}$ of the domain of definition, the global minimum $x_{\min}$ of f and the acceptable error $\varepsilon > 0$ of the result of DEM* from the true $x_{\min}$. Note, that for this *learning* or *training* phase we need to know location(s) of minima to derive kernels later to be *transferred* to new problem instances.

The term to optimize is of the form $\mathbb{R}^n \times [0,1]^n \rightarrow \mathbb{N}$ plus the convexity constraint of Definition 2 for a fixed $n \in \mathbb{N}$. One easily sees that it is difficult to map this problem into a typical continuously differentiable parameter optimization problem because basic properties like continuous differentiability cannot be derived easily. Foremost, because the application of DEM* is not an analytic procedure beyond simple cases.

Therefore, we have chosen a genetic algorithm, which is applicable to a wider class of even discontinuous problems to tackle our combinatorial optimization problem of well performing Generalized DEM Kernels. Genetic algorithms only require that deviations in parameter values – in our context α and μ defining a Generalized DEM Kernel – bound deviations in fitness values – in our context Eq. 3 [14, p. 158,159]. This is a fair assumption because all of the necessary intermediate computation steps provide good smoothness properties. The following subsections define the mapping of Eq. 3 onto the genetic algorithm setting.

4.1   Individuals

An individual in the population $\mathscr{P}$ is a Generalized DEM Kernel. Its chromosomes are implemented as a vector of pairs of weights and frequency shifts as real variables. Any such pair represents a Gaussian peak within the kernel. The implementation takes care of the weights fulfilling the convexity constraint of Definition 2.

4.2   Fitness Function and Selection

As described above the fitness function is evaluated for each individual, i.e., each Generalized DEM Kernel. The domain is discretized into a finite set of test points $\mathscr{I}$ to make the definition of convergence operational from a computational/experimental point of view. DEM* is executed for each of these points using f and $\hat{f}$, respectively. Then the fitness is computed as the fraction of the sampling points that converged to the correct global minimum within the predefined ε-precision over the total number of points $(= |\mathscr{I}|)$. The numerator of this fraction is the term to be maximized in Eq. 3.

We use an adapted classical roulette wheel selection: (a) an individual's probability of being selected to advance into the next generation is proportional to its fitness function value $p(i) = \frac{\text{fitness}(i)}{\sum_{j \in \mathscr{P}} \text{fitness}(j)}$. In contrast to Elitism Selection it allows functions with little fitness to survive and take part in the recombination which is often necessary to find the optimal solution [14,19]. Furthermore, (b) the best individual always advances to the next generation to include some greedy behavior in the Generalized DEM kernel search.

4.3   Crossover

Our crossover operator takes two individuals of the population, cuts both of them at a randomly selected point k and performs a crossover recombination of the four partial sequences of the genome. A frequency shift μ_j and its corresponding weight α_j are treated as one unit in the crossover operation. A pair $(\alpha_j; \mu_j)$ is passed on together because the two of them define one Gaussian of a Generalized DEM Kernel in Fourier space and thus a particular band-pass filter. Due to additive commutativity the order of the Gaussians in a Generalized DEM Kernel is irrelevant for the computation of the fitness function. Consequently, the sequence of Gaussian kernel elements in the individual is shuffled before the crossover operation.

The probability for a crossover between any two individuals was set to 75 % (empirically determined, sufficiently well working). To this end, the crossover routine loops over all individuals in the current population. Now, after a crossover operation the weights of the frequency shifted Gaussian kernels typically do not meet the convexity constraint anymore. Therefore, they are also normalized by $\alpha_i^* = \frac{\alpha_i}{\sum_{j=1}^n \alpha_j}$. Note, that this implementation guarantees that now individual is broken and always an eligible kernel.

4.4   Mutation

The mutation operator is applied on each member of the population after the crossover operator. Weight and frequency are treated as distinct entries of genotypes in context of the mutation operator. Both of them are real valued. Therefore, a Gaussian mutation as proposed in [14, p. 184] is implemented, i.e., a normally distributed random number is added[2]. This random number is bounded in order to avoid negative weights.

The probability for a mutation of an individual was set to 5 %. Again, after mutation the weights of the frequency shifted Gaussian kernels might not meet the convexity constraint any more. They are again renormalized (see Sect. 4.3).

4.5   Local Minimization

For the local minimization of $\text{Re} \, \mathscr{F}^{-1}(\hat{k}_{\alpha,\mu,t} \hat{g})$ we use the Polak-Ribiere version of the conjugate gradient method. This variant is known to perform superior to the original Fletcher-Reeves version for non-quadratic functions [20, p. 122].

---

[2] For α's the width of the normal distribution was set to 0.2 and for the ν's to 1.0.

4.6   Genetic Algorithm Kernel Search as Pseudocode

We implemented and used the GA in Algorithm 3 to search for suitable and efficient Generalized DEM kernels.

5   DEM* for a Family of Simple Test Functions

First, we applied the DEM* algorithm and the GA-based kernel search to a simple, but tunable set of test functions. We therefore introduce the following definition:

Definition 3. *A function* $f : \mathbb{R} \to \mathbb{R}$ *is called a* **Real Cosine Function** *if* $\exists \beta, \gamma, \eta \in \mathbb{R}^m$ *such, that* f *can be written as:*

$$f(x) = \sum_{j=1}^{m} \beta_j \cos(\gamma_j \cdot x - \eta_j)$$

A Real Cosine Function is obviously 2π-periodic if all γ_j are integers.

Algorithm 3. DEM* Adaption

Require: Objective function f and its Fourier transform $\hat{f}$
Require: Number of samples for the fitness function N
Require: Start time the DEM* process T
Require: Time stride of the DEM* process h
Require: $x_{\min} := \text{GlobalMinimum}(f)$
⊳ Note: f and its gl. min. are known, f is a training instance
    $\mathscr{I} := \text{Discretization}(N)$
    $\mathscr{P} := \text{InitialPopulation}()$
    fitness $:= (0, \dots, 0)$
    $c := 0$
    $p_{\text{best}} := \{\}$
    while $\max_{p \in \mathscr{P}} \text{fitness}(p) < 1$ and $c < 100$ **do**
       for $p \in \mathscr{P}$ **do**
          fitness$(p) :=$
             $\frac{1}{|\mathscr{I}|}|\{x \in \mathscr{I} \ : \ |\text{DEM}^*(f, p, x, T, h) - x_{\min}| < \varepsilon\}|$
       end for
       $p_{\text{best}} := \text{argmax}_{p \in \mathscr{P} \cup p_{\text{best}}} \text{fitness}(p)$
       $\mathscr{P} := \text{RouletteWheelSelection}(\mathscr{P}, \text{fitness})$
       $\mathscr{P} := \text{Crossover}(\mathscr{P}, 0.75)$
       $\mathscr{P} := \text{Mutation}(\mathscr{P}, 0.05)$
       $c := c + 1$
    end while
    return p_{best}

5.1   A DEM Counter Example from the Real Cosine Function Class

Hamacher [7] introduced the test function $g(x) = \cos(x - 3.4) + 6\cos(2x + 1) + 3\cos(3x + 2.5)$ which falls into the class of Real Cosine Functions. When discretizing the interval $\mathscr{I} = [0, 2\pi]$ into 100 distinct points, the classical DEM fails to return the global minimum of g from any starting point.

Now, our GA is able to evolve kernel functions for DEM* which yield the global minimum of g from 50 % of the starting points. One evolved kernel function and the DEM standard kernel are shown in Fig. 2(a).

In Fig. 2(b) we show the counter-example g of [7] and the starting points from which the classical DEM fails to converge to the global optimum (100 %).

In contrast, Fig. 2(c) shows the points from which an GA-evolved Generalized DEM kernel for this test function indeed converged to the global optimum.

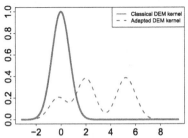

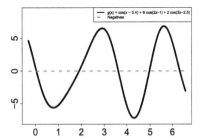

(a) Standard DEM and GA evolved kernel in Fourier Space (at $t = 1$) for the test instance g of [7]. We can clearly observe the band-pass filter like characteristics of the evolved kernel vis à vis the low-pass filter behavior of the standard DEM kernel.

(b) Starting points (indicated as a red line) of g (black) using classical DEM that did not converge to the known global optimum.

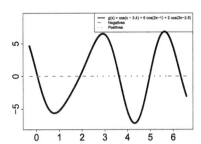

(c) Starting points of g (black) using DEM* and the evolved Generalized DEM kernel for which DEM* found (blue) the global optimum and failed to do so (red).

Fig. 2. Comparing DEM and DEM* for a difficult g. (Color figure online)

5.2   Testing DEM* on General Real Cosine Functions

Real Cosine Functions have a clear interpretation in Fourier/frequency space: they show distinct peaks and identifiable patterns due to their 2π periodicity and their construction via linear combination of periodic functions. This could potentially imply a "rule" on what band-pass filters need to be chosen to DEM*-optimize General Real Cosine Functions.

In order to evaluate if such a (probabilistic) dependency between the objective function and the evolved kernel function can be established, 10,000 Real Cosine Functions were generated randomly, each containing 10 terms. The β_i and γ_i were drawn from a Poisson distribution with mean 2.5. The frequency shift values η_i were drawn from a uniform distribution on $[-2\pi, 2\pi]$. The Poisson distribution returns integer values for γ_i which ensures the 2π-periodicity of the generated functions.

To use the GA to obtain DEM* kernels, it is necessary to know the global minimum of these functions beforehand. The location of the global minimum is approximated by starting Conjugate Gradient minimizations from 10,000 equidistant points in $[0, 2\pi]$. Then, DEM* was used with a start time of 10 and a time window of $h = 0.01$. Each Generalized DEM Kernel consisted of four peaks. The population size $|\mathscr{P}|$ of the GA was set to 728 as a boundary for the combinatorics of starting parameters μ_j, α_j. We initialized the population to Gaussians on a grid of frequency shifts, each of equal weight.

5.3   Results

For all of the 10,000 test instances, the individuals of the last GA generation with best fitness were put into a matrix. The result of the adaption was a matrix with 38 columns, consisting of the 30 independent parameters ($\beta_i, \gamma_i, \eta_i$ with $i \in [1 \ldots 10]$) that define one Real Cosine Function and the 8 dependent parameters that define the adapted Generalized DEM Kernel (α_j, μ_j with $j \in [1 \ldots 4]$), and a total of 1,655,128 rows.

We then *transfer* the "learned" band-pass filter-like Generalized DEM kernels to new test instances and predict their respective frequency shifts and weights based solely on the properties of the Real Cosine Functions. To this end, 8 Support Vector Machines (SVMs) were trained with the resulting data, one for each dependent parameter. 2,000 additional Real Cosine Functions were used to test the prediction quality of the trained SVM. The R [21] package e1071 [22] was used as a freely available SVM implementation.

The fitness of the unmodified DEM Kernel was evaluated for all of the 2,000 Real Cosine Functions in the test set. DEM finds the global optimum for 68.75 % of these tests. However, the unmodified DEM fails completely for 19.85 % of them – namely, for *all* starting points. For DEM*, the number of test Real Cosine Functions for which the DEM* fails completely reduces to 5.35 %. However, the number of functions for which the DEM* works perfectly (100 % of starting points) is reduced to 14.95 %.

So, we observe a trade-off between general applicability (DEM* finds for *more* test instances the global minimum) vs. easiness of use (for those tests where DEM converges at all, one does not need to sample so many starting points).

Therefore, we proceed to combine classical DEM (which fails frequently, but is successful for more starting points whenever it converges) and DEM* (which is more successful overall, but might fail for a particular starting point). The combined fitness for an objective function is then computed as the fraction of starting points for which at least one of the methods obtains the real minimum, i.e. the synergy effect one obtains when executing both DEM and DEM* with the predicted kernel. One obtains the best possible result for 68.75 % and a complete fail for only 3.25 % of the objective functions.

Those few failures could arise basically for three reasons: (a) our Generalized DEM kernels are not suitable, (b) the GA did not find suitable ones, or (c) the SVM did not generalize enough and predicted weights and frequency shifts incorrectly. Typically, machine learning techniques cannot find "exact" patterns, therefore, a failure rate of 3.25 % could possibly be attributed to the SVM.

Whatever the source, the improvements by DEM* are striking. Therefore, these results encouraged us to apply DEM* to a challenging problem in high-dimensional search spaces: ground states of atoms and particles that interact via a Lennard-Jones potential. Our findings are described in the following section.

6   DEM* for Potential Energy Surfaces

6.1   Lennard-Jones Potential

The Lennard-Jones potential models the pairwise interactions of atoms and molecules including so-called dispersion effects. The potential function reads

$$V = \sum_{i<j}^{N} 4\varepsilon \left(\frac{\sigma^{12}}{r_{ij}^{12}} - \frac{\sigma^6}{r_{ij}^6} \right) \tag{4}$$

where r_{ij} is the euclidean distance between the N interacting molecules i and j in 3D-space. The minimum potential well that can be reached per interacting pair is given by the pair equilibrium depth well ε. The length scale is defined by σ [23,24]. The "surface" in the $3N$ dimensional space is frequently called Potential Energy Surface (PES) in the natural sciences.

Now, the ground state of such a collection of N particles (such as noble gases or the partial charges in atoms of biomolecules) is the global optimum of the full potential V. Thus, V is our objective function and we eventually predict the native state of these set of particles.

Without loss of generality we set $\varepsilon = \sigma = 1$ as in [24]. This allows us to use known ground states [24] as reference results for the GA-based training of Generalized DEM kernels in our DEM*.

The potential consists of $\frac{N(N-1)}{2} \sim O(N^2)$ terms for N interacting particles. Furthermore, the number of variables, namely the coordinates of the interacting atoms, increases linearly with the number of particles $(3 \cdot N)$.

Our goal is to use derive *transferable* Generalized DEM Kernels with the approach laid out above. In particular, it is desirable to derive parameters or rules for Generalized DEM Kernels for our DEM* for just a few particles and use them in larger problem instances with more particles.

However, the naive approach of using the same kernels is not possible as the dimensionality of the kernels differ for varying number of particles. Therefore our DEM* is instead applied for one-dimensional kernels (in the respective r_{ij} on each of the terms in Eq. 4.

6.2   Computational Details

We applied the GA learning procedure with a population size of $1,000$ individuals and 100 starting configurations in the discretization $\mathscr{I}$. We restricted our study to those Generalized DEM Kernels with just five Gaussians and therefore five peaks.

The Lennard Jones potential is unbound for $r \to 0$. Now, the limit $r \to 0$ is (a) physically unrealistic and (b) leads to singularities in Eq. 4. We therefore use a truncated potential: DEM and DEM* operate on a function $\tilde{V}$ that truncates V of Eq. 4 within the interval $r \in [0, 0.85)$. For $r < 0.85$, $\tilde{V}$ has the form $\alpha \exp(-\beta r^2)$ and for $r \geq 0.85$ we use the full Lennard-Jones potential of Eq. 4. The parameters α, β were chosen to provide for continuity $\left(\tilde{V}(0.85) = V(0.85) \right)$ and continuous differentiability $\left(\tilde{V}'(0.85) = V'(0.85) \right)$ and thus physically realistic, finite forces.

Now, the analytic IFT $\mathscr{F}^{-1} \left(\hat{k}_{\alpha,\mu,t} \; \mathscr{F} \left(\tilde{V}(r) \right) \right)(r)$ and its real part cannot be given in closed form. Due to this and for computational efficiency the FT of the term and the IFT of the product were approximated by a Radix-2-Fast-Fourier-Transform (FFT) and used via a lookup table – a traditional memory-CPU trade-off.

To avoid "exploding" communities of particles, that is solutions that – for mainly numerical reasons – separate particles by infinite distances, we also added a confining potential for each of the distances r_{ij}. The on-set of such a confinement was chosen to be much larger than the typical size of a cluster of optimized particles.

6.3   Results

Our performance results for DEM and DEM* are summarized in Fig. 3. First, we found the classical DEM to be able to correctly find the global minima of V of Eq. 4 from all starting configurations in the respective $\mathscr{I}$ for small systems – namely for $2 \leq N \leq 6$. Hence, there is no need for the more complex DEM* for such small instances. For $N \geq 7$ the performance of DEM deteriorates. Eventually, the classical DEM found the global optimum only for $N = 10$ (probability 32%) and $N = 14$ (probability 1%) for the 100 starting points in $\mathscr{I}$. For all other $N \in [7, \dots, 30] \setminus \{10\}$ it failed *always* to find a global optimum.

Now, DEM* was used to cope with the poor performance of the classical DEM. To this end, we used the GA procedure laid out above to train Generalized DEM kernels for instances $N = \{7, 8, 17\}$ for later usage in DEM* in the cases of $N \in [2, \ldots, 30]$. The performance results for those three different DEM* trainings are also shown in Fig. 3. Furthermore, Table 1 gives the exact values for the instances trained on.

Table 1. Results of the GA trainings. We show the number of particles the respective Generalized DEM kernel was trained on; the kernels we were able to find with maximal fitness for those respective cases; and the fitness obtained for the best performing kernels on the training examples.

# particles	Kernels with max. fitness	Max. fitness
7	21	1.0
8	1	0.56
17	4	0.16

In general and in agreement with intuition, the overall fitness decreases with an increasing number of particles tested for. But there are also counter-intuitive improvements, e.g., for $N = 12$ and $N = 13$ tests where the performance increases to some 80 % to 100 % of starting points and thus fitness.

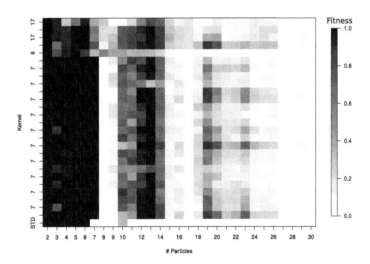

Fig. 3. Performance of the DEM and the DEM* for Lennard-Jones potentials of Eq. 4. "STD" denotes the classical DEM; 7,8 and 17 on the y-axis denote the size of *training* examples (number of particles in the Lennard-Jones potential). The x-axis shows the size of the *test* examples (the respective number of particles) the evolved kernels are applied to.

A remarkable number of the trained kernels in DEM* obtain the global minimum for even large instances as in the $N = 29$ case. One kernel, which resulted from the 17 particle training, is even transferable to the system with 30 particles.

Interestingly the DEM* modifications adapted for 7 particles, as well as the unmodified DEM, fail to obtain the global minimum from any starting configuration for 8,17,27 and 30 particles. At present, we have no analytic verifiable explanation for this empirical finding. The overall pattern of transferability is as follows: for almost all but some few system sizes N we can use kernels trained for smaller N and use them in DEM* to find with a high likelihood the global optimum. The two trainings for $N = 17$ are superior to the ones for $N = 7$ but do not overall suffer from poor performance on isolated cases. Still, $N = 8$ trained kernels are not transferable. But $N = 8$ as a test case is solvable by a DEM* kernel trained upon $N = 17$.

7 Conclusions and Future Work

We have shown that it is possible to GA-evolve Generalized DEM Kernels that improve the convergence of the DEM to the global minimum for Real Cosine Functions. The results show that a regression tool based on standard machine learning techniques such as SVMs can be found that *predicts* kernels for previously not visited Real Cosine Functions for which the classical DEM approach fails always.

Furthermore, the Genetic Algorithm from Sect. 4 is able to evolve Generalized DEM Kernels for mid-sized Lennard-Jones-Potential functions that yield the global minimum for bigger instances.

Both findings taken together suggest that the DEM* kernels are *transferable* to larger, high-dimensional and thus more challenging optimization problems. Note, that the overall computational costs need to be investigated in a future study – our goal here was to derive the *conceptual framework* of the DEM*, that is a proof-of-concept: a genetic algorithm supports the meta-optimization of approximate optimization procedures such as the DEM. Future developments in this regard require improved and adapted mutation and recombination operators to more efficiently diffuse through the space of linear combinations of Gaussians.

In the case of Lennard-Jones potentials, this finding translates into that optimization problems of 15,18 and 45 degrees of freedom can be used to solve optimization problems with 84 degrees of freedom while the unmodified DEM is only able to obtain the correct minimum for up to 12 degrees of freedom[3].

For future research, we note in passing that it is possible for the DEM* to omit approximations for functions with a multidimensional domain if the multidimensional FTs are analytically available. This is the case for separable objective functions f which are of the form $f(x) = \prod_{i=1}^{N} f_i(x_i)$ as their Fourier transform reads $\hat{f}(\omega) = \prod \hat{f}_{i=1}^{N}(\omega_i)$ [11, p. 331].

---

[3] Here degrees of freedom $= 3(N - 2)$ due to a free choice of the point of origin and orientation, thus translational and rotational invariance of the potential V. The prefactor 3 is the relation between no. of particles and no. of Eucledian coordinates.

A multidimensional extension of the Generalized DEM Kernel $\sum_{i=1}^{n} \alpha_i \exp(-t\|\omega - \mu_i\|^2)$ with the Euclidean norm ($\|y\|^2 = \sum_{j=1}^{N} |y_j|^2$) is obviously trivially Fourier transformable as the FT is linear [11, p. 331].

Our DEM* idea modifies the Fourier Transform of the fundamental solution Φ_t of a linear differential operator while fulfilling the convergence to the original objective function for $t \to 0$. Hence, the modification is transferable to all diffusion processes of the objective function that are described by a linear differential operator and have a fundamental solution. The Malgrange-Ehrenpreis theorem guarantees the existence of fundamental solutions for all linear differential operators with constant coefficients [25].

Note, that our approach is quite different from previous studies on, e.g., evolving kernels for SVMs [26,27] as we evolve and transfer kernels in an optimization procedure called DEM and not a machine learning method like SVMs. Here, we have merely used SVMs and are not concerned with them beyond their usefulness to transfer patterns of Generalized DEM Kernels.

Acknowledgements. KH gratefully acknowledges funding by the LOEWE project compuGene of the Hessen State Ministry of Higher Education, Research and the Arts.

References

1. Piela, L., Kostrowicki, J., Scheraga, H.A.: On the multiple-minima problem in the conformational analysis of molecules: deformation of the potential energy hypersurface by the diffusion equation method. J. Phys. Chem. **93**(8), 3339–3346 (1989)
2. Wawak, R.J., Pillardy, J., Liwo, A., Gibson, K.D., Scheraga, H.A.: Diffusion equation and distance scaling methods of global optimization. J. Phys. Chem. A **102**(17), 2904 (1998)
3. Mobahi, H., Fisher III, J.W.: On the link between Gaussian homotopy continuation and convex envelopes. In: Tai, X.-C., Bae, E., Chan, T.F., Lysaker, M. (eds.) EMMCVPR 2015. LNCS, vol. 8932, pp. 43–56. Springer, Heidelberg (2015)
4. Mobahi, H., Fisher III, J.W.: A theoretical analysis of optimization by Gaussian continuation. In: AAAI 2015: 29th Conference on Artificial Intelligence (2015)
5. Kostrowicki, J., Piela, L., Cherayil, B.J., Scheraga, H.A.: Performance of the diffusion equation method in searches for optimum structures of clusters of Lennard-Jones atoms. J. Phys. Chem. **95**(10), 4113–4119 (1991)
6. Wawak, R., Wimmer, M., Scheraga, H.: Application of the diffusion equation method of global optimization to water clusters. J. Phys. Chem. **96**(12), 5138–5145 (1992)
7. Hamacher, K.: Optimization of high-dimensional functions with respect to protein-structure-prediction. Dipl. thesis, Dortmund University (1998)
8. Galassi, M.: GNU Scientific Library Reference Manual. Network Theory Ltd. (2009)
9. Miklavcic, M.: Applied Functional Analysis and PDEs. World Scientific, Singapore (1998)
10. Abels, H.: Pseudodifferential and Singular Integral Operators. Walter de Gruyter, Berlin (2011)
11. Marks, R.J.: Handbook of Fourier Analysis & Its Applications. Oxford University Press, Oxford (2009)

12. Aluffi-Pentini, F., Parisi, V., Zirilli, F.: Global optimization and stochastic differential equations. J. Optim. Theor. Appl. **47**(1), 1–16 (1985)
13. Blum, C., Roli, A.: Hybrid metaheuristics: an introduction. In: Blum, C., José Bleasa Aguilera, M., Roli, A., Sampels, M. (eds.) Hybrid Metaheuristics, pp. 1–30. Springer, Heidelberg (2008)
14. Kruse, R., Borgelt, C., Klawonn, F., Moewes, C., Steinbrecher, M., Held, P.: Computational Intelligence: Eine methodische Einführung in Künstliche Neuronale Netze, Evolutionäre Algorithmen, Fuzzy-Systeme und Bayes-Netze. Springer, Heidelberg (2011)
15. Mitchell, M.: An Introduction to Genetic Algorithms. Bradford, Denver (1998)
16. Yu, X., Gen, M.: Decision Engineering. Springer, Heidelberg (2010)
17. Nagel, R.: One-Parameter Semigroups for Linear Evolution Equations. Graduate Texts in Mathematics, vol. 194. Springer, Heidelberg (2000)
18. Engel, K.J., Nagel, R.: One-Parameter Semigroups for Linear Evolution Equations. Semi-group Forum, vol. 63. Springer, Heidelberg (2001)
19. Mayer, B.E., Hamacher, K.: Stochastic tunneling transformation during selection in genetic algorithm. In: Proceedings of the 2014 Annual Conference on Genetic and Evolutionary Computation, GECCO 2014, pp. 801–806. ACM, New York (2014)
20. Nocedal, J., Wright, S.: Numerical Optimization. Operations Research and Financial Engineering. Springer, New York (2006)
21. R Development Core Team.: R: A Language and Environment for Statistical Computing. R Foundation for Statistical Computing, Vienna, Austria (2008)
22. Dimitriadou, E., Hornik, K., Leisch, F., Meyer, D., Weingessel, A.: Misc functions of the Department of Statistics (e1071), TU Wien. R package 1–5 (2008)
23. Hopfinger, A.: Conformational Properties of Macromolecules. Academic Press, London (1973)
24. Wales, D.J., Doye, J.P.: Global optimization by basin-hopping and the lowest energy structures of Lennard-Jones clusters containing up to 110 atoms. J. Phys. Chem. A **101**(28), 5111–5116 (1997)
25. Wagner, P.: A new constructive proof of the Malgrange-Ehrenpreis theorem. Am. Math. Month. **116**(5), 457–462 (2009)
26. Diosan, L., Rogozan, A., Pecuchet, J.P.: Evolving kernel functions for svms by genetic programming. In: Sixth International Conference on Machine Learning and Applications, ICMLA 2007, 19–24 December 2007
27. Sullivan, K.M., Luke, S.: Evolving kernels for support vector machine classification. In: Proceedings of the 9th Annual Conference on Genetic and Evolutionary Computation, GECCO 2007, pp. 1702–1707. ACM, New York (2007)

EvoPAR

Implementing Parallel Differential Evolution on Spark

Diego Teijeiro[1], Xoán C. Pardo[1], Patricia González[1(✉)], Julio R. Banga[2],
and Ramón Doallo[1]

[1] Grupo de Arquitectura de Computadores, Universidade da Coruña,
A Coruna, Spain
{diego.teijeiro,xoan.pardo,patricia.gonzalez,doallo}@udc.es
[2] BioProcess Engineering Group, IIM-CSIC, Pontevedra, Spain
julio@iim.csic.es

Abstract. Metaheuristics are gaining increased attention as an efficient way of solving hard global optimization problems. Differential Evolution (DE) is one of the most popular algorithms in that class. However, its application to realistic problems results in excessive computation times. Therefore, several parallel DE schemes have been proposed, most of them focused on traditional parallel programming interfaces and infrastructures. However, with the emergence of Cloud Computing, new programming models, like Spark, have appeared to suit with large-scale data processing on clouds. In this paper we investigate the applicability of Spark to develop parallel DE schemes to be executed in a distributed environment. Both the master-slave and the island-based DE schemes usually found in the literature have been implemented using Spark. The speedup and efficiency of all the implementations were evaluated on the Amazon Web Services (AWS) public cloud, concluding that the island-based solution is the best suited to the distributed nature of Spark. It achieves a good speedup versus the serial implementation, and shows a decent scalability when the number of nodes grows.

Keywords: Metaheuristics · Differential evolution · Cloud computing · Spark · Amazon web services

1 Introduction

Global optimization problems arise in many areas of science and engineering [1–3]. Most of these problems are NP-hard, so many research efforts have focused on developing metaheuristic methods which are able to locate the vicinity of the global solution in reasonable computation times. Moreover, in order to reduce the computational cost of these methods, a number of researchers have studied parallel strategies for metaheuristics [4,5]. However, all these efforts are focused on traditional parallel programming interfaces and traditional parallel infrastructures.

With the advent of Cloud Computing effortless access to large number of distributed resources has become more feasible. But developing applications that

© Springer International Publishing Switzerland 2016
G. Squillero and P. Burelli (Eds.): EvoApplications 2016, Part II, LNCS 9598, pp. 75–90, 2016.
DOI: 10.1007/978-3-319-31153-1_6

execute at so big scale is hard. New programming models are being proposed to deal with large scale computations on commodity clusters and Cloud resources. Distributed frameworks like MapReduce [6] or Spark [7] provide high-level programming abstractions that simplify the development of distributed applications including implicit support for deployment, data distribution, parallel processing and run-time features like fault tolerance or load balancing. We wonder how much benefit can we expect from implementing parallel metaheuristics using these new programming models because, besides the many advantages, they also have some shortcomings. Cloud-based distributed frameworks prefer availability to efficiency, being the speedup and distributed efficiency frequently lower than in traditional parallel frameworks due to the underlying multitenancy of virtualized resources.

The aim of this paper is to explore this direction further considering a parallel implementation of Differential Evolution (DE) [8], probably one of the most popular heuristics for global optimization, to be executed in the Cloud. The main contribution of the proposal is an analysis of different alternatives in implementing parallel versions of the DE algorithm using Spark and a thoroughly evaluation of their feasibility to be executed in the Cloud using a real testbed on the Amazon Web Services (AWS) public cloud.

The organization of this paper is as follows. Section 2 briefly presents the background and related work. Some new programing models in the Cloud are described in Sect. 3. Section 4 describes the proposed implementations of the Differential Evolution algorithm using Spark. The performance of the proposal is evaluated in Sect. 5. Finally, Sect. 6 concludes the paper and discusses future work.

2   Related Work

The parallelization of metaheuristics methods has received much attention to reduce the run time for solving large-scale problems [9]. Many parallel algorithms have been proposed in the literature, most of them being parallel implementations based on traditional parallel programming interfaces such as MPI and OpenMP. However, research on cloud-oriented parallel metaheuristics based mainly on the use of MapReduce has also received increasing attention in recent years. MRPSO [10] uses the MapReduce model to parallelize the Particle Swarm Optimization (PSO). MRPGA [11] attempts at combining MapReduce and genetic algorithms (GA). They properly claim that GAs cannot be directly expressed in MapReduce due to their specific characteristics. So they extend the model featuring a hierarchical reduction phase. A different approach is followed in [12], that tries to hammer the GAs into the MapReduce model. In [13] the applicability of MapReduce to distributed simulated annealing (SA) was also investigated. They design different algorithmic patterns of distributed SA with MapReduce and evaluate their proposal on the AWS public cloud. Recently, in [14], a practical framework to infer large gene networks through a parallel hybrid GA-PSO optimization method using MapReduce has also been proposed.

Some proposals are more specific on studying how to apply MapReduce to parallelize the DE algorithm to be used in the Cloud. In [15] the fitness evaluation in the DE algorithm is performed in parallel using Hadoop (the well-known open-source MapReduce framework). However, the experimental results reveal that the extra cost of Hadoop DFS I/O operations and the system bookkeeping overhead significantly reduces the benefits of the parallelization. In [16], a concurrent implementation of the DE based on MapReduce is proposed, however, it is a parallelized version of a neoteric DE based on the steady-state model instead of on the generation alternation model. While the generational model holds two populations and generates all individuals for the second population from those of the current population, the steady-state model holds only one population and each individual of the population is updated one by one. Comparing with the generational model, the parallelization of the steady-state model is simpler because it does not require synchronization for replacing the current population by newborn individuals simultaneously. On the other hand, the experiments reported in that paper were conducted on a multi-core CPU, thus, their implementation take advantage of the shared-memory architecture, sharing the population among the different threads, which is not possible in a distributed cloud environment. In [17] a parallel implementation of DE based clustering using MapReduce is also proposed. This algorithm was implemented in three levels, each of which consists of different DE operations.

To the best of our knowledge, there is no previous work that explores the use of Spark for evolutionary computation. Also, previous works using MapReduce have rarely evaluated their proposals in a real testbed on a public cloud.

3   New Programming Models in the Cloud

From the new programming models that have been proposed to deal with large scale computations on cloud systems, MapReduce [6] is the one that has attracted more attention since its appearance in 2004. In short, MapReduce executes in parallel several instances of a pair of user-provided *map* and *reduce* functions over a distributed network of *worker* processes driven by a single *master*. Executions in MapReduce are made in batches, using a distributed filesystem (typically HDFS) to take the input and store the output. MapReduce has been applied to a wide range of applications, including distributed pattern-based searching, distributed sorting, graph processing, document clustering or statistical machine translation among others.

When it comes to iterative algorithms as those that are typical in areas like machine learning or evolutionary computation, MapReduce has shown serious performance bottlenecks. Computations in MapReduce can be described as a *directed acyclic data flow* where a network of stateless mappers and reducers process data in single batches (see Fig. 1). All input, output and intermediate data is stored and accessed via the file system and map/reduce tasks are created in every single batch. Having several of these single batches executed inside a loop has shown to introduce considerable performance overhead [18] mainly because there is no way of reusing data or computation from previous iterations efficiently.

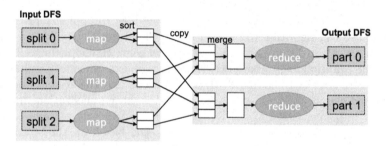

Fig. 1. MapReduce dataflow.

Although some extensions have been proposed to improve the support to iterative algorithms in MapReduce like Twister [18], iMapReduce [19], or HaLoop [20], they still perform poorly on the kind of algorithms we are interested in, mainly due to those systems inability to exploit the (sparse) computational dependencies present in these tasks [21]. New proposals, not based on MapReduce, like Spark [7] or Fink (which has its roots on Stratosphere [22]), are designed from the very beginning to provide efficient support for iterative algorithms.

Spark provides a language-integrated programming interface to *resilient distributed datasets* (RDDs), a distributed memory abstraction for supporting fault-tolerant and efficient in-memory computations. According to authors [7] the performance of iterative algorithms can be improved by an order of magnitude when compared to MapReduce (using Hadoop).

Formally, an RDD is a read–only fault–tolerant partitioned collection of records. Users can manipulate them using a rich set of operators, control their partitioning to optimize data placement and explicitly persist intermediate results (in memory by default but also to disk). RDDs are created from other RDDs or from data in stable storage by applying coarse-grained *transformations* (e.g., *map*, *filter* or *join*) that apply the same operation to many data items. Once created, RDDs are used in *actions* (e.g. *count*, *collect* or *save*) which are operations that return a value to the application or export data to a storage system.

RDDs are computed lazily the first time they are used in an action, so transformations can be pipelined to form a *lineage*. By storing enough information about their lineages, RDDs do not need to be materialized at all times, as every RDD can recompute its partitions from previously persisted RDDs or data in stable storage at any time. This feature is the one used to provide fault-tolerance in case of an RDD partition lost.

Spark runtime is composed of a single *driver* program and multiple *workers* which are long-lived processes launched by the driver. Workers read data blocks from a distributed file system and persist RDD partitions in RAM across operations. Developers write the driver program where they define one or more RDDs and invoke actions on them. Whenever an action is executed on an RDD, the Spark job scheduler uses its lineage to compute a *directed acyclic graph* (DAG)

Algorithm 1. Differential Evolution algorithm (seqDE)

input : A population matrix P with size D x NP
output: A matrix P whose individuals were optimized

repeat

    for *each element x of the P matrix* **do**

        $\vec{a}$, $\vec{b}$, $\vec{c}$ ← different random individuals from P matrix

        for $k \leftarrow 0$ **to** D **do**

            if *random point is less than* CR **then**

                $\overrightarrow{Ind}(k) \leftarrow \vec{a}\,(k) + F(\,\vec{b}\,(k) - \vec{c}\,(k))$

            end

        end

        if *Evaluation($\overrightarrow{Ind}$) is better than Evaluation($\overrightarrow{P(x)}$)* **then**

            Replace_Individual($P,\overrightarrow{Ind}$)

        end

    end

until *Stop conditions*;

of stages. The scheduler then launches tasks to compute missing partitions from each stage until it has computed the target RDD. Assignment of tasks to workers takes into account data locality. Tasks end up being assigned to workers that already hold the RDD partitions of interest in memory.

An example of how Spark computes job stages is showed in Fig. 2. In the figure, boxes with solid outlines are RDDs. Partitions are shaded rectangles, darker if they are persisted in memory. Each stage contains as many pipelined

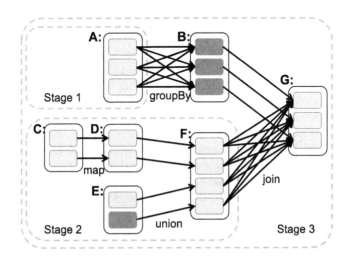

Fig. 2. Example of how Spark computes job stages.

transformations with *narrow* (one-to-one or one-to-many) dependencies as possible. The boundaries of the stages (boxes with doted outlines in the figure) are the shuffle operations required for *wide* (many-to-one or many-to-many) dependencies or the presence of an already computed RDD in the lineage. In the example, to run an action on RDD G, as output RDD from stage 1 is already in RAM only stages 2 and 3 need to be executed.

4   Implementing Differential Evolution on Spark

Differential Evolution is an iterative mutation algorithm where vector differences are used to create new candidate solutions. Starting from an initial population matrix composed of NP D-dimensional solution vectors (individuals), DE attempts to achieve the optimal solution iteratively through changes in its vectors. Algorithm 1 shows the basic pseudocode for the DE algorithm. For each iteration, new individuals are generated in the population matrix through operations performed among individuals of the matrix (mutation - F), with old solutions replaced (crossover - CR) only when the fitness value of the objective function is better than the current one.

A population matrix with optimized individuals is obtained as output of the algorithm. The best of these individuals are selected as solution close to optimal for the objective function of the model. However, in some real applications, such as parameter estimation in dynamic models, the performance of the classical sequential DE is not acceptable due to the large number of objective function evaluations needed. As a result, typical runtimes for realistic problems are in the range from hours to days. Parallelism can help improving both computational time and number of iterations for convergence. In the literature, different parallel models can be found [9], being the most popular ones the *master-slave* model and the *island-based* model. In the *master-slave* model the inner-loop of the algorithm is parallelized. A master processor distributes computation operations between the slave processors. Therefore, the parallel algorithm has the same behavior of the sequential one. In the *island-based* model the population matrix is divided in subpopulations (*islands*) where the algorithm is executed isolated. Sparse individual exchanges are performed among islands to introduce diversity into the subpopulations, preventing search from getting stuck in local optima.

With the aim of better understanding Spark intricacies and assess the performance of different alternatives when implementing DE, we have developed three different versions of the algorithm: (1) the classic sequential algorithm (seqDE) which has been implemented for comparative purposes and it is the only that does not make use of Spark; (2) three different variants of the master-slave parallel implementation (SmsPDE); and (3) an island-based parallel implementation (SiPDE). All of them have been coded using the Scala language [23] which is the one used to implement Spark itself although APIs for Python and Java also exist. The rest of this section features relevant facts about each of the implementations. As it will be demonstrated, the main conclusion is that the island-based parallel implementation is best suited to the distributed nature of Spark and obtains the best performance results.

4.1 Master-Slave DE

To implement a master-slave parallel version of the DE algorithm using Spark, some previous insight into the way data is distributed and processed by Spark is needed. Spark uses the RDD abstraction to represent fault-tolerant distributed data. RDDs are inmutable sets of records that optionally can be in the form of key-value pairs. Spark driver (the master in Spark terminology) partitions RDDs and distributes the partitions to workers (the slaves in Spark terminology), which persist and transform them and return results to the driver. There is no communication among workers. Shuffle operations (i.e. join, groupBy) that need data movement among workers through the network are expensive and should be avoided.

Our Spark-based master-slave DE implementation (SmsPDE) follows the scheme shown in Fig. 3. A key-value pair RDD has been used to represent the population where each individual is uniquely identified by its key. Some DE algorithm steps have been selected as appropriate to be executed in a distributed fashion:

- The random generation and initial evaluation of individuals that form the population, implemented as a Spark map transformation.
- The mutation strategy including random pick of individuals and replacement of old individuals with new improved ones, implemented using three different variants that are explained later.
- The checking of the termination criterion, implemented as a Spark reduce action (a distributed OR operation).

The two last steps are arranged into a loop that is executed until the termination criterion is met. After that the final selection of the best individual is also executed as a Spark reduce action (a distributed MIN operation).

The main issue found was the implementation of the mutation strategy because the population is partitioned and distributed among workers. For the

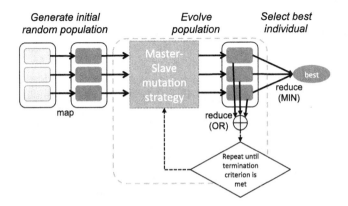

Fig. 3. Spark-based Master-Slave implementation of the DE algorithm (SmsPDE).

mutation of each individual, random different individuals have to be selected from the whole population. How to access to individuals of other partitions from a given worker, having the constraint that only the driver has access to the complete population, was the main difficulty to be tackled. Three different variants have been considered for solving this problem:

- The driver distributes the random generation of keys to the workers, collects them, selects from the whole population the individuals corresponding to the generated random keys and distributes selected individuals to the workers that perform mutations and replacements.
- The driver itself makes the random pick of individuals for each member of the population and distribute them to the workers that perform mutations and replacements.
- The driver broadcasts the whole population to every worker using Spark *broadcast variables*. This Spark feature allows workers to have access to a local memory-cached read-only copy of the complete population. Therefore each worker can perform mutations picking the needed random individuals from its local copy of the population.

After benchmarking the performance of the three variants, broadcasting the population showed to be by far the best option. This is not surprising because the broadcasting feature of Spark is highly optimized and it is the recommended method for iterative algorithms to distribute data to workers that has to be reused by different iterations. Only the size of data to be broadcasted could discourage its use, but this is not the case with DE where the size of populations is small (usually in the range of 5D and 10D being D the problem dimension).

4.2   Island-Based DE

When testing each one of the previous approaches, even using the version that has shown best benchmarking results, the penalty due to broadcast the whole population to workers in each iteration was unaffordable. For instance, using one of the benchmark functions used later on in Sect. 5, the f_{15} function, and a stopping criterion based on a predefined effort of 800,000 evaluations, the execution time of seqDE was 30 s, while the execution time of SmsPDE using 4 nodes was 263 s. The main conclusion of our experience with the master-slave implementation of the DE algorithm was that this approach does not fit well with the distributed model of Spark. Therefore we decided to implement a new parallel version of the algorithm using an island-based approach which in advance seemed to be a more promising one.

Figure 4 shows the scheme of the Spark-based island DE (SiPDE) implementation. As it can be seen it has some steps in common with the master-slave implementation: i.e. generation of the population, checking of the termination criterion and selection of the best individual. The main difference resides in the way the population evolves. Every partition of the population RDD has been considered to be an island, all with the same number of individuals. Islands evolve isolated during a number of evolutions. This number can be configured and is the same for all islands. During these evolutions every worker calculates mutations picking random individuals from its local partition only. To introduce diversity a migration strategy that exchanges selected individuals among islands is executed every time the number of evolutions is reached. This evolution-migration loop is repeated until the termination criterion is met.

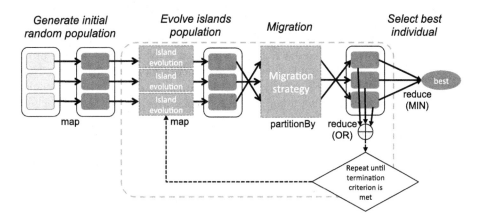

Fig. 4. Spark-based island implementation of the DE algorithm (SiPDE).

For implementing the migration strategy a Spark feature known as *partitioner* has been used. In Spark the partitioner is responsible for assigning key-value pair RDD elements to partitions based on their keys. Default partitioner implements a hash-based partitioning using the Java hash code of the key. For this work we have implemented a custom partitioner that randomly and evenly shuffles elements among partitions. It must be noted that this partitioner leads to a migration strategy that randomly shuffles individuals among subpopulations without replacement. This partitioner proposal is intended to evaluate the migration communications overhead and not to improve the searching quality of the algorithm. Adding migration strategies with that purpose in mind are left for future work.

5   Experimental Results

In order to evaluate the efficiency of the Spark-based parallel implementation of the island DE algorithm (SiPDE), different experiments have been carried out. Its behavior, in terms of convergence and total execution time, has been compared with the sequential implementation (seqDE). For the experimental testbed Spark was deployed with default settings in the AWS public cloud using virtual clusters formed by 2, 4, 8 and 16 nodes communicated by the AWS standard network (Ethernet 1 GB). For the nodes the m3.medium instance (1 vCPU, 3.75 GB RAM,

4 GB SSD) was used. Each experiment was executed a number of 10 independent runs on every virtual cluster, and the average and standard deviation of the execution time are reported in this section. It must be noted that, since Spark runs on the Java Virtual Machine (JVM), usual precautions (i.e. warm-up phase, effect of garbage collection) has been taken into account to avoid distortions on the measures.

The performed experiments used two sets of benchmark problems: a set of problems out of an algebraic black-box optimization testbed, the Black-Box Optimization Benchmarking (BBOB) data set [24]; and a challenging parameter estimation problem in systems biology [25]. On the one hand, the experiments over the BBOB data set were carried out to evaluate the efficiency of the proposed parallelization in a popular and accessible benchmarking testbed. On the other hand, the aim of the experiments with the parameter estimation in systems biology is to demonstrate the potential of the proposed techniques for improving the convergence and execution time of very hard problems. In these benchmarks, the execution of seqDE can take hours or even days to complete one only test. Four well known benchmarks problems from the BBOB data set were evaluated: Rastringin function (f_{15}), Schaffers function (f_{17}), Schwefel function (f_{20}), and Gallagher's Gaussian 21-hi Peaks function (f_{22}). The considered benchmark from the domain of computational system biology was a parameter estimation problem in a dynamic model of the *circadian* clock in the plant *Arabidopsis thaliana*, as presented in [25]. It must be noted that, as already available implementations in C/C++ and/or FORTRAN existed for all the benchmarks, we have wrapped them in our Scala code by using Java/Scala native interfaces (i.e. JNI, JNA, SNA).

There are many configurable parameters in the classical DE algorithm, such as the mutation scaling factor (F), the crossover constant (CR) or the mutation strategy (MSt), whose selection may have a great impact in the algorithm performance. Since the objective of this work is not to evaluate the impact of these parameters, only results for one configuration are reported here. For the selection of the settings in these experiments, in general, the suggestions in [8] have been followed. Previous tests have been done to select those parameters that lead to reasonable computation times. Table 1 shows the selected configuration for each benchmark.

Comparing the sequential and the parallel metaheuristics is not an easy task, therefore, guidance of [24, 26] has been followed when analyzing the results of these experiments. On the one hand, the behavior of the proposed solution was compared with the sequential classic version of DE (seqDE), therefore, speedups calculated as T_{seqDE}/T_{SiPDE} are reported in this section. On the other hand, both vertical and horizontal views can be used when evaluating a parallel metaheuristic. A vertical view assesses the performance of a fix number of evaluations, i.e., a pre-defined effort; while an horizontal view assesses the performance by measuring the time needed to reach a given target value. Thus, two different stopping criteria were considered in these experiments: maximum effort, for the vertical view,

Table 1. Benchmark functions. Parameters: dimension (D), population size (NP), crossover constant (CR), mutation factor (F), mutation strategy (MSt), value-to-reach/*ftarget* (VTR).

B	Function	D	NP	CR	F	MSt	VTR
BBOB benchmarks							
f_{15}	Rastrigin Function	5	800	.8	.9	DE/rand/1	1000
f_{17}	Schaffers F7 Function	6	1024	.8	.9	DE/rand/1	−16.94
f_{20}	Schwefel Function	6	1024	.8	.9	DE/rand/1	−546.5
f_{22}	Gallagher's Gaussian	10	1600	.8	.9	DE/rand/1	−1000
Systems Biology benchmark							
circadian	Circadian model	13	640	.8	.9	DE/rand/1	1e-5

and solution quality (using as stopping criterion a Value-To-Reach), for the horizontal view.

Results from both views are shown in Table 2. For each experiment, the number of nodes ($\#n$) used, the mean execution time and the standard deviation (in seconds) of the 10 independent runs in each experiment, the average number of migrations ($\#m$) performed in the SiPDE method, and the speedup (sp) achieved are shown. It should be noted that the stopping criterion is evaluated during each island evolution but, when it is met by one or more islands, the algorithm only stops after the reduce operation at the end of the stage (see Fig. 4). Thus, because no communication among workers is possible in Spark, the parallel SiPDE implementation cannot stop just right when the stopping criterion is reached (as the serial one does).

The figures for the vertical view (predefined effort) show that the proposed SiPDE method accelerates the computation of seqDE by performing the same number of evaluations in parallel. The figures for the horizontal view (quality solution) also demonstrate that the proposed SiPDE method reduces the computation time needed to achieve the VTR of the seqDE by improving the convergence of the algorithm.

The speedup achieved when using the predefined effort as stopping criterion deviates from the ideal one because of the overhead introduced by the communications. This fact can be observed in Fig. 5 where both the speedup and efficiency are shown, the latter calculated as *speedup/np*. In the experiments with the BBOB benchmarks, due to their short execution times, only two migrations were performed among islands before the stopping criterion is met, while for the *circadian* 20 migrations were performed. The efficiency results show that the overhead of the migrations barely affects on the performance when the execution time between two of them is significant (case of the *circadian* benchmark, where the efficiency is above 0.9), but it may greatly impact if it is small (case of BBOB benchmarks, where the efficiency is below 0.8 and significatively decreases when the number of nodes grows).

Table 2. Execution time in seconds, number of migrations ($\#m$), and speedup (sp) results for predefined effort (stopping criterion: $Nevals_{f15} = 1,000,000$; $Nevals_{f17} = 1,048,576$; $Nevals_{f20} = 1,040,000$; $Nevals_{f22} = 3,200,000$; $Nevals_{circadian} = 1,280,000$) and for a given solution quality (*Value-To-Reach* reported in Table 1), using different number of nodes ($\#n$). Average results from 10 independent runs in each experiment.

	method	#n	Predefined Effort time±std	#m	sp	Quality Solution time±std	#m	sp
f_{15}	seqDE	1	35.71 ± 0.20	-	-	47.89 ± 1.67	-	-
	SiPDE	2	27.86 ± 0.24	2	1.28	34.26 ± 1.47	2	1.40
		4	12.34 ± 0.21	2	2.89	15.88 ± 0.45	2	3.02
		8	6.39 ± 0.21	2	5.59	9.09 ± 0.25	2	5.27
		16	4.40 ± 0.40	2	8.12	3.73 ± 0.52	1.7	12.85
f_{17}	seqDE	1	36.60 ± 0.08	-	-	65.42 ± 2.13	-	-
	SiPDE	2	31.63 ± 0.27	2	1.16	53.74 ± 1.01	2	1.22
		4	13.31 ± 0.17	2	2.75	22.72 ± 0.28	2	2.88
		8	6.80 ± 0.24	2	5.38	12.79 ± 0.43	2	5.11
		16	4.25 ± 0.36	2	8.62	6.11 ± 0.35	2	10.71
f_{20}	seqDE	1	34.48 ± 0.07	-	-	55.12 ± 1.86	-	-
	SiPDE	2	27.00 ± 0.15	2	1.28	45.48 ± 2.19	2	1.21
		4	12.64 ± 0.16	2	2.73	17.78 ± 1.02	2	3.10
		8	6.47 ± 0.26	2	5.33	9.90 ± 1.05	2	5.57
		16	4.16 ± 0.41	2	8.28	4.08 ± 0.54	1.7	13.49
f_{22}	seqDE	1	112.14 ± 0.95	-	-	598.38 ± 478.31	-	-
	SiPDE	2	101.18 ± 0.94	2	1.11	703.97 ± 465.52	12.6	0.85
		4	44.03 ± 0.46	2	2.55	155.32 ± 156.19	7.8	3.85
		8	19.00 ± 0.40	2	5.90	84.60 ± 74.77	9.4	7.07
		16	10.25 ± 0.39	2	10.94	48.07 ± 37.64	11	12.45
circadian	seqDE	1	6267.97 ± 76.26	-	-	84482.53 ± 2369.75	-	-
	SiPDE	2	3111.14 ± 33.80	20	2.01	41736.17 ± 2114.45	26.7	2.03
		4	1575.26 ± 15.56	20	3.98	19029.74 ± 1042.31	24	4.44
		8	799.65 ± 9.23	20	7.84	8247.04 ± 558.07	20.4	10.24
		16	412.90 ± 4.93	20	15.18	2799.53 ± 377.43	13.4	30.39

Figure 5 also shows the speedup and efficiency when using the quality of the solution as stopping criterion. Speedups are larger than the ones obtained for predefined effort because the cooperation among islands in the parallel searches modifies the systemic properties of the algorithm, improving its convergence and outperforming the serial one. In the figures reported in Table 2 it can be seen that for the f_{15}, f_{17} and f_{20} benchmarks, again due to their short execution times, most of them converged after two migrations. However, when the number of nodes grew, some benchmarks required only one migration (so giving an average of 1.7 in some experiments), thus, improving the efficiency obtained for 16 nodes. The f_{22} benchmark is a highly multimodal function that frequently fall into an undesired stagnation condition. Thus, the dispersion of the experimental results

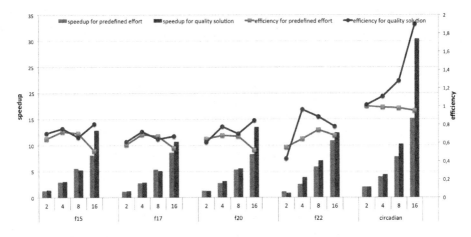

Fig. 5. Speedup and efficiency for results in Table 2

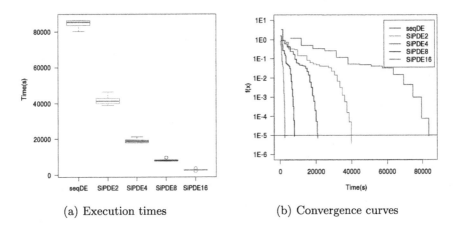

(a) Execution times   (b) Convergence curves

Fig. 6. Box plot of the execution times and convergence curves for the *circadian* benchmark with quality solution stopping criterion.

is very large (see the standard deviation in Table 2) and the number of migrations also varies from a minimum of 3 to a maximum of 19 in different runs.

For complex problems, like the *circadian* benchmark, the number of migrations clearly decreases with the number of nodes, demonstrating the potential of the parallel algorithm for improving the convergence of the DE method. The harder the problem is, the most improvement is achieved by the parallel algorithm, since the diversity introduced by the migration phase, although using a naive strategy as explained in Sect. 4.2, actually improves the effectiveness of the DE algorithm. Thus, for the *circadian* benchmark superlinear speedups are obtained, as well as efficiency above 1.

In this kind of stochastic problems it is also important to evaluate the dispersion of the experimental results. Figure 6(a) illustrates how the proposed SiPDE method reduces the variability of the DE execution time. This is an important feature that can be used to more accurately predict the boundaries in the cost of resources when using a public cloud like AWS.

Finally, to better illustrate the improvement of the proposed SiPDE method versus the seqDE method, Fig. 6(a) shows, for the circadian benchmark, the convergence curves for different number of nodes. As expected, convergence time is considerably reduced by SiPDE with respect to seqDE. It must be noted that these results could be further improved using more skilled mutation and migration strategies and adding enhancements, like local search [27], which have not been considered in this work.

6   Conclusions and Future Work

In order to explore how parallel metaheuristics could take advantage of the recent advances in Cloud programming models, in this paper Spark-based implementations of two different parallel schemes of the Differential Evolution (DE) algorithm, the master-slave and the island-based, are proposed and evaluated. Early benchmarking results showed that the island-based solution is by far the best suited to the distributed nature of Spark. Thus a thorough evaluation of this implementation was conducted on the AWS public cloud using a real testbed consisting on virtual clusters of different sizes.

Both synthetic and real biology-inspired benchmarks were used for the evaluation. The experimental results show that the proposal achieves not only a competitive speedup against the serial implementation, but also a good scalability when the number of nodes grows. The paper can be useful for those interested in the potential of Spark in computationally intensive nature-inspired methods in general and DE in particular. To the best of our knowledge, this is the first work on using Spark to parallelize DE.

Future work will be focus on developing new migration strategies and including further optimizations to improve the convergence of the Spark-based island parallel DE proposed.

Acknowledgements. This research received financial support from the Spanish Ministerio de Economía y Competitividad (and the FEDER) through the Project SYN-BIOFACTORY (grant number DPI2014-55276-C5-2-R). It has been also supported by the Spanish Ministerio de Ciencia e Innovación (and the FEDER) through the Project TIN2013-42148-P, and by the Galician Government (Xunta de Galicia) under the Consolidation Program of Competitive Research Units (Network Ref. R2014/041 and Project Ref. GRC2013/055) cofunded by FEDER funds of the EU.

References

1. Floudas, C.A., Pardalos, P.M.: Optimization in Computational Chemistry and Molecular Biology: Local and Global Approaches, vol. 40. Springer Science and Business Media, Heidelberg (2013)

2. Banga, J.R.: Optimization in computational systems biology. BMC Syst. Biol. **2**(1), 47 (2008)
3. Grossmann, I.E.: Global Optimization in Engineering Design, vol. 9. Springer Science and Business Media, Heidelberg (2013)
4. Crainic, T.G., Toulouse, M.: Parallel Strategies for Meta-Heuristics. Springer, Heidelberg (2003)
5. Alba, E.: Parallel Metaheuristics: a New Class of Algorithms. Wiley-Interscience, New York (2005)
6. Dean, J., Ghemawat, S.: MapReduce: simplified data processing on large clusters. In: Proceedings of the 6th USENIX Symposium on Operating Systems Design and Implementation, OSDI 2004 (2004)
7. Zaharia, M., Chowdhury, M., Das, T., Dave, A., Ma, J., McCauly, M., Franklin, M.J., Shenker, S., Stoica, I.: Resilient distributed datasets: a fault-tolerant abstraction for in-memory cluster computing. In: Proceedings of the 9th USENIX Symposium on Networked Systems Design and Implementation, NSDI 2012 (2012)
8. Storn, R., Price, K.: Differential evolution - a simple and efficient heuristic for global optimization over continuous spaces. J. Global Optim. **11**(4), 341–359 (1997)
9. Alba, E., Luque, G., Nesmachnow, S.: Parallel metaheuristics: recent advances and new trends. Int. Trans. Oper. Res. **20**(1), 1–48 (2013)
10. McNabb, A.W., Monson, C.K., Seppi, K.D.: Parallel PSO using MapReduce. In: IEEE Congress on Evolutionary Computation, CEC2007, IEEE, pp. 7–14 (2007)
11. Jin, C., Vecchiola, C., Buyya, R.: MRPGA: an extension of MapReduce for parallelizing genetic algorithms. In: IEEE Fourth International Conference on eScience, eScience 2008, IEEE, pp. 214–221 2008)
12. Verma, A., Llora, X., Goldberg, D.E., Campbell, R.H.: Scaling genetic algorithms using MapReduce. In: Ninth International Conference on Intelligent Systems Design and Applications, ISDA 2009, IEEE, pp. 13–18 (2009)
13. Radenski, A.: Distributed simulated annealing with MapReduce. In: Di Chio, C., Agapitos, A., Cagnoni, S., Cotta, C., de Vega, F.F., Di Caro, G.A., Drechsler, R., Ekárt, A., et al. (eds.) EvoApplications 2012. LNCS, vol. 7248, pp. 466–476. Springer, Heidelberg (2012)
14. Lee, W.P., Hsiao, Y.T., Hwang, W.C.: Designing a parallel evolutionary algorithm for inferring gene networks on the cloud computing environment. BMC Syst. Biol. **8**(1), 5 (2014)
15. Zhou, C.: Fast parallelization of differential evolution algorithm using MapReduce. In: Proceedings of the 12th Annual Conference on Genetic and Evolutionary Computation, ACM, pp. 1113–1114 2010)
16. Tagawa, K., Ishimizu, T.: Concurrent differential evolution based on MapReduce. Int. J. Comput. **4**(4), 161–168 (2010)
17. Daoudi, M., Hamena, S., Benmounah, Z., Batouche, M.: Parallel differential evolution clustering algorithm based on MapReduce. In: 6th International Conference of Soft Computing and Pattern Recognition (SoCPaR), IEEE, pp. 337–341 (2014)
18. Ekanayake, J., Li, H., Zhang, B., Gunarathne, T., hee Bae, S., Qiu, J., Fox, G.: Twister: a runtime for iterative MapReduce. In: The First International Workshop on MapReduce and Its Applications (2010)
19. Zhang, Y., Gao, Q., Gao, L., Wang, C.: IMapReduce: a distributed computing framework for iterative computation. In: Proceedings of the 1st International Workshop on Data Intensive Computing in the Clouds (DataCloud), p. 1112 (2011)
20. Bu, Y., Howe, B., Balazinska, M., Ernst, M.D.: HaLoop: efficient iterative data processing on large clusters

21. Ewen, S., Tzoumas, K., Kaufmann, M., Markl, V.: Spinning fast iterative data flows. CoRR abs/1208.0088 (2012)
22. Alexandrov, A., Bergmann, R., Ewen, S., Freytag, J.C., Hueske, F., Heise, A., Kao, O., Leich, M., Leser, U., Markl, V., Naumann, F., Peters, M., Rheinländer, A., Sax, M., Schelter, S., Höger, M., Tzoumas, K., Warneke, D.: The stratosphere platform for big data analytics. VLDB J. **23**(6), 939–964 (2014)
23. Odersky, M., Micheloud, S., Mihaylov, N., Schinz, M., Stenman, E., Zenger, M., et al.: An overview of the Scala programming language. Technical report (2004)
24. Hansen, N., Auger, A., Finck, S., Ros, R.: Real-parameter black-box optimization benchmarking 2009: experimental setup. Technical report RR-6828, INRIA (2009)
25. Locke, J., Millar, A., Turner, M.: Modelling genetic networks with noisy and varied experimental data: the circadian clock in arabidopsis thaliana. J. Theor. Biol. **234**(3), 383–393 (2005)
26. Alba, E., Luque, G.: Evaluation of parallel metaheuristics. In: PPSN-EMAA 2006, Reykjavik, Iceland, 9–14 September 2006
27. Penas, D., Banga, J., González, P., Doallo, R.: Enhanced parallel differential evolution algorithm for problems in computational systems biology. Appl. Soft Comput. **33**, 86–99 (2015)

ECJ+HADOOP: An Easy Way to Deploy Massive Runs of Evolutionary Algorithms

Francisco Chávez[1]([✉]), Francisco Fernández[1], César Benavides[2], Daniel Lanza[3], Juan Villegas[4], Leonardo Trujillo[5], Gustavo Olague[6], and Graciela Román[2]

[1] Department of Computer Science, University of Extremadura,
C/. Santa Teresa de Jornet, 38, CP 06800 Mérida, Spain
Chavez.fchavez@unex.es

[2] Departamento de Ing. Eléctrica, Universidad Autónoma Metropolitana,
Iztapalapa, San Rafael Atlixco 186, Vicentina, CP 09340 Mexico, D.F., Mexico

[3] CERN, European Organization for Nuclear Research, Meyrin, Switzerland

[4] Departamento de Electrónica, Universidad Autónoma Metropolitana,
Azcapotzalco, Av. San Pablo Xalpa No.180, Col Reynosa Tamaulipas,
CP 02200 Mexico, D.F., Mexico

[5] Instituto Tecnológico de Tijuana, Calzada Del Tecnológico S/N,
Fraccionamiento Tomas Aquino, CP 22414 Tijuana, BC, Mexico

[6] CICESE, Carretera Ensenada-Tijuana 3918, Zona Playitas, 22860
Ensenada, BC, Mexico

Abstract. This paper describes initial steps towards allowing Evolutionary Algorithms (EAs) researchers to easily deploy computing intensive runs of EAs on Big Data infrastructures. Although many proposals have already been described in the literature, and a number of new software tools have been implemented embodying parallel versions of EAs, we present here a different approach. Given traditional resistance to change when adopting new software, we try instead to endow the well known ECJ tool with the *MapReduce* model. By using the Hadoop framework, we introduce changes in ECJ that allow researchers to launch any EA problem on a big data infrastructure similarly as when a single computer is used to run the algorithm. By means of a new parameter, researchers can choose where the run will be launched, whether in a Hadoop based infrastructure or in a desktop computer. This paper shows the tests performed, how the whole system has been tuned to optimize the running time for ECJ experiments, and finally a realworld problem is shown to describe how the MapReduce model can automatically deploy the tasks generated by ECJ without additional intervention.

Keywords: Multi-objective evolutionary algorithm · Face recognition · Hadoop · ECJ

1 Introduction

Real life complex optimization problems are perfect candidates for Evolutionary Algorithms (EA), and new parallel hardware provide the means for addressing

© Springer International Publishing Switzerland 2016
G. Squillero and P. Burelli (Eds.): EvoApplications 2016, Part II, LNCS 9598, pp. 91–106, 2016.
DOI: 10.1007/978-3-319-31153-1_7

those problems. Yet, most of researchers still rely on sequential versions of the algorithms, mainly because of the difficulty of migrating software tools to these new parallel and distributed environments and also due to the typical resistance to change when adopting new software [29]. There are available a plethora of parallel programming languages, libraries and even paradigms. Moreover, open source solutions supporting big data approaches can be easily found, but resistance to change keep them frequently out of the path followed by researchers, who still continue relaying on sequential algorithms that are run on desktop computers.

During the last decades a number of new software tools have been developed embodying latest proposals for distributed EAs, including pool based models [23], p2p communication protocols allowing new interaction patterns among individuals within the populations [27], among others. Thus old parallel models already studied for GAs [19] and GP [20] as well as new ones are available for researchers, matching any standard parallel or distributed infrastructure or technology: clusters and message passing libraries, such as PVM and MPI [11,12], GRID based tools [21], also including Desktop grids [22] and pool based approaches [23]. Nevertheless, few have considered providing a tool that makes use of MapReduce models as described first by Google [28].

Cloud models dominate the distributed computing landscape of today, mainly because of their easy-to-use virtualization based approach: researchers can still rely in their preferred programming language and operating system while running the algorithms in third-party hardware infrastructure. Yet, for optimally profiting from those cloud resources, parallel versions of the algorithms must be employed. The EA community is aware of that need, and different research groups continue offering solutions, such as FlexGP [24] or EvoSpace [23] both providing EAs following the SaaS (Software as a Service) approach that can be run in Paas (Platform as a Service).

Even though those models and tools are useful, we are still far from the goal: avoiding steep learning curves for new software tools and infrastructures; what many would like is to be able to continue using a well known tool that progressively incorporates the latest technologies. This would allow researchers to continue using it without changes on any new available hardware, and this is the goal we pursue here.

ECJ [16] is one of the best known and employed tool for running EAs, and already offer the possibility of using multicore systems, but still lack the capability for running on Grids nor new models typically employed when *Big Data* applications are addressed. Although some projects have recently tried to include Desktop Grid Models [25], new trends in Big Data show us the path towards improving the tool. We think that new solutions that allow users to transparently launch EA runs with ECJ on Big Data infrastructures will be welcome.

We present below a first proposal for allowing ECJ to deploy massive version of EAs by means of the MapReduce model, making it transparent for users, who will simply have to select the way of running the algorithm, sequentially or in the distributed model, by means of a single parameter. Our proposal relies on Hadoop [13], which uses the MapReduce parallel paradigm. We include below the performance tests that we run, issues found and possible solutions, as well as a real life problem implemented and run with this improved version of ECJ, a computationally intensive face recognition problem.

1.1 Evolutionary Approaches to Face Recognition

Although face recognition is an easy task for humans, it becomes very complex when addressed by means of computers. Different approaches have been already described [5–7], but the algorithms typically incur in a high computational cost [2]. Algorithms typically explore a big database containing hundreds of photographs, which are employed for both training and testing the algorithm capabilities.

Some of the best known approaches to the problem begin with a previously established set of points located over the human face [8–10]. The computing intensive task required to address the problem have already led other researchers to introduce MapReduce methodologies [3,4]. But the difficulty of the problem increases if we want the algorithm to also decide which are the preferred points in the face for the training and testing process. This meta-learning process notably increases the difficulty and the time required for any machine learning algorithm, although if solved will provide useful information. Given the computing resources required to address this problem, we decided to use it for testing the scalability of ECJ+Hadoop. In the work presented below we use the technique based on Content-Based Image Retrieval (CBIR) [1], and the EA will be in charge of deciding which are the specific points of interest (POI) that should be employed.

If we use a population size 100 individuals, and given that the algorithm needs around 4 min per individual -as we will see below-, it will take around 6 h evaluating a single generation. Depending on the number of generations, the total time will be weeks or months. Thus, this is a perfect problem to be deployed using ECJ+Hadoop as described below.

The rest of the paper is structured as follows: Sect. 2 describes the tools and technologies. Section 3 presents the hardware infrastructures employed to test the system. Results obtained are shown in Sect. 4. Finally, Sect. 5 draws our conclusions.

2   ECJ and Hadoop

ECJ is one of the best known EA tools, developed in java. ECJ offers a number parallel models, such as the island model, that can be run on top of desktop multicore systems and clusters. But when somebody is interested in using a different infrastructure, a number of changes must be applied to the tool. Such was the case for the recently published Desktop Grid version of the tool [25].

In this work we follow an alternative path, also trying to improve the ECJ project: we try to adapt it to the MapReduce model [17,18]. However another important goal is that internal changes should not affect the way researchers use the tool, thus avoiding resistance to change that may arise.

MapReduce is a model deriving from the functional programming paradigm, well known for its inherent parallel way of describing actions to be performed. MapReduce allows to apply *map* operations massively in a parallel way to a big amount of data, and then employ the *reduce* operation to unify and extract conclusions.

The interest that this model has arisen, has allowed the development of a number of software tools and frameworks embodying it. One of the best known open source initiative has been developed by the Apache foundation, giving rise to Hadoop [13,14].

Apache Hadoop includes an ecosystem that allows to massively process big amounts of data by means of a simple programming model. Hadoop easily scales, and it can be run in a single computer as well as several thousands of processors. The framework is fault tolerant, thus providing high availability features. Hadoop provides: (1) a high performance distributed file system, HDFS, so that big amount of data can be easily shared and accessed by every computer node; (2) tools for managing and balancing work units, and finally (3) the MapReduce programming model.

In this work we present a modification on the ECJ tool so that it can run using MapReduce over clusters, so that the EAs community can profit from that technology. Although several parallel and distributed models of EAs could be considered, we focus here on the simplest one, that aims at evaluating simultaneously a number of individuals from the population. Therefore, the idea is that the fitness function can be considered as a Hadoop task, so that every generation all of the fitness evaluations can be managed by the tool and run by means of the MapReduce model.

2.1   Modifying ECJ

As described above, we decided to employ an embarrassingly parallel model when connecting ECJ and Hadoop, so that fitness evaluations are deployed as working units in the MapReduce model. In order to improve performances, we consider the possibility of including not just a single individual per working unit but a number of them, so that latencies related to communication processes can be reduced. According to the MapReduce model, the fitness function will be the task to be distributed, so that the main evolutionary loop must be changed accordingly.

Once we checked ECJ source code, we noticed that several functions were already available providing check pointing facilities: the population may thus be stored when required, freezing the process that can continue later by simply loading the *serialized file* that was previously saved.

This has thus been the main operation on which we build the new functionality, given that the saved file can be accessed on the HDFS by as many *mapper* processes as required, and all of them will share the state of the evolutionary process and know the set of individuals to be evaluated. We must nevertheless take into account that the serialized information does not include the population of individuals, that must also be saved in a different file. Figure 1 graphically depicts how Hadoop works with the *Master* process, in charge of managing the main loop, and then the *slave* layer is in charge of launching each of the map works, known here as *TaskTrackers*.

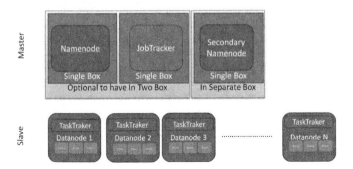

Fig. 1. Layers of Hadoop.

Once the check pointing process has saved the required information, a file containing all of the individuals to be evaluated is also produced, that will be the input for the first task that hadoop will launch. This first task will then produce as many new *mapper* tasks as we require to evaluate all the population. Every *mapper* will then read info from the problem previously saved and shared through the file system and load the set of individuals to be evaluated. We delegate on Hadoop the load balancing strategy that benefits the way work units are distributed along the computing nodes. Figure 2 shows the relationship between map and reduce processes for ECJ.

A map process requires a series of input/output parameters, namely a key-value pair as input and another one as output. In our specific case the key-value pair will be the index for the individual to be evaluated and the value the individual itself. For the output, the value will be thus provided by the fitness function. Once the work is done, all of the fitness values are gathered thus finishing the map step, and the algorithm can continue within the master process.

All of the above described changes have been applied in a new version we name *ECJ+Hadoop*, sometimes providing new functions required, such as *ec.hadoop.HadoopEvaluator*. This new process will be in charge of computing fitness values. The code 1 includes the detail for the map process that will be run simultaneously in every processor.

Any ECJ user knows that for any new problem to be solved by means of the tool, it is typically enough to implement the fitness function and select parameter values. For the new model described, and given that we want all the process to be transparent for users, the idea must be the same: The user will only have to provide the fitness function, and then to set the *eval* parameter to the new available value, *ec.hadoop.HadoopEvaluator*. Other parameters involved are the following ones:

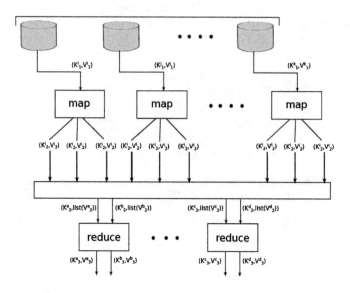

Fig. 2. Map and Reduce processes

```
import java.io.IOException;

import org.apache.hadoop.mapreduce.Mapper;

import ec.EvolutionState;
import ec.Individual;
import ec.hadoop.writables.FitnessWritable;
import ec.hadoop.writables.IndividualWritable;
import ec.hadoop.writables.IndividualIndexWritable;
import ec.simple.SimpleProblemForm;

public class EvaluationMapper extends
  Mapper<IndividualIndexWritable,
      IndividualWritable, IndividualIndexWritable,
    FitnessWritable> {

  public static EvolutionState state;

  @Override
  protected void map(IndividualIndexWritable key,
    IndividualWritable value, Context context)
        throws IOException, InterruptedException {

    state = value.getEvaluationState();
    Individual ind = value.getIndividual();

    //Evaluate individual
    SimpleProblemForm problem =
      ((SimpleProblemForm) state.evaluator.p\_problem);
    problem.evaluate(
        state,
        ind,
        key.getSubpopulation(),
        0);

    //Emit key and value
    context.write(
        key,
        new FitnessWritable(state, ind.fitness));}}
```

Algorithm 1. EvaluationMapper

- Main folder for the serialized file. Parameter name: eval.hdfs-prefix, Default value: ecj_work_folder_hc.
- Address for the JobTracker. Parameter name: eval.jobtracker-address, Default value: localhost.
- JobTracker port. Parameter name: eval.jobtracker-port, Default value: 8021.
- HDFS address. Parameter name: eval.hdfs-address, Default value: localhost.
- HDFS port. Parameter name eval.hdfs-port, Default value: 8020.

This way, anybody interested can easily implement and deploy an ECJ+Hadoop based optimizacion problem. Of course, it is desirable to run this new version of the tool on a cluster of computers. In the experimental stage described in the next sections, we have employed *Cloudera*.

3   System Deployment

Among the different possibilities for setting up a Hadoop infrastructure, we decided to use Cloudera [15]. Cloudera is a single platform for Big Data projects integrating Hadoop among other tools. It allows to centrally manage and monitor the hardware infrastructure on which it is installed. It allows both to run parallel processes and also share distributed storage system. Cloudera provides a quick and easy installation of all the required elements to use Hadoop, and also includes an entire ecosystem of tools that facilitate managing tasks, scheduling and monitoring jobs. Figure 3 shows the toolkit that accompanies Cloudera Hadoop when installed.

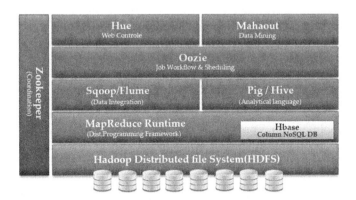

Fig. 3. Tools integrated in Cloudera

Among the tools included in Cloudera we find:

- Pig: high level data-flow language to facilitate MapReduce programming.
- Hive: provides a SQL interface with data stored in HDFS.
- Oozie: workflow planner to manage Hadoop jobs.

- Hue: Web interface to simplify the use of Hadoop.
- HBase: distributed non-relational database (NoSQL) running on HDFS (inspired by Google BigTable).
- Sqoop: allows efficient transfer of data between Hadoop and relational databases
- ZooKeeper: centralized configuration, named, distributed and synchronization services groups for large distributed file systems.
- Flume: collection, aggregation and movement of large log files to HDFS.
- Mahout: Scalable machine learning algorithms and data mining on Hadoop.

We configured Cloudera on a hardware infrastructure consisting of 6 blades servers, 2 blades with 4 cores each and 4 blades with 8 cores each. The main features for each of the servers are described in Table 1

Table 1. Cloudera based infrastructure

Node	RAM (Gb)	HD (Gb)	Cores	Roles Cloudera
1	16	75.5	8	13
2	16	74.7	4	9
3	16	74.7	4	10
4	16	141.4	8	7
5	16	75.5	8	5
6	16	75.5	8	6

Once Cloudera is deployed on the cluster, a shared cache memory and a distributed storage through HDFS are shared among all the computing nodes. This allows us to view the system as a single unit, where data is distributed and can be accessed by all processes deployed in the cluster.

4 Experiments and Results

As described below, we have performed a series of experiments that have allowed us to both apply a tuning process to the *ECJ+Hadoop* tool and also check the advantages of the approach when it is applied to solve an optimization problem. Both a benchmark problem and also a real-world problem have been considered.

4.1 Analysing Performance

We have first analyzed the performance of this MapReduce version of ECJ. The first series of experiments have consisted in running one of the benchmark problems already available: *Even Parity*. This problem was presented first by J. Koza [26] and has since then been considered one of the benchmarks for Genetic Programming. We have run evenp-12, instead of the more standard five bits version, with the idea of making it difficult enough to be able to work with large populations during a number of generations.

As described above, we are not here mainly interested in the problem itself. We simply employ it to study how large population sizes might be used considering the latencies that arise due to the handling of the population during the process of creating packages of individuals to be distributed, fitness values collection, application of centralized genetic operators, etc. Sumarizing, we analyze the performance of the approach to identify ideal situations where ECJ+Hadoop is of interest.

Although many possibilities exist to analyze results, we decided here to compare with the standard ECJ tool which allows us to run the problem using multithread capabilities. The base case comparison is thus a single computer with up to 8 threads.

Therefore, we started running the problem with different population sizes: from 30,000 to 1,500,000 individuals. Larger sizes of the population were only employed with larger number of cores. Table 2 displays a summary of the experiments run on a single computer.

Table 2. Experiments in a standard PC with 8 threads. Fitness evaluation time in seconds for a generation

Threads/Individuals	30000	50000	100000	300000	500000	1000000	1500000
1	19	31	65				
2	10	16	32				
4	5	8	16	47	82		
8	3	4	8	25	41	87	130

As we can see, the larger the number of cores, the shorter the time required to evaluate a single generation.

The idea now is to run similar experiments using EJC+Hadoop and compare results with those shown below for the ECJ tool. When running Hadoop based processes we have to take into account some parameters that highly influence performance, and some values whose analysis allow us to understand the behaviour of the whole system:

- HDFS block size (sz-HDFS) is a configurable parameter that allows us to manage the number of data blocks in HDFS, allowing greater parallelization of *mappers* created by the *JobTracker*.
- Time to store the population in HDFS (t-wrt): This measure tells how long it will take to write the population in the distributed file system.
- Average task time (t-task) shows the average execution time for each task.
- Tasks are the number of tasks launched for each job. This number depends on the workload given by the popultion size and of the used HDFS block size.
- Evaluation time (t-eval) is the total time to evaluate a population.

Table 3 shows an experiment summary with different configurations, where the parameters and meassures described above are shown.

Table 3. ECJ+Hadoop results with different HDFS blocks and populations sizes, using 40 cores

sz-HDFS	sz-Pob	t-wrt (s)	t-task (s)	Tasaks	t-eval (s)
14	3,000,000	22	45	55	141
	2,500,000	18	44	46	132
	2,000,000	14	46	36	123
	1,500,000	10	45	27	101
	1,000,000	18	40	18	84
10	2,500,000	18	40	64	174
	2,000,000	14	38	50	111
	1,500,000	11	38	38	94
	1,000,000	7	38	25	64
	750,000	6	30	19	58
	500,000	4	30	12	52
5	2,500,000	18	25	127	132
	2,000,000	14	22	101	101
	1,500,000	11	22	75	92
	1,000,000	8	24	49	73
	750,000	6	22	37	60
	500,000	4	21	25	40
3	2,500,000	19	20	211	140
	2,000,000	14	18	167	109
	1,500,000	11	20	124	103
	1,000,000	8	17	82	61
	500,000	4	17	41	44
1.25	1,000,000	8	11	196	70
	500,000	4	11	97	43
	300,000	2	11	58	31
	100,000	1.5	8	19	24
	50,000	1	7	10	19
	30,000	0.6	7	6	19
1	30,0000	2	10	72	34
	10,0000	1.5	8	24	25
	5,0000	1	6	12	19
	3,0000	0.6	6	7	19

Considering small number of individuals this first series of experiments using ECJ+Hadoop were discouraging, providing poor performance when compared with the standard multicore approach, so we decided to further explore some of the elements that could influence the computing time.

We decided thus to avoid storing the whole population within the checkpoints. Instead individuals are serialized and directly saved within HDFS. Previously text format was used but a serialized model saves time and resources when saving individuals due to the shorter length of the serialized model. The impact of this improvement is large due to the way the HDFS distributed file system must save the information on a number of devices. Moreover, we employed compression processes to save read/write operation time.

Second, given the big size of the populations we are managing, we decided to enlarge the buffers employed for transferring information, which also influence the transfer rates.

We must also take into account that different processes running on nodes are held in Java virtual machines (JVM), and with each new generation these Virtual Machines are created and then destroyed again, which is time consuming. However Hadoop has a parameter to avoid this behaviour and allow the *JVM* to be reused, saving time. This feature was enabled and an improvement was observed in processing times.

All of the above described improvements were applied in the next round of experiments. Figure 4 shows the time required for different population sizes when using hadoop versus the standar runs of ECJ using differet number of cores. Graphs shows computing time in seconds required for evaluating a whole generation for each of the population sizes tested, while the color refers to different configurations both of ECJ and ECJ+Hadoop using Hadoop.

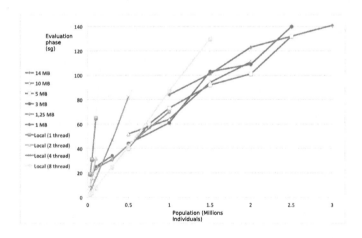

Fig. 4. Evaluation times in Hadoop and multithreads (Color figure online).

The red, green, purple and yellow lines show the result of the experiments on a single computer using 1, 2, 4 and 8 threads respectively, using the standard ECJ tool. We can see in these results that the time linearly scales with the size of the population making it easy to predict the behaviour with any number of individuals. Also, the relationship with the number of cores employed can be clearly noticed.

On the other hand, different blue lines corresponds with Hadoop-based runs using up to 40 cores available, configured with different block sizes each, which affects the total number of individuals that will be included in every working unit. All of the runs employ the total number of cores available. Although many operations are performed by Hadoop when launching and running map/reduce operations, we may clearly notice the difference obtained with ECJ standard multicore performances.

The most interesting part of the graph is where blue lines overpass the yellow one, corresponding to ECJ with 8 threads; this is the time when ECJ+Hadoop begins to provide better performance. Therefore, we could firstly conclude that with more standard population sizes, which are always below 500,000 individuals, it is not useful to employ ECJ+Hadoop, particularly in problems whose fitness evaluation time is short, as was the case for the even parity problem. We must take into account that for the even parity problem, the computing time is so short, that only when huge population sizes are required ECJ+Hadoop will be of interest. On the other hand, if we are addressing a population with large fitness evaluation time, the critical point when ECJ+Hadoop is of interest will be reached with much smaller population sizes.

Therefore, in our next experiment we have tried to test performances of the tool in a specific real life problem that employees a computationally intensive fitness function.

4.2   A Real Life Problem: Face Recognition

A number of databases are available with images for the learning step. In this work we have employed BioID, MUCT and MEDS II [30]. Each image from the database includes meta information, known as *POI* coordinates that are located in different locations of the face.

The problem we want to run consists of applying an EA capable of deciding which are the best POIs for the learning process. Thus, this meta-learning algorithm will employ a Genetic Algorithm (GA) with a binary chromosome: each of the bits describes whether a given POI will be employed during the training and testing process of the algorithm. The fitness function will then check whether the algorithm is capable of learning and then correctly classifying faces only based on the POIs selected by each of the GA individuals, while the learning algorithm is CBIR, as described in [31]. Given the training-testing process each of the individuals must perform for assing its fitness value, the problem is computationally intensive, and this is precisely the reason for selecting it to test ECJ+Hadoop. At this stage we are not still trying to solve the proposed problem, so other GA parameters are not tuned here, and we are only interested in the size of the population; we simply use the problem as a test for understanding the performances that may be obtained with the new version of ECJ. We thus focus on the step that needs more computing power, the individual's evaluation. We describe below what is required for a proper evaluation of the individuals in the population.

Each of the individuals provide a set of POIs. The learning algorithm must then employ each of the faces belonging to the training set from the database to extract the points and train. Then, the images belonging to the test set are used to check the accuracy of the algorithm. Thus the fitness function must apply a number of image processing operations to convert images to HSI color space, compute statistical values and generate a descriptor vector, which will be the output for the image. Given the way hadoop works, images will be processed by a different Map operation, and then a final Reduce process will take outputs produced by every Map process and generate a normalised matrix that is finally used by a *k-Means* algorithm to compute the final matrix.

This matrix is then employed by a query job, where the algorithm tries to recognise the faces. Results obtained are checked with classes stored into image database, and the true positives percentage is used as the final fitness value, which is then stored into the HDFS. Once all jobs have finished, the evaluation phase of the EA concludes, and the algorithm can continue with the next step.

In order to test the advantages of the new ECJ+Hadoop, we launched a series of experiments trying to analyse when the new implementation outperforms the results obtained with the sequential approach, and also the results obtained in a multithread system. The following table shows results obtained with different population sizes. We must take into account that Hadoop incurs in a computational overhead due to the extra processes and operations necessary to deploy and manage a distributed platform.

Table 4. Time (minutes) for evaluating a population of individuals

ECJ tool/Individuals	10	20	30	40	50	60
Standard sequential	44.5	89	133.5	178	222.5	267
Standard multithreading	10.7	21.4	32.1	42.8	53.5	64.2
Hadoop	6.4	12.8	19.2	25.6	32	38.4

Table 4 shows running times obtained experimentally with up to 60 individuals using the standard sequential version of ECJ, the multithreading version running on a single Pc with 8 cores, and then the Hadoop based model with up to 40 cores. We see that for this specific problem the new version of the tool makes sense once we simply reach the population size of 10 individuals. Given the difficulty of the faced problem, we think that probably a much larger population will be required in future work. Therefore, we have computed the running times that may be obtained when enlarging the population with up to 100 individuals. Given that ECJ+Hadoop can make use of as many computer nodes as required, we understand that the running time can be keep fixed by simply adding more nodes when required, while the standard ECJ tool cannot use more than 8 threads available in the multicore desktop PC where it is run. Figure 5 allows to graphically see that differences increase as we manage larger populations.

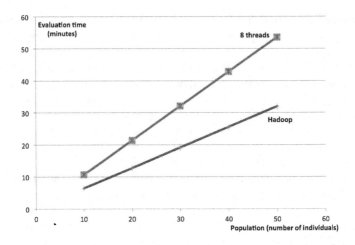

Fig. 5. Comparing ECJ 8 threads vs. ECJ-Hadoop

As we have seen, ECJ+Hadoop is of interest when long computing times are required for assessing the quality of individuals in the population or when large populations are required to solve the problem, or both.

5   Conclusions

This preliminary work presents a new version of ECJ tool, which allows to run EA over BigData infrastructure making use of Map/Reduce model. We provide an embarrassingly parallel version of the model that allows to run experiments on Hadoop based cluster of computers.

The implementation makes use of the checkpointing facility already provided by ECJ. The system has been tuned and some hints for important parameters affecting running times for the experiments are provided.

We have performed a series of tests considering both traditional benchmark problems already available within the ECJ tool and also a real life problem. Results show that the new version of ECJ allows to save computing time when the fitness function is computationally intensive and or when large population sizes are required to solve a given problem. Moreover, the implementation and running of the experiments does not change the way ECJ tool is traditionally used, and a single parameter allows to specify whether the run will be launch in a single computer on in the Hadoop based cluster available.

Much work must be done yet: allowing other models to be run within the map/reduce approach, such as the Island Model, better optimize the launching time for the tasks, thus allowing to save time even when small population sizes are considered. Yet, the tests presented show the usefulness of the approach, that allow researchers to profit from hadoop based infrastructures without changing their way of working. We also would like to make ECJ automatically decide which is the best parameter to be used (traditional or Hadoop based ECJ), depending on the application characteristics.

Acknowledgements. This work has been supported by FP7-PEOPLE-2013-IRSES, Grant 612689 ACoBSEC, Spanish Ministry of Economy, Project UEX:EPHEMEC (TIN2014-56494-C4-2-P); Junta de Extremadura, and FEDER, project GR15068. Also it has been supported by CONACyT México by the project 155045 – "Evolución de Cerebros Artificiales en Visión por Computadora" and TESE by the project DIMI-MCIM-004/08.

References

1. González, A., Prieto, F.: Extracción de puntos característicos del rostro para medidas antropométricas. Revista Ingenierías Universidad de Medellín, **9**(17), 139–150 (2010)
2. Jain, A.K.: Automatic face recognition: state of the art, Distinguished Lecture Series, 0–44, Septiembre (2010)
3. Sun, K., Kang, H., Park, H.-H.: Tagging and classifying facial images in cloud environments based on KNN using MapReduce. Optik Int. J. Light Electron Opt. **126**(21), 3227–3233 (2015). ISSN 0030–4026. http://dx.doi.org/10.1016/j.ijleo.2015.07.080
4. Zhang, Z., Li, W., Jia, H.: A fast face recognition algorithm based on MapReduce. In: 2014 Seventh International Symposium on Computational Intelligence and Design (ISCID), vol. 2, pp. 395–399, 13–14 December 2014. doi:10.1109/ISCID.2014.195
5. Zhao, W., Chellappa, R., Phillips, P.J., Rosenfeld, A.: Face recognition: a literature survey. ACM Comput. Surv. (CSUR) **35**(4), 399–458 (2003)
6. Yang, M.H., Kriegman, D.J., Ahuja, N.: Detecting faces in images: a survey. IEEE Trans. Pattern Anal. Mach. Intell. **24**(1), 34–58 (2002)
7. Hsu, R.L., Abdel-Mottaleb, M., Jain, A.K.: Face detection in color images. IEEE Trans. Pattern Anal. Mach. Intell. **24**(5), 696–706 (2002)
8. Zhu, X., Ramanan, D.: Face detection, pose estimation, and landmark localization in the wild. In: 2012 IEEE Conference on Computer Vision and Pattern Recognition (CVPR), pp. 2879–2886, 16–21 June 2012. doi:10.1109/CVPR.2012.6248014
9. Zhang, L., Yang, M., Feng, X.: Sparse representation or collaborative representation: which helps face recognition? In: 2011 IEEE International Conference on Computer Vision (ICCV), pp. 471–478, 6–13 November 2011. doi:10.1109/ICCV.2011.6126277
10. Wagner, A., Wright, J., Ganesh, A., Zhou, Z., Mobahi, H., Ma, Y.: Toward a practical face recognition system: robust alignment and illumination by sparse representation. IEEE Trans. Pattern Anal. Mach. Intell. **34**(2), 372–386 (2012). doi:10.1109/TPAMI.2011.112
11. Fernández, F., Sánchez, J.M., Tomassini, M., Gómez, J.A.: A parallel genetic programming tool based on PVM. In: Margalef, T., Dongarra, J., Luque, E. (eds.) PVM/MPI 1999. LNCS, vol. 1697, pp. 241–248. Springer, Heidelberg (1999)
12. Tomassini, M., Vanneschi, L., Bucher, L., Fernandez de Vega, F.: An MPI-based tool for distributed genetic programming. In: IEEE International Conference on Cluster Computing (CLUSTER 2000) pp. 209–209. IEEE Computer Society (2013)

13. Shvachko, K., Hairong, K., Radia, S., Chansler, R.: The hadoop distributed file system. In: 2010 IEEE 26th Symposium on Mass Storage Systems and Technologies (MSST), pp. 1–10 (2010)
14. White, T.: Hadoop: the definitive guide (2009)
15. Cloudera. http://www.cloudera.com/
16. ECJ: A Java-based Evolutionary Computation Research System. http://cs.gmu.edu/eclab/projects/ecj/
17. Dean, J., Ghemawat, S.: MapReduce: a flexible data processing tool. Commun. ACM **53**(1), 72–77 (2010)
18. Dean, J., Ghemawat, S.: MapReduce: simplified data processing on large clusters. Commun. ACM **51**(1), 107–113 (2008)
19. Cantu-Paz, E.: Efficient and Accurate Parallel Genetic Algorithms. Genetic Algorithms and Evolutionary Computation, vol. 1. Springer, New York (2011)
20. Fernandez, F., Tomassini, M., Vanneschi, L.: An empirical study of multipopulation genetic programming. Genet. Program. Evolvable Mach. **4**(1), 21–51
21. Melab, N., Cahon, S., Talbi, E.G.: Grid computing for parallel bioinspired algorithms. J. Parallel Distrib. Comput. **66**(8), 1052 1061
22. Gonzalez, D.L., et al.: Increasing gp computing power for free via desktop grid computing and virtualization. In: 17th Euromicro International Conference on Parallel, Distributed and Network-based Processing, pp. 419–423. IEEE (2009)
23. García-Valdez, M., Trujillo, L., Fernández de Vega, F., Merelo Guervós, J.J., Olague, G.: EvoSpace: a distributed evolutionary platform based on the tuple space model. In: Esparcia-Alcázar, A.I. (ed.) EvoApplications 2013. LNCS, vol. 7835, pp. 499–508. Springer, Heidelberg (2013)
24. Sherry, D., Veeramachaneni, K., McDermott, J., O'Reilly, U.-M.: Flex-GP: genetic programming on the cloud. In: Di Chio, C., et al. (eds.) EvoApplications 2012. LNCS, vol. 7248, pp. 477–486. Springer, Heidelberg (2012)
25. Fernndez de Vega, F., Chvez, F., Trujillo, L., Mediero, E., Muoz, L.: A Hybrid ECJ+Boinc tool for distributed evolutionary algorithms. In: Research in Computing Science, pp. 120–130 (2014)
26. Koza, J.R.: Genetic Programming: On the Programming of Computers by Means of Natural Selection, vol. 1. MIT Press, Cambridge (1992)
27. Laredo, J.L.J., Eiben, A.E., Steen, M., Merelo, J.J.: Evag: a scalable peer-to-peer evolutionary algorithm. Genet. Program. Evolvable Mach. **11**(2), 227–246 (2010)
28. Du, X., Ni, Y., Yao, Z., Xiao, R., Xie, D.: High Performance parallel evolutionary algorithm model based on mapreduce framework. Int. J. Comput. Appl. Technol. **46**(1), 290–295 (2013)
29. Stelzer, D., Mellis, W.: Success factors of organizational change in software process improvement. Softw. Process Improv. Pract. **4**(4), 227–250 (1998)
30. Fei-Fei, L., Fergus, R., Perona, P.: Learning generative visual models from few training examples: an incremental bayesian approach tested on 101 object categories. Comput. Vis. Image Underst. **106**(1), 59–70 (2007)
31. Benavide, C., Villegas, J., Román, G., Avilés, C.: Face recognition using CBIR techniques (Spanish). In: Proceedings MAEB 2015, pp. 733–740 (2015)

Addressing High Dimensional Multi-objective Optimization Problems by Coevolutionary Islands with Overlapping Search Spaces

Pablo García-Sánchez$^{(\boxtimes)}$, Julio Ortega, Jesús González, Pedro A. Castillo, and Juan J. Merelo

Department of Computer Architecture and Technology,
University of Granada, Granada, Spain
pablogarcia@ugr.es

Abstract. Large-scale multi-objective optimization problems with many decision variables have recently attracted the attention of researchers as many data mining applications involving high dimensional patterns can be leveraged using them. Current parallel and distributed computer architectures can provide the required computing capabilities to cope with these problems once efficient procedures are available. In this paper we propose a cooperative coevolutionary island-model procedure based on the parallel execution of sub-populations, whose individuals explore different domains of the decision variables space. More specifically, the individuals belonging to the same sub-population (island) explore the same subset of decision variables. Two alternatives to distribute the decision variables among the different sub-populations have been considered and compared here. In the first approach, individuals in different sub-population explore disjoint subsets of decision variables (i.e. the chromosomes are divided into disjoints subsets). Otherwise, in the second alternative there are some overlapping among the variables explored by individuals in different sub-populations. The analysis of the obtained experimental results, by using different metrics, shows that coevolutionary approaches provide statistically significant improvements with respect to the base algorithm, being the relation of the number of islands (subpopulations) to the length of the chromosome (number of decision variables) a relevant factor to determine the most efficient alternative to distribute the decision variables.

Keywords: Multi-objective algorithms · NSGA-II · Island model · Distributed evolutionary algorithms

1 Introduction

Evolutionary Algorithms (EAs) are inherently parallelizable, as each individual can be considered as an independent unit [1], and therefore, the computational performance can be improved over their non-parallel versions. This can also be applied to Multi-Objective Evolutionary Algorithms (MOEAs). Besides, as this kind of algorithms may be computationally expensive, several parallelization

© Springer International Publishing Switzerland 2016
G. Squillero and P. Burelli (Eds.): EvoApplications 2016, Part II, LNCS 9598, pp. 107–117, 2016.
DOI: 10.1007/978-3-319-31153-1_8

methods have been proposed [2]. However, as MOEAs deal with whole solution sets called Pareto Fronts (PFs), different distribution and sharing mechanisms need to be addressed, as there exist a tension between the speedup achievable from parallelization and the need to globally recombine results to accurately identify the PF [3].

Classic approaches, such as the global parallel EAs (Master-Slave), or the spatially structured algorithms (Island model or Cellular EAs) have been applied successfully in the past [4,5].

MOEAs have gained attention in the last years, mostly because their application in real-world problems [2,6]. Usually these problems require a higher number of decision variables, and these larger individuals require a significant extra time for crossover, mutation and transmission. Therefore, dividing the decision space (that is, the chromosome) may improve high performance and solution quality. In this aspect, the co-evolution model is a dimension-distributed model where a high-dimensional problem is divided into lower dimensional problems [7], and evolved separately. Also, applying parallelism techniques in EAs not only implies a reduction in the execution time, but developing new and more efficient search models [2]. The idea of dividing the decision space has been studied in [8]. In that work, different workers evolve sub-populations created and recombined by a master process, which performs different recombination alternatives from the parts returned by worker processes. A high dimensional problem was used to compare these alternatives. In [9], a distributed coevolutionary island model was used. In both previous works significant speedups were attained, although only a low number of nodes were used (8). However, when increasing the number of islands, the division of each section of the chromosome becomes smaller, and the scalability may be affected by obtaining lower quality solutions in the same amount of time.

The hypothesis of this paper is that using overlapped sections of the chromosome in a coevolutionary multi-objective island-based algorithm can improve the quality of the solutions in the same computing time when the number of islands increases. To demonstrate this, a new overlapping islands scheme will be compared with a baseline version of a distributed NSGA-II [10] and with a coevolutionary disjoint section approach, in different benchmarks problems and with different large number of islands. Results show that this new technique can improve different quality indicators in the same amount of time.

The rest of the paper is structured as follows: after a background in parallelization in MOEAs, the compared algorithms and the used methodology (Sect. 3) are presented. Then, the results of the experiments are shown (Sect. 4), followed by conclusions and suggestions for future work lines.

2   State of the Art

Distributed and parallel Evolutionary Algorithms have been used since the early 2000s [11] to leverage the capabilities of new common parallel computer architectures, such as clusters or grids. However, distributing MOEAs is not as easy as it is in single-objective EAs, as they have to deal with a whole set of dependent solutions, the Pareto Front, that should be managed in several steps of the algorithm. Some authors have focused on Master-Slave approaches to parallelize this kind of EAs. For example, Nebro and Durillo [12] compared different master-slave approaches to save time when running the EA. The method proposed by Hiroyasu et al. [13] generates offspring depending on the computation power.

First approaches for distributed MOEAs (dMOEAs) were studied in [14]. In that work, the dominance of solutions is divided in the islands using a transformation of coordinates. Authors concluded that dividing the search space is a good idea, but achieve this is not trivial. Other ideas for dividing the search space include objective separation in different processors [15], or divide in two populations: one for elite and other search [16]. Martens, in [17] generated a Barabasi-Albert network as the island topology, and selection based on migrating individuals from a not-crowded area and acceptance based on diversity.

More similar approaches to the one presented here were studied in the next two works, also using cooperative coevolution for high-dimensional problems. Dorronsoro et al. in [9] obtained super-linear performance in several instances using cooperative coevolution, focusing each island on a portion of the chromosome. However, in some instances the parallel version not always improved the sequential version. Kimovski et al. [8] proposed a master-slave method that splits the population into several processors, each one running in parallel a MOEA that only affects some portion of the individuals. After a certain number of generations, the master process receives all the sub-populations to be combined. Different combination alternatives in this step are compared. Up to 8 processors were used in the experimental setup. Our approach in this paper does not broadcast all solutions to all islands for recombination, as previous works, but only one solution to a random island, needing less communication time. Besides, the maximum number of islands of Dorronsoro or Kimovsky approaches was 8, while in this paper we have used up to 128 islands.

Lately, some researchers [18] have focused on working on solution space, dividing the Pareto front in different clusters in order to model it. This *divide and conquer* approach is readily paralellizable and, in the sense of giving the responsibility of different parts of the search space to different modules, is similar to the one presented in this paper. We will describe how we have tested our approach in the next Section.

3   Methodology and Experimental Setup

This section describes the quality indicators used and the methods compared. The chosen quality indicators, are:

- Hypervolume (HV): measures the area formed by all non-dominated solutions found with, respect to a reference point. Higher values imply better quality of the PF.
- Inverted Generational distance (IGD): calculates the distance of the obtained set of solutions to the optimal PF. Therefore, this metric requires the optimal PF found in the literature, or the theoretical one. In this metric, the lower the better.

– Spread (S): Measures the spread between solutions, taking into account the euclidean distance between consecutive solutions. As in previous metric, the lower value, the better, as it implies solutions distributed along all the PF.

These metrics have been used extensively, especially in some of the papers presented previously in Sect. 2.

3.1 Baseline Algorithm

This algorithm is a regular NSGA-II algorithm distributed along a number P of islands. After a fixed number of generations, one individual is migrated to another random island. At the end of the run, all PFs of all islands are aggregated in a new one to compute the quality measures.

3.2 Coevolutionary Algorithms Compared

Two different coevolutionary methods have also been used. In both methods, each island only performs crossover and mutation in specific sections of the individuals.

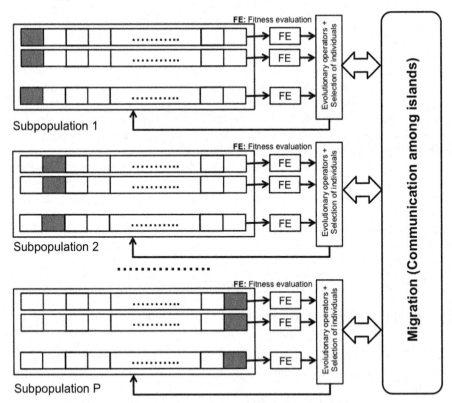

Fig. 1. Disjoint algorithm: every island p only modifies the p_{th} components (in grey) of the individuals.

As in the baseline, every certain number of generations, an individual is migrated to another random island. In the new island, this individual will be considered as one of the others in the island, crossed and mutated in the same way, depending of the island identifier. Note that, on the contrary of other works such as the ones described [11] all the islands deal with complete chromosomes for fitness calculation, so our approach can deal with no decomposable problems.

Coevolution with Disjoint Islands. In this approach, each individual of size L is split into P chunks of size L/P. Every island p only performs crossover and mutation on the p_{th} part of the individuals. Figure 1 describes this approach.

Coevolution with Overlapping Islands. This approach is similar to the previous one, but every island also uses the $p+1$ and $p-1$ (module size) chunks of the individual for crossover and migration. Therefore, some kind of overlapping of the crossed and mutated parts exists between islands. Figure 2 shows the affected parts of the individuals in each island.

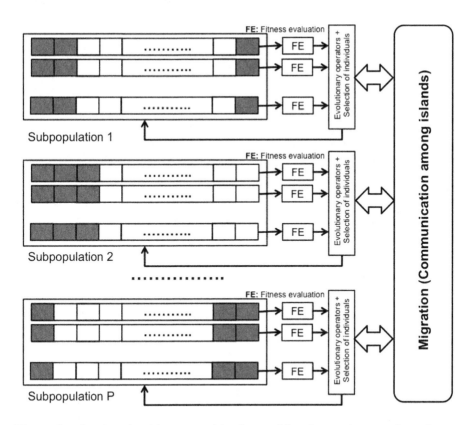

Fig. 2. Overlapping algorithm: every island p modifies the $p+1$, p_{th} and $p-1$ components (in grey) of the individuals.

4   Experiments and Results

The three approaches have been run with two different chromosome lengths (L): 512 and 2048. Different number of islands (P) have also been compared: 8, 32 and 128. This maximum number of islands have also been used in previous work in the literature [17]. The crossover and mutation chosen, SBX and polynomial, have been used previously by other authors in [12].

ZDT [19] has been chosen as a benchmark, since it is the most widely used in this area [12,14,16,17]. The optimal PF distribution used for comparison has 1000 solutions[1].

The termination criterion used is the execution time: 25 s for dimension 512 and 100 (four times more) for 2048. We have used the time instead the number of evaluations firstly because our hypothesis argue that the time saved in crossover in mutation can be spent on improving the sub-populations and more operations and migrations can be achieved. Also, we are using different number of islands (with different sub-population sizes) that could lead to different execution times, so it would be difficult to compare different times and quality solutions at the same time. A summary of the parameters used in the experiments is shown in Table 1.

Table 1. Parameters and operators used in the experiments.

Parameter Name	Value
Number of islands (P)	8, 32 and 128
Chromosome size (L)	512 and 2048
Execution time (s)	25 (for 512) and 100 (for 2048)
Global population size (N)	1024
Selection type	Binary tournament selection
Replacement type	Generational
Crossover type	SBX
Mutation type	Polynomial
Mutation probability	$1/L$
Generations between migration	5
Individuals per migration	1
Selection for migration	Binary Tournament
Runs per configuration	30

The ECJ framework [20] has been used to run the experiments. Specific operators have been developed as new modules for ECJ, and they can be downloaded

---

[1] Optimal PFs are available at: http://www.tik.ee.ethz.ch/sop/download/supplemen tary/testproblems/.

Table 2. Quality metrics obtained after 30 runs per configuration, for the three methods compared: Baseline, Disjoint and Overlapping. An asterisk (*) means that there is not significant difference with respect to Baseline, and a bullet (•) indicates there is not difference between Disjoint and Overlapping. Best values are marked in bold font.

	HV			Spread			IGD		
512 dimensions									
#Islands	Baseline	Disjoint	Overlapping	Baseline	Disjoint	Overlapping	Baseline	Disjoint	Overlapping
ZDT1									
8	0.870	**0.961**	0.943	0.775	**0.546**	0.720 *	0.019	**0.0004**	0.004
32	0.803	0.887	**0.957**	0.815	0.837 *	**0.621**	0.033	0.014	**0.001**
128	0.744	0.687	**0.827**	0.850	0.874 *	**0.839** *	0.045	0.054	**0.024**
ZDT2									
8	0.787	**0.919**	0.878	0.913	**0.646**	0.907 *	0.035	**0.001**	0.011
32	0.744	0.890	**0.917**	0.950	0.754	**0.627**	0.048	0.007	**0.002**
128	0.693	0.592	**0.765**	0.971	0.944	**0.917** •	0.062	0.089	**0.039**
ZDT3									
8	0.894	**0.976**	0.965	0.895	**0.832**	0.922 *	0.012	**0.001**	0.003
32	0.829	0.902	**0.972**	0.877	0.851	**0.835**	0.019	0.011	**0.001**
128	0.770	0.726	**0.865**	0.891	0.845	**0.845** •	0.026	0.029	**0.015**
ZDT6									
8	0.224	**0.503**	0.331	0.992	1.008 *	**0.993** * •	0.192	**0.070**	0.145
32	0.192	0.396	**0.454**	0.998	0.998 *	**0.980** * •	0.206	0.118	**0.091**
128	0.157	0.114	**0.214**	1.004	**0.985**	0.979 •	0.221	0.238	**0.195**
	HV			Spread			IGD		
2048 dimensions									
#Islands	Baseline	Disjoint	Overlapping	Baseline	Disjoint	Overlapping	Baseline	Disjoint	Overlapping
ZDT1									
8	0.754	**0.948**	0.889	0.844	**0.587**	0.807 *	0.044	**0.003**	0 015
32	0.679	0.928	**0.942**	0.854	0.714	**0.646**	0.061	0.006	**0.004**
128	0.660	0.765	**0.907**	0.865	0.860 *	**0.802**	0.065	0.036	**0.010**
ZDT2									
8	0.641	**0.877**	0.794	0.965	**0.942**	0.969 * •	0.078	**0.011**	0.033
32	0.605	**0.898**	0.885•	0.980	**0.704**	0.755 •	0.088	**0.006**	0.009 •
128	0.560	0.689	**0.830**	0.984	0.929	**0.870**	0.101	0.060	**0.021**
ZDT3									
8	0.773	**0.969**	0.922	0.910	**0.865**	0.919 •	0.026	**0.002**	0.009
32	0.713	0.937	**0.966**	0.903	**0.796**	0.839	0.032	0.007	**0.002**
128	0.691	0.817	**0.935**	0.902	**0.839**	0.812 •	0.035	0.020	**0.007**
ZDT6									
8	0.136	**0.334**	0.231	0.995	1.002	**0.996** * •	0.230	**0.144**	0.189
32	0.113	**0.394**	0.327	0.996	0.975	**0.990** * •	0.240	**0.118**	0.147
128	0.101	0.165	**0.267**	0.997	**0.985**	0.979 •	0.245	0.216	**0.172**

from our GitHub repository under a LGPL V3 License[2]. The island model has been executed synchronously, using the ECJ Internal Island Model, in a CentOS 5.4 machine with Intel(R) Xeon(R) CPU E5520 @2.27GHz, 16 GB RAM, with Java Version 1.6.0_16. As in this paper we only focus on the behaviour of our model in a single machine scenario (the islands share the same memory

---

[2] https://github.com/hpmoon/hpmoon-islands.

Table 3. Number of generations and average number of solutions per island obtained after 30 runs per configuration, for the three methods compared: Baseline, Disjoint and Overlapping. An asterisk (*) means that there is not significant difference with respect to Baseline, and a bullet (•) indicates there is not difference between Disjoint and Overlapping.

	Generations			Avg. solutions per island		
512 dimensions						
#Islands	Baseline	Disjoint	Overlapping	Baseline	Disjoint	Overlapping
ZDT1						
8	112.933	453.333	225.233	118.733	137.000	141.467 •
32	111.333	851.933	561.667	69.433	33.067	73.833 *
128	126.433	607.200	606.533	62.367	18.200	25.567
ZDT2						
8	110.333	426.033	222.667	42.233	122.500	101.367
32	113.300	801.800	595.267	28.800	26.600 *	62.367
128	134.300	615.467	595.733 •	20.300	11.433	13.200 •
ZDT3						
8	106.967	449.533	227.467	143.933	138.167 *	160.967 *
32	113.500	819.533	554.767	90.033	38.200	69.767
128	122.633	630.533	596.300	70.167	17.933	25.133
ZDT6						
8	109.767	435.800	215.833	16.933	35.533	17.367 *
32	113.267	720.233	518.733	15.800	11.233	15.100 * •
128	131.333	597.200	569.500	15.367	8.467	8.533 •
	Generations			Avg. solutions per island		
2048 dimensions						
#Islands	Baseline	Disjoint	Overlapping	Baseline	Disjoint	Overlapping
ZDT1						
8	118.333	574.100	262.700	136.167	126.267	136.267 * •
32	115.567	1195.800	721.733	103.867	35.000	60.267
128	141.067	1266.700	1170.767	98.000	20.167	30.467
ZDT2						
8	111.833	584.233	265.933	32.100	86.600	78.067 •
32	121.367	1278.700	757.667	25.600	41.833	45.133 •
128	143.567	1306.700	1133.367	22.967	12.767	20.800 *
ZDT3						
8	111.700	589.800	265.567	167.633	129.333	144.000
32	117.633	1154.133	772.633	129.167	36.667	72.767
128	136.267	1379.067	1128.467	105.767	19.367	29.400
ZDT6						
8	116.867	585.333	267.133	22.867	27.267 *	32.600
32	118.533	1082.633	684.500	22.133	12.067	16.633 •
128	162.267	1296.733	1102.867	20.167	7.400	9.267 •

and processor), the migration time will not be as important factor as in a real distributed system. This task will be addressed in future works.

Different metrics, explained in the previous section, have been used to calculate the quality of the obtained PFs in each configuration. As some of the metrics require a reference point to be calculated (such as HV), the point (1,9) has been chosen as reference, as none of the generated PFs in all runs are dominated by it. Metrics are then normalized with respect to that point. A Kruskall-Wallis significance test has been performed to the metrics of all runs of the configurations, as the Kolmogorov-Smirnov test detected non-normal distributions. The average results for each configuration are shown in Table 2.

At a first glance results show that dividing the chromosome produces an improvement in all the quality indicators in all the configurations, with respect to the baseline. Also, results show that all the quality indicators lowers when the number of islands is increased, as the subpopulations are smaller. This is consistent with the claim by Durillo in [9], who found that cooperative coevolutionary MOEAs work better on bigger populations (more than 100 individuals). Paying attention to the disjoint and overlapping methods, only when using 8 islands the disjoint method always attain better metrics, and even more times with the 2048 chromosome length. This can be explained because only $1/8$ of the individual is changed in each island, while in the overlapping, $3/8$ of the island may be more similar to the baseline configuration. Therefore, there exist some kind of limit point where one method will be preferable to another, depending the number of islands, population size and length.

This can be explained also comparing the number of generations and average solutions per island. Table 3 shows these values obtained in each method. It is clear that in the same amount of time, the disjoint and overlapping method needs more generations than the baseline. This is even clearer with $L = 2048$, where almost 10 times the number of generations is attained, and therefore, more migrations and crossovers/mutations can be achieved. Surprisingly, the average number of solutions of PFs per island in the baseline configuration is significantly higher in some configurations (mostly with 128 islands), but their combination in the final PF never attains higher values than Disjoint and Overlapping methods.

5   Conclusions

High-performance problems that require a large number of decision variables can leverage the division of the decision space that parallel and distributed algorithms imply. This can be done in dMOEAs by dividing the chromosome into different parts, each one modified by a different island. This paper compares a baseline distributed NSGA-II with two different strategies to separate the chromosome (disjoint or overlapping parts), using a high number of islands. Results show that these methods can achieve better quality metrics than the baseline in the same amount of time. This can be explained because of the time reduction in crossover and mutation in the chromosomes, producing a higher number of generations and more modifications of the solutions of the PF in each subpopulation, and therefore improving quality of the global PF.

Results also show that when increasing the number of islands, the overlapping method significantly improves the results with respect to the disjoint method.

Therefore, the relation between the number of islands used, the global population size, and the length of the chromosome may be a key factor to decide if using the overlapping method or the disjoint one. Studying this factor with more types of problems, and new configurations of population size and chromosome lengths will be addressed in the future.

Also, distributed implementations in several systems (such as a GRID or cluster) with more different quantities of islands/processors, will be used to perform a scalability study of the different methods, being the transmission time between islands a relevant issue to address. New benchmarks and real problems will also be used to validate our approach.

Acknowledgments. This work has been supported in part by projects TIN2014-56494-C4-3-P and TIN2012-32039 (Spanish Ministry of Economy and Competitivity), V17-2015 of the Microprojects program 2015 from CEI BioTIC Granada, PROY-PP2015-06 (Plan Propio 2015 UGR), PETRA (SPIP2014-01437, funded by Dirección General de Tráfico), and MSTR (PRY142/14, Fundación Pública Andaluza Centro de Estudios Andaluces en la IX Convocatoria de Proyectos de Investigación).

References

1. Alba, E., Luque, G., Nesmachnow, S.: Parallel metaheuristics: recent advances and new trends. Int. Trans. Oper. Res. **20**(1), 1–48 (2013)
2. Luna, F., Alba, E.: Parallel multiobjective evolutionary algorithms. In: Kacprzyk, J., Pedrycz, W. (eds.) Springer Handbook of Computational Intelligence, pp. 1017–1031. Springer, Berlin (2015)
3. Branke, J., Schmeck, H., Deb, K., Maheshwar, R.S.: Parallelizing multi-objective evolutionary algorithms: cone separation. In: Proceedings of the IEEE Congress on Evolutionary Computation, CEC 2004, pp. 1952–1957, Portland, OR, USA. IEEE, 19–23 June 2004
4. Folino, G., Pizzuti, C., Spezzano, G.: A scalable cellular implementation of parallel genetic programming. IEEE Trans. Evol. Comput. **7**(1), 37–53 (2003)
5. Alba, E., Tomassini, M.: Parallelism and evolutionary algorithms. IEEE Trans. Evol. Comput. **6**(5), 443–462 (2002)
6. Mukhopadhyay, A., Maulik, U., Bandyopadhyay, S., Coello, C.A.C.: A survey of multiobjective evolutionary algorithms for data mining: part I. IEEE Trans. Evol. Comput. **18**(1), 4–19 (2014)
7. Gong, Y., Chen, W., Zhan, Z., Zhang, J., Li, Y., Zhang, Q., Li, J.: Distributed evolutionary algorithms and their models: a survey of the state-of-the-art. Appl. Soft Comput. **34**, 286–300 (2015)
8. Kimovski, D., Ortega, J., Ortiz, A., Baños, R.: Parallel alternatives for evolutionary multi-objective optimization in unsupervised feature selection. Expert Syst. Appl. **42**(9), 4239–4252 (2015)
9. Dorronsoro, B., Danoy, G., Nebro, A.J., Bouvry, P.: Achieving super-linear performance in parallel multi-objective evolutionary algorithms by means of cooperative coevolution. Comput. OR **40**(6), 1552–1563 (2013)
10. Deb, K., Agrawal, S., Pratap, A., Meyarivan, T.: A fast elitist non-dominated sorting genetic algorithm for multi-objective optimisation: NSGA-II. In: Deb, K., Rudolph, G., Lutton, E., Merelo, J.J., Schoenauer, M., Schwefel, H.-P., Yao, X. (eds.) PPSN 2000. LNCS, vol. 1917, pp. 849–858. Springer, Heidelberg (2000)

11. Talbi, E.-G., Mostaghim, S., Okabe, T., Ishibuchi, H., Rudolph, G., Coello Coello, C.A.: Parallel approaches for multiobjective optimization. In: Branke, J., Deb, K., Miettinen, K., Słowiński, R. (eds.) Multiobjective Optimization. LNCS, vol. 5252, pp. 349–372. Springer, Heidelberg (2008)

12. Nebro, A.J., Durillo, J.J.: A study of the parallelization of the multi-objective metaheuristic MOEA/D. In: Blum, C., Battiti, R. (eds.) LION 4. LNCS, vol. 6073, pp. 303–317. Springer, Heidelberg (2010)

13. Hiroyasu, T., Yoshii, K., Miki, M.: Discussion of parallel model of multi-objective genetic algorithms on heterogeneous computational resources. In: Lipson, H. (ed.) Genetic and Evolutionary Computation Conference, GECCO 2007, Proceedings, p. 904, London, England, UK. ACM, 7–11 July 2007

14. Deb, K., Zope, P., Jain, S.: Distributed computing of pareto-optimal solutions with evolutionary algorithms. In: Fonseca, C.M., Fleming, P.J., Zitzler, E., Deb, K., Thiele, L. (eds.) EMO 2003. LNCS, vol. 2632, pp. 534–549. Springer, Heidelberg (2003)

15. Xiao, N., Armstrong, M.P.: A specialized island model and its application in multi-objective. In: Cantú-Paz, E., Foster, J.A., Deb, K., Davis, L., Roy, R., O'Reilly, U.-M., Beyer, H.-G., Kendall, G., Wilson, S.W., Harman, M., Wegener, J., Dasgupta, D., Potter, M.A., Schultz, A., Dowsland, K.A., Jonoska, N., Miller, J., Standish, R.K. (eds.) GECCO 2003. LNCS, vol. 2724, pp. 1530–1540. Springer, Heidelberg (2003)

16. Zhi-xin, W., Ju, G.: A parallel genetic algorithm in multi-objective optimization. In: Control and Decision Conference, CCDC 2009, pp. 3497–3501, Chinese (2009)

17. Märtens, M., Izzo, D.: The asynchronous island model and NSGA-II: study of a new migration operator and its performance. In: Blum, C., Alba, E. (eds.) Genetic and Evolutionary Computation Conference, GECCO 2013, pp. 1173–1180, Amsterdam, The Netherlands. ACM, 6–10 July 2013

18. Cheng, R., Jin, Y., Narukawa, K.: Adaptive reference vector generation for inverse model based evolutionary multiobjective optimization with degenerate and disconnected pareto fronts. In: Gaspar-Cunha, A., Henggeler Antunes, C., Coello, C.C. (eds.) EMO 2015. LNCS, vol. 9018, pp. 127–140. Springer, Heidelberg (2015)

19. Zitzler, E., Deb, K., Thiele, L.: Comparison of multiobjective evolutionary algorithms: empirical results. Evol. Comput. 8(2), 173–195 (2000)

20. Luke, S., et al.: ECJ: a Java-based Evolutionary Computation and Genetic Programming Research System (2009). http://www.cs.umd.edu/projects/plus/ec/ecj

Compilable Phenotypes: Speeding-Up the Evaluation of Glucose Models in Grammatical Evolution

J. Manuel Colmenar[1,2], J. Ignacio Hidalgo[1(✉)], Juan Lanchares[1],
Oscar Garnica[1], Jose-L. Risco[1], Iván Contreras[1], Almudena Sánchez[1],
and J. Manuel Velasco[1]

[1] Adaptive and Bioinspired Systems Research Group,
Universidad Complutense de Madrid, Madrid, Spain
hidalgo@ucm.es
http://absys.dacya.ucm.es
[2] Universidad Rey Juan Carlos, Madrid, Spain

Abstract. This paper presents a method for accelerating the evaluation of individuals in Grammatical Evolution. The method is applied for identification and modeling problems, where, in order to obtain the fitness value of one individual, we need to compute a mathematical expression for different time events. We propose to evaluate all necessary values of each individual using only one mathematical Java code. For this purpose we take profit of the flexibility of grammars, which allows us to generate Java compilable expressions. We test the methodology with a real problem: modeling glucose level on diabetic patients. Experiments confirms that our approach (compilable phenotypes) can get up to 300x reductions in execution time.

Keywords: Grammatical evolution · Model identification · Diabetes mellitus

1 Introduction

Evolutionary Algorithms (EAs) are search methods which can be applied for solving a great variety of optimization, engineering, search, machine learning and identification problems among others. As the reader surely knows, EAs work with a set of solutions, instead of an unique solution as the vast majority of optimization algorithms do. Most EAs follow a similar flow for obtaining a final solution after an iterative process over groups of solutions. This process, although with a random component, is directed by an evaluation function.

The evaluation of the solutions (individuals) is usually the most costly part in the execution of an EA, in terms of the computation time. In most of the

J.M. Colmenar—Support from Spanish Government Grant TIN2014-54806-R is acknowledged by ABSyS group.

G. Squillero and P. Burelli (Eds.): EvoApplications 2016, Part II, LNCS 9598, pp. 118–133, 2016.
DOI: 10.1007/978-3-319-31153-1_9

cases, the evaluation process is complex, since we not only need to compute a value of a mathematical function but also, to decode the individual in order to evaluate the solution which it is representing. Hence, the evaluation complexity depends on (1) the mathematical process for the computation of the fitness and, (2) the kind of the EA, which determines the decoding degree of complexity. Standard GAs, for instance, usually have direct representations, while Genetic Programming (GP) or Grammatical Evolution (GE) need more computational effort for decoding a solution.

In addition, there exist some kind problems, such as model characterization, where each individual represents a model through a mathematical expression. Here, it is necessary to compute the values of the mathematical expression for a set of time values, to obtain a complete evaluation of one individual. What we propose here, is to reinterpret other approaches for Grammatical Evolution (GE), by reducing the complexity of the evaluation of different models over the same time series. To this aim, we take profit of the flexibility of grammars, which allows us to generate compilable expressions, in this case, for Java.

We test the methodology with a real problem which originally motivated our investigation: modeling the glucose level on diabetic patients. Experiments confirm that our approach (compilable phenotypes) can get up to 300x improvement in execution time for this difficult problem.

The rest of the paper is organized as follows. Some previous approaches are revised on Sect. 2. A description of the problem is made on Sect. 3. On Sect. 4 we describe the application of GE to generate models represented by time dependent expressions. The main proposal to generate compilable phenotypes is described in Sect. 5. On Sect. 6.1 we validate our proposal measuring the speedup on a test experiment. Section 6.2 gives the details of the results for the real problem that motivated our investigation, glucose models in humans. Finally, Sect. 7 summarizes the conclusions of this work.

2 Related Work

In order to obtain results in affordable times, some parallel solutions have been proposed in the past for accelerating the execution time of EAs. There are several ways to parallelize an evolutionary algorithm [1]. The first and more intuitive consist on parallelizing the evaluation of individuals maintaining a unique population, which is called global parallelization. There is another form of global parallelization that is to perform a sequential execution of various EAs simultaneously. Since robust EAs need several independent executions, we can run those on a set of processors. The final approach divides the population into subpopulations, that evolve independently for several generations and exchange individuals periodically. In this article we will focus on the parallelization of the evaluation of individuals for GE.

Global parallelization affects only the computational performance and rarely changes the evolutionary search process, as opposed to approaches with several populations that tend to produce benefits also in terms of quality of the solutions. We can find in the literature several approximations. For instance, we can find parallel evaluation approaches which use several processors for evaluating one generation [2]. In these cases, we can save computing time by simply sending simultaneously a part of the population, to a set of different processors. Those processors will return the fitness of the evaluated individuals. By sharing the workload among N processors we can expect the system to work N times faster than with a single processor, allowing treatment of large and complicated problems. Things are not so simple, because there are several factors of overloading, which decrease the expected performance. The solution proposed in this work do not need more than one processor.

Recently the use of Graphics Processing Units (GPU) has been also proposed to accelerate the evaluation of individuals, we refer to the review work made by Langdon in [3]. Unfortunately, the implementation of GE on GPU is not efficiently solved up to date [4].

Another idea proposed by Hu et al. [5] is to reduce the number of evaluations, by using a population that varies its size over generations. They investigated the effect of variable population size in a parallel algorithm and obtained important conclusions about the evolution dynamics, fitness progression and population diversity. They observed that dramatic changes in population size allow the acceleration of the evolution. When working with GE, the reduction of the population size does not guarantee the reduction in the decoding time.

There also some papers proposing volunteer computing, or collaborative evaluations. Arenas et al. [6] for instance released the so called DREAM (Distributed Resource Evolutionary Algorithm Machine) framework. It allows the automatic distribution of EA processing through virtual machines connected by standard Internet protocols. Although sometimes this is a good alternative, it is not the best option seeking for a high speed-up.

Another option is to convert the population into a single program and compile it. The complete population program contains the necessary code to run each of its component individuals and compute their fitness. In [7], the authors showed that compiled code runs much faster than interpreting it. Following with this line, in this paper we investigate a particular parallelization of the evaluation of individuals under a GE paradigm. In particular we study identification of mathematical expressions. The objective is to produce compilable expressions from the decoded solutions therefore reducing the evaluation time for the whole population.

3   Evaluation of Time Dependent Mathematical Expressions

When using an EA, or other search method, for obtaining a mathematical expression of a time series, it is necessary to evaluate the solutions several times. Let us describe the process by an example represented by the information reported on Table 1. First, look only to the first five columns, labeled as y, t, x_1, x_2 and x_3 respectively. The problem can be defined as follows: *knowing the values of the output variable y for six different values of the time t and the values of 3 input*

Table 1. An example of the input data of a mathematical time-dependent expression.

t	x_1	x_2	x_3	y	y_1	y_2	$|y - y_1|$	$|y - y_2|$
0	2	1	1	0	0	3	0	3
1	0	5	1	2	2	7	0	5
2	5	12	2	2	12	5	10	3
3	7	0	2	18	23	23	5	5
4	2	2	3	34	4	41	30	7
5	5	7	4	8	9	3	1	1
Fitness							46	24

variables x_1, x_2 and x_3 at the same time values, find a mathematical expression which represents the value of y as a function of the 3 input variables.

$$y(t) = F(x_1(t), x_2(t), x_3(t))$$

For the values of the table the solution is

$$y(t) = x_1(t - 1) + 2 \cdot x_2(t - 2) + 3 \cdot x_3(t - 3) \tag{1}$$

Let us now consider that we want to evaluate the quality of the expression y_1 (a decoded individual):

$$y_1(t) = x_1(t - 1) + 2 \cdot x_2(t - 2) - x_3(t - 3)$$

For that, we need to compute the values of $y_1(t)$ for $t = 0$ to $t = 5$. Once calculated, we compute the differences with $y(t)$ (column $|y - y_1|$). Adding the values of this column, we obtain the fitness of y_1 which is 46. Columns y_2 and $|y - y_2|$ contains the values needed for computing the fitness of y_2, which is 24

$$y_2(t) = x_1(t - 1) + 2 \cdot x_2(t - 2) + 3$$

As seen, for each time value, it is required to recall previous values of the input variables, which depend on the concrete t we were calculating. Notice that if a negative index is required for a variable, then we use the value 0.

4 Grammatical Evolution for Mathematical Expressions, Curve Fitting and Regression

Grammatical Evolution (GE) [8] is a grammar-based form of Genetic Programming (GP) [9]. GE manages a population of individuals that consist of chromosomes of integer values. The phenotype of the individuals are obtained by applying a grammar in the process of decodification. Hence, a suitable Backus Naur Form (BNF) grammar definition must initially be defined. In an optimization run, GE can theoretically evolve programs in any language described by a BNF. For the sake of space, we refer the reader to the bibliograpy to obtain more details of the GE technique. Due to the flexibility of the grammars, one of the more useful applications of GE is the generation of mathematical expressions. In such a case, it is usually required to evaluate the expression in order to calculate the fitness value of each individual.

However, the evaluation of a mathematical expression generated by a grammar may be time costly in high-level languages like Java. In the case of recursive grammars, the length of the expression is not known, as well as the number of variables involved. Hence, a straightforward approach is to begin with a string of characters that represents an expression and then try to use an evaluator like JEval [10] to substitute each variable with its corresponding numerical value and, after that, compute the result of the mathematical operators. Then, the evaluation of the expression is completed.

The key point is not the mathematical calculus when all the terms are numbers in a given string of characters. There are many fast ways in Java to perform this task, like JEval or the Apache Java EXpression Language [11]. The problem arises when the expression represents a time dependent model that depends on previous data values. The following example will try to clarify this situation.

In the particular case of an algorithm where a model has to be found, the phenotype of an individual is an expression that must be evaluated in order to compute the fitness of this individual (the goodness of the model). In a time dependent model, the evaluation has to be performed for all individuals in all the generations and also for time $t = 0$ to $t = T$. Notice that, in the case of GE, a grammar may generate heterogeneous individuals. This means that it is not mandatory that all the variables were present in all the phenotypes.

Now, if we select JEval as the evaluation library, we need to slightly change the expression to fit its sintax rules regarding arrays as input values. In GE, this is accomplished by tunning the grammar. In particular, Fig. 1 shows the grammar that will work with JEval in this example.

```
<func> ::= <expr> <op> <expr> <op> <expr>

<expr> ::= <preop> (<expr>)|(<cte> <op> <var>)|<var>

<op> ::= +|-|/|*
<idx> ::= <dgt>|<dgtNoZero><dgt>
<cte> ::= <dgt><dgt>.<dgt><dgt>
<dgt>::= 0|1|2|3|4|5|6|7|8|9
<dgtNoZero>::= 1|2|3|4|5|6|7|8|9

<var> ::= #{X1[t_<idx>]}|#{X2[t_<idx>]}|#{X3[t_<idx>]}|<cte>
<preop>::= sin|cos|tan|exp
```

Fig. 1. Grammar to be used with JEval.

Hence, following the rules of the grammar we presented, Expression (1) will be represented with following code:

$$\#\{X1[t\_1]\} + 2 * \#\{X2[t\_2] + 3 * \#\{X3[t\_3]\} \tag{2}$$

The evaluation process for each time value will take the model expression and, for each input variable, will substitute the string with the real value of the input array. That is, $\#X1[t\_1]$ will be changed by the value of the array x_1 ($x_1[t-1]$), and so on. Algorithm 1 shows the process that evaluates just one generation considering that the maximum time is T.

Algorithm 1. Evaluation process of one generation using compilation of the individuals with JEval. Arrays x1, x2 and x3 store real values.

```
// evaluator is an object of the class net.sourceforge.jeval.Evaluator
for (int indiv = 0; indiv < POPULATION\_SIZE; indiv++) {

    String ph = obtainPhenotype(indiv).toString();
    double res = 0.0;

    // Evaluation of one individual (cuadratic complexity):
    for (int t=INPUT_SIZE; t<T; t++) {
        // Data on the evaluator should be cleared
        evaluator.clearVariables();

        for (int i=0; i<INPUT_SIZE; i++) {
            // Update variables in the evaluator depends on t !!
            String si = String.valueOf(i);
            evaluator.putVariable("X1[t_"+si+"]", String.valueOf(x1[t-i]));
            evaluator.putVariable("X2[t_"+si+"]", String.valueOf(x2[t-i]));
            evaluator.putVariable("X3[t_"+si+"]", String.valueOf(x3[t-i]));
        }

        res += Double.valueOf(evaluator.evaluate(ph));
    }

    // Store objective function value
    setObjective(indiv,res);
}
```

As seen in the algorithm, there is a constant value named INPUT_SIZE which corresponds to the size of the input arrays. As seen in this case, the model is evaluated starting in INPUT_SIZE because, given that the input variables must be accessed backwards, $t - i$ will be a non-negative value only if $t \geq i$. In other case, the access to a negative position of an array will generate an error.

So, considering the computational complexity of this algorithm, it is easy to see that the evaluation of one individual presents a time $O(T*INPUT\_SIZE)$, which is almost quadratic. Besides, the inner for loop is repeated for all individuals on each value for t.

The aim of this work is to present an alternative linear method to evaluate one individual removing that inner for loop. This method is faster than the straightforward approach of the evaluation libraries.

5   Compilable Phenotypes

The main problem of the method using an evaluation library, is that an inner loop is needed to recall the previous values of the input variables because the access to the input arrays depends on the time variable. We propose the evaluation of each individual by using mathematical Java code. This is faster than the use of an external library, and it is possible in GE given that we can define a grammar to produce model expressions that could be computed in Java. In addition, we incorporate the input arrays into the evaluation by allowing a direct access to the input arrays from the compiled model expression.

However, model expressions that correspond to individuals are generated in run time and must be evaluated on each generation. Therefore, we need to be able to make this process happen.

In brief, the process we perform conducts the following steps:

1. For a given generation, collect the phenotype of the individuals (model expression).
2. Generate a Java source code file where each individual is a branch of a `case` instruction.
3. Compile the Java file and load it as an evaluator object.
4. Loop to evaluate all the individuals of the generation by calling the evaluator object.
5. Proceed to the next generation.

In Java it is possible to compile an object in run-time, and dynamically load this object to the execution thread. Therefore, the only requirement is to define a Java abstract class to be inherited by the evaluator object. This way, the object evaluation method will be defined and this process could be completed in run time. Algorithm 2 shows the Java code of our abstract class. As seen, a method is defined that receives the identificator of the individual to be evaluated (`indiv`), the instant of time for the evaluation (`t`), and the input arrays (`x1`, `x2` and `x3`). Notice that a similar process could be performed by using a Java interface. The main difference is that an abstract class allows the inclusion of attributes, while the implementation of a Java interface does not. Instead, the implementation of a Java interface allows inheritance from a different class.

Algorithm 2. Abstract class for evaluation objects.

```java
public abstract class AbstractPopEvaluator {

    public abstract double evaluateExpression(int indiv, int t,
                double[] x1, double[] x2, double[] x3);

}
```

Figure 1 presented a grammar that produces expressions to be managed with JEval. With our proposal, the mathematical expressions generated by the grammar are going to be part of a Java file. Hence, they have to be compilable. Therefore, the grammar must be tunned in order to accomplish this requirement. In this way, we have to modify the grammar used for JEval into a grammar to produce compilable expressions. Figure 2 shows this new grammar. It is remarkable to note the changes we performed in relation to the previous grammar.

We modified the first rule, which now produces an assignation sentence where variable **res** takes a value coming from an expression. In addition, the final two rules have been also modified. The first one of them now manages the input arrays x1, x2 and x3 (matching the header of the evaluation method), and the final rule adds the mathematical prefix Math for the mathematical functions.

```
<func> ::= res = <expr> <op> <expr> <op> <expr>;
...
<var> ::= x1[t-<idx>]|x2[t-<idx>]|x3[t-<idx>]|<cte>
<preop>::= Math.sin|Math.cos|Math.tan|Math.exp
```

Fig. 2. Grammar used with the compilation technique.

Let us assume a toy example with 3 individuals. In this case, the first generation of our experimentation provides 3 different phenotypes. Hence, the grammar will provide compilable expressions like the following ones:

```
res = x1[t-5] + Math.sin(x2[t-12]) * x3[t-64];
res = 4.02 * (45.76 + x2[t-23]) - (Math.exp(Math.exp(x1[t-1])) + 5);
res = x2[t-31] / 23.5 - x3[t-2] + Math.cos(x1[t-74]);
```

Now, the idea is to generate a compilable Java code to include those expressions. For this example, the code is shown in Algorithm 3. Notice that the class is fully compilable and the method always returns a value as a result. In this case, we return a positive infinity if the execution does not select any of the case branches because we assume a minimization problem. The process of evaluation will follow the same for the rest of the generations in the optimization procedure. That is, for each generation, one Java file is generated including all the phenotypes of this generation, then it is compiled and loaded into the execution. Algorithm 4 shows the complete process.

As stated before, the evaluation of one individual is now a process that is performed in a linear time. Hence, the algorithm should be faster. However, the generation of the file and the compilation are slow processes compared to regular Java instructions. Hence, an empirical analysis has to be performed.

6   Experiments

In order to perform the experiments, we have used our own Java implementation of the Grammatical Evolution, called ABSys JECO [12]. This implementation

Algorithm 3. Class for an evaluator object including the individuals of a given generation.

```
public class PopEvaluator extends AbstractPopEvaluator {

    public abstract double evaluateExpression(int indiv, int t,
                        double[] x1, double[] x2, double[] x3) {
        double res;

        switch(indiv) {
            case 0: res = x1[t-5] + Math.sin(x2[t-12]) * x3[t-64];
                break;
            case 1: res = 4.02 * (45.76 + x2[t-23]) - (Math.exp(Math.exp(x1[t-1])) + 5);
                break;
            case 2: res = x2[t-31] / 23.5 - x3[t-2] + Math.cos(x1[t-74]);
                break;
            default:
                res = Double.POSITIVE_INFINITY;
        }

        return res;

    }

}
```

Algorithm 4. Evaluation process of one generation using compilation of the individuals. Assume x1, x2 and x3 are available in this context.

```
// Obtain the phenotypes and put them in a list.
ArrayList<String> phens = new ArrayList<>();
for (int indiv = 0; indiv < POPULATION\_SIZE; indiv++) {
    phens.add(obtainPhenotype(indiv).toString());
}
// Produce the file and compile it.
PopEvaluator evaluator = generateAndCompileFile(phens);
// Evaluate population
for (int indiv = 0; indiv < POPULATION\_SIZE; indiv++) {
    double res = 0.0;
    // Evaluation of one individual (linear complexity):
    for (int t=INPUT_SIZE; t<T; t++) {
        res += evaluator.evaluateExpression(indiv,t,x1,x2,x3);
    }
    // Store objective function value
    setObjective(indiv,res);
}
```

is similar to GEVA [13], the original code for GE. However, JECO presents many modifications that were implemented for solving our research work. We can for instance avoid the use of tree objects, which allows an improvement in the memory footprint of the optimization engine.

All the experiments reported in this paper were run on a computer provided by an Intel Core2 Quad CPU (Q8300) running at 2.50 GHz, with 6 Gb of RAM, and using the Java 7u79 JRE.

6.1   Validation of the Proposal

We will conduct a set of experiments that will compare the performance of both evaluation alternatives. Hence, given that we want to measure the performance of the evaluation, we have stated a similar scenario for both evaluation alternatives under the GE optimization technique. This scenario consists on the generation of a random set of individuals using a grammar, then evaluate all of them, compute the time required for this evaluation, and repeat the process again. This loop iterates 30 times, giving us the evaluation time of 30 random populations of individuals. In order to check how the evaluation scales, we have tested 15 populations with sizes ranging from 100 to 1500 individuals.

In the case of the evaluation with JEval, we have used the grammar described in Fig. 1, and the evaluation procedure is the one described in Algorithm 1. The length of the time interval for each expression is $T = 1440$, which corresponds to the number of evaluations of each phenotype. We have considered, as described in the example, three input variables whose size is $INPUT\_SIZE = 100$.

Figure 3 shows the execution time of the evaluation of the 15 different population sizes through a box-plot. The y-axis indicates the time in seconds, while the x-axis corresponds to the size of the population being evaluated. As seeing in the figure, there is a very clear trend towards linear dependence on the size of the population. In addition, the mean and quartile values are relatively close to each other on every population size. Hence, the behavior is quite clear.

The same experiment was run for the compiled phenotypes we propose. In this case, the grammar we used is analogous to the case of JEval, but adjusted to produce compilable phenotypes. Hence, the grammar for these experiments is the one shown in Fig. 2, and the evaluation procedure is the one shown in Algorithm 4. The number of individuals, and the measurement of computation times is the same than in the previous experiments with JEval.

Figures 4 shows the results for the evaluation of compiled phenotypes. As seen, there is a huge difference in the computation time, being shorter by far in this case (lower than 1 s on average, for all the cases). Comparing the average computation time, the compilation technique is approximately from $210x$ to $450x$ times faster than JEval.

Again, we can see a trend where computation time depends on the size of the population in the compilation technique. However, the slope is not so high than in the case of JEval. This fact occurs because the most of the computation time is due to the compilation process, which is very similar in all the population sizes. Besides, notice that there are outsider points in the plot. They appear once and correspond to the first compilation of a population. Given that each population

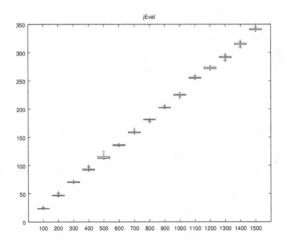

Fig. 3. Execution time with JEval. Time in y-axis is indicated in seconds; x-axis indicates the size of the evaluated population.

size is repeated 30 times, the compilation environment is loaded in the compilation of the first population and then is cached for the other 29 populations. This point is interesting because the compilation environment needs to be loaded just once on an optimization run, while the individuals of a population may change a lot along the whole optimization process.

Once we saw that the compilation time is important in the compiled phenotypes, we wonder where is the frontier between JEval and compilation. In this regard, we run experiments with a population on 100 individuals. However, we stated T to take values of 2, 5, 10, 25, 50 and 100 evaluations per each individual, and a size of $INPUT\_SIZE = 1$ values for the incoming variables. Therefore, once an individual is compiled, we try to discover how many evaluations are required to state that the compilation of the phenotype is better than JEval.

Figure 5 shows the comparison of computation times. Approximately, if more than 10 evaluations are required per individual, it is better to use compilation instead of JEval, using a size of only one input data ($INPUT\_SIZE = 1$).

Therefore, the compilation of phenotypes has been proven to be a faster technique in the case that one expression has to be evaluated several times (more than 10) for different input data. Next, we will show an application of this idea in a real case of experimentation for glucose model extraction.

6.2   Case Study: Glucose Model Extraction

Diabetes Mellitus is a disease that affects to hundreds of millions of people worldwide. Maintaining a good control of the disease is critical to avoid severe long-term complications. In recent years, several artificial pancreas systems have been proposed and developed, which are increasingly advanced. However there is still a lot of research to do. One of the main problems that arises in the (semi)

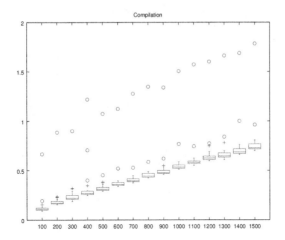

Fig. 4. Execution time with compilation technique. Time in y-aixs is indicated in seconds; x-axis indicates the size of the evaluated population.

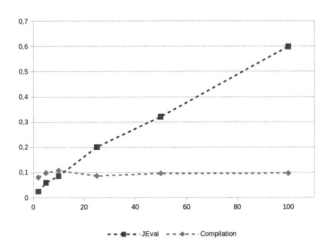

Fig. 5. Execution time comparison. The number of evaluations per individual (T) is shown in the x-axis. Time (in seconds) is shown in y-axis. Notice there are only 6 data values for each technique.

automatic control of diabetes, is to get a model explaining how glycemia (glucose levels in blood) varies with insulin, food intakes and other factors, fitting the characteristics of each individual or patient. In [14] we proposed the application of GE, to obtain customized models of patients. Although we have progressed in this procedure, the focus of this paper is to evaluate the compilable phenotypes proposal, and not to compare the model extraction with other techniques.

The glucose model of a patient is an expression that depends on several input variables. In the case of our research, we consider the previous glucose values (GL), the carbohydrates ingested in the meals (CH) and the insulin units injected, which may belong to two different types: short effect (IS) and long effect (IL). So, there are 4 different variables which, considered along one day of observation, provided 1440 data values corresponding to each minute of a day.

The idea then, is to find out a mathematical expression that will produce glucose values as close as possible to the real glucose values of a patient. There-fore, each model expression will be evaluated for a whole day producing 1440 values of "estimated glucose", one for each minute of a day. These values will be compared with the real glucose values we have for this day, and the correspond-ing root mean square error (RMSE) value will be the fitness of the model. In this experimentation we will use data from a synthetic patient generated with the AIDA simulator [15]. The crossover and mutation probabilities were stated to 0.6 and 0.1 respectively.

Given that we are trying to compare two evaluation techniques, we have stated an experimentation setup based on Grammatical Evolution where 250 generations were run. For the case of JEval, we will run the experiment 30 times considering a population of 100 individuals. We have considered the grammar shown in Fig. 6, which has been adapted to our four input variables: GL, CH, IS and IL. The size of this input variables is 1440 because their values were collected along a day. However, we will consider a window of 100 previous values for each model expression, as stated in the rule which defines idx.

```
<func> ::= <exprGL> <op> <exprCH> <op> <exprIS> <op> <exprIL>

<exprGL> ::= <preop>(<exprGL>)|(<cte> <op> <varGL>)|<varGL>
<exprCH> ::= <preop>(<exprCH>)|(<cte> <op> <varCH>)|<varCH>
<exprIS> ::= <preop>(<exprIS>)|(<cte> <op> <varIS>)|<varIS>
<exprIL> ::= <preop>(<exprIL>)|(<cte> <op> <varIL>)|<varIL>

<op> ::=+|-|/|*
<idx> ::= <dgtNoZero>|<dgtNoZero><dgt>
<cte> ::= <dgt><dgt>.<dgt><dgt>
<dgt>::=0|1|2|3|4|5|6|7|8|9
<dgtNoZero>::=1|2|3|4|5|6|7|8|9

<varGL> ::= #{GL[k_<idx>]}|<cte>
<varCH> ::= #{CH[k_<idx>]}|<cte>
<varIS> ::= #{IS[k_<idx>]}|<cte>
<varIL> ::= #{IL[k_<idx>]}|<cte>

<preop>::=sin|cos|tan|exp
```

Fig. 6. Grammar for glucose model extraction to be used with JEval.

In the case of the experiments where we compile the phenotypes, the execution time will be faster. Hence, with the same number of generations (250), we will run five different population sizes: 100, 250, 500, 750 and 1000 individuals. The phenotypes will be generated by a grammar similar to the one presented before. However, in order to produce compilable phenotypes, the last four rules of the JEval grammar have to be changed, resulting in those shown in Fig. 7.

```
...
<varGL> ::= GL[k-<idx>]|<cte>
<varCH> ::= CH[k-<idx>]|<cte>
<varIS> ::= IS[k-<idx>]|<cte>
<varIL> ::= IL[k-<idx>]|<cte>

<preop>::=Math.sin|Math.cos|Math.tan|Math.exp
```

Fig. 7. Extract of the grammar for glucose model extraction to be used when compiling phenotypes.

Firstly, we will compare the execution time between the six sets of experiments. Figure 8 shows the execution time in percentage, where JEval represents the slowest time (100 %) and the five population sizes of compilable phenotypes present execution times lower than 2.1 %. Hence, it can be shown that for the same population, the compilation technique is $300x$ faster. In addition, considering the progression followed by the execution time, it can be predicted that with the same number of generations, a population of 48633 compiled individuals could be evaluated in the same time spent by JEval evaluating 100 individuals.

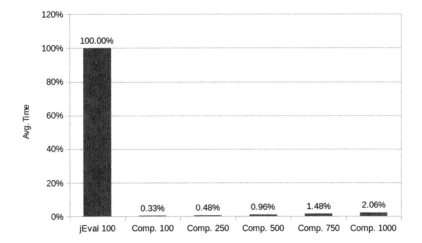

Fig. 8. Execution time (%) comparison between JEval execution (100 individuals) and executions with compilable phenotypes (from 100 individuals to 1000 individuals, Comp. 100 to Comp. 1000).

In addition to the execution time comparison, we have analyzed the behavior of the fitness along each execution. Therefore, we plot this evolution (the average best fitness for the 30 runs of each experiment), in Fig. 9. As seen, the behavior of JEval and compilation evaluations is similar for the case of 100 individuals. Hence, the conclusion is that the improved evaluation method does not influence on the effectiveness of the optimization process. However, the bigger the population size, the better the fitness evolution which, in the case of 1000 individuals presents the best behavior.

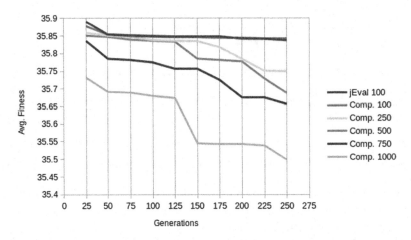

Fig. 9. Fitness comparison between JEval execution (100 individuals) and executions with compilable phenotypes (from 100 to 1000 indivs., Comp. 100 to Comp. 1000).

7   Conclusions and Future Work

In this paper we have presented a method to boost the evaluation process in Grammatical Evolution. Given the flexibility of the grammar approach, it is possible to easily produce phenotypes that could be inserted into compilable source code. In our proposal, we have tested this approach using the Java language, where the inheritance and the dynamic load of methods allow the compilation and the use of new objects in run time. Hence, the phenotypes of one generation can be compiled once as an object, and evaluated many times, as required in the case of expressions whose input values depend on variables like the time.

We have compared our approach with JEval, a well-known Java library for expression evaluation. Our compilable phenotypes improve the computation times up to $450x$ in the case of one generation of individuals. Besides, we have proven that if an expression has to be evaluated more than 10 times, the compilation approach is faster than the JEval library.

In addition, we have tested this approach in a real case of study: the glucose model extraction. In this scenario, the compilable phenotypes have shown an improvement of $300x$ in execution time, without influencing the exploration of the optimization process. In fact, we have shown that the compilable phenotypes technique allows greater populations which, in this case, are much more convenient to obtain better results and avoid premature convergence and starvation of the algorithm.

References

1. Cantú-Paz, E.: Efficient and Accurate Parallel Genetic Algorithms. Kluwer Academic Publishers, Norwell, MA, USA (2000)
2. Hidalgo, J.I., Lanchares, J., Ibarra, A., Hermida, R.: A hybrid evolutionary algorithm for multi-FPGA systems design. In: Proceedings of the Euromicro Symposium on Digital System Design, pp. 60–67 (2002)
3. Langdon, W.B.: Graphics processing units and genetic programming: an overview. Soft Comput. **15**, 1657–1669 (2011)
4. Pospichal, P., Murphy, E., O'Neill, M., Schwarz, J., Jaros, J.: Acceleration of GE using GPUs: computational intelligence on consumer games and graphics hardware. In: Companion Proceedings of the 13th GECCO, pp. 431–438 (2011)
5. Hu, T., Harding, S., Banzhaf, W.: Variable population size and evolution acceleration: a case study with a parallel evolutionary algorithm. Genet. Program Evolvable Mach. **11**(2), 205–225 (2010)
6. Arenas, M., Collet, P., Eiben, A.E., Jelasity, M., Merelo, J.J., Paechter, B., Preuß, M., Schoenauer, M.: A framework for distributed evolutionary algorithms. In: Guervós, J.J.M., Adamidis, P.A., Beyer, H.-G., Fernández-Villacañas, J.-L., Schwefel, H.-P. (eds.) PPSN 2002. LNCS, vol. 2439, pp. 665–675. Springer, Heidelberg (2002)
7. Harding, S.L., Banzhaf, W.: Distributed genetic programming on GPUs using CUDA. In: Workshop on Parallel Architectures and Bioinspired Algorithms, Raleigh, USA (2009)
8. O'Neill, M., Ryan, C.: Grammatical Evolution: Evolutionary Automatic Programming in an Arbitrary Language. Kluwer Academic Publishers, Norwell (2003)
9. Poli, R., Langdon, W.B., McPhee, N.F.: A field guide to genetic programming (2008). Published via http://lulu.com and freely available at http://www.gp-field-guide.org.uk
10. Breidecker, R.: JEval. Java library for functional expression parsing and evaluation (2007). http://sourceforge.net/projects/jeval/
11. JEXL: Java EXpression Language (2015). http://commons.apache.org/proper/commons-jexl/
12. Adaptive and Bioinspired Systems Group: ABSys JECO (Java Evolutionary COmputation) library (2015). https://github.com/ABSysGroup/jeco
13. O'Neill, M., Hemberg, E., Gilligan, C., Bartley, E., McDermott, J., Brabazon, A.: GEVA - Grammatical Evolution in Java. Technical report, Natural Computing Research and Applications Group - UCD Complex and Adaptive Systems Laboratory, University College Dublin, Ireland (2008)
14. Hidalgo, J.I., Colmenar, J.M., Risco-Martin, J.L., Cuesta-Infante, A., Maqueda, E., Botella, M., Rubio, J.A.: Modeling glycemia in humans by means of Grammatical Evolution. Appl. Soft Comput. **20**, 40–53 (2014)
15. AIDA: AIDA diabetic software simulator (2011). http://www.2aida.org/

GPU Accelerated Molecular Docking Simulation with Genetic Algorithms

Serkan Altuntaş[1], Zeki Bozkus[1(✉)], and Basilio B. Fraguela[2]

[1] Department of Computer Engineering,
Kadir Has Üniversitesi, Istanbul, Turkey
serkan.altuntas@stu.khas.edu.tr,
zeki.bozkus@khas.edu.tr
[2] Depto. de Electrónica e Sistemas,
Universidade da Coruña, A Coruña, Spain
basilio.fraguela@udc.es

Abstract. Receptor-Ligand Molecular Docking is a very computationally expensive process used to predict possible drug candidates for many diseases. A faster docking technique would help life scientists to discover better therapeutics with less effort and time. The requirement of long execution times may mean using a less accurate evaluation of drug candidates potentially increasing the number of false-positive solutions, which require expensive chemical and biological procedures to be discarded. Thus the development of fast and accurate enough docking algorithms greatly reduces wasted drug development resources, helping life scientists discover better therapeutics with less effort and time.

In this article we present the GPU-based acceleration of our recently developed molecular docking code. We focus on offloading the most computationally intensive part of any docking simulation, which is the genetic algorithm, to accelerators, as it is very well suited to them. We show how the main functions of the genetic algorithm can be mapped to the GPU. The GPU-accelerated system achieves a speedup of around $\sim 14x$ with respect to a single CPU core. This makes it very productive to use GPU for small molecule docking cases.

Keywords: GPU · OpenCL · Molecular docking · Genetic algorithm · Parallelization

1 Introduction

The binding of small molecule ligands to large protein targets is central to numerous biological processes. For example the accurate prediction of the binding modes between a ligand and a protein, which is known as the docking problem, is of fundamental importance in modern structure-based drug design [1]. Docking algorithms try to generate different poses (binding modes) throughout possible three-dimensional conformations, which can be seen in Fig. 1, with the purpose of evaluating them so as to choose the best possible pose.

Molecular dynamic applications use a highly sophisticated force field while searching only a small portion of the conformational space. This approach uses physically based energy functions combined with full atomic level simulations to yield

© Springer International Publishing Switzerland 2016
G. Squillero and P. Burelli (Eds.): EvoApplications 2016, Part II, LNCS 9598, pp. 134–146, 2016.
DOI: 10.1007/978-3-319-31153-1_10

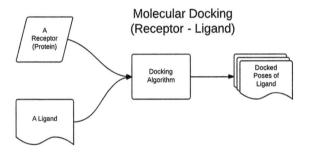

Fig. 1. Molecular docking (receptor - ligand)

accurate estimates of the energetics of molecular processes. However, these methods are too computationally time consuming to allow blind docking of a ligand to a protein. On the other hand, molecular docking simulations often use simpler force fields and explore a wider region of the conformational space [2].

Docking simulations require two basic components:

- A search method for exploring the conformational space available to the system.
- A force field to evaluate the energetics of each conformation (scoring function).

The extensive search performed by docking algorithms involves the sampling of many high-energy unfavorable states. This can restrict the success of an optimization algorithm. Therefore the computational expense is limited by applying constraints, restraints and approximations to sample such a large search space. In practice a limited search may reduce the dimensionality of the problem in an attempt to locate the global minimum as efficiently as possible [1].

Stochastic methods such as Monte Carlo (MC) or genetic algorithms (GA) are general optimization techniques with a limited physical basis, and are able to explore the search. Evolutionary algorithms are generic iterative stochastic optimization procedures mimicking the adaptive process of natural evolution, classified as artificial intelligence techniques [3].

We develop a docking algorithm whose search method is a GA. Based on genotypes of parents. Each generation has a number of offspring, which replaces the worst solutions of the population since population size is limited. The latter is then exposed again to the scoring function, and the evolution goes on for the next iteration (generation). As the number of generations increases, the average fitness of the population of solutions is supposed to increase, and several highly fit solutions are expected to appear.

The fact that almost all subunits of GAs have an embarrassingly parallel nature makes these algorithms very suitable for massively parallel accelerators such as graphics processing units (GPUs). While GPUs were originally designed to process graphics with a massive number of threads, in recent years, they have evolved to accelerate broader types of applications. This has been favored by the availability of new programming models such as Compute Unified Device Architecture (CUDA) [4] and Open Computing Language (OpenCL) [5], which allow the development of general purpose, non graphics programs for GPUs. We have implemented our application using the Heterogeneous Programming Library (HPL) [6, 7] because it largely improves the programmability of heterogeneous platforms with respect to CUDA and OpenCL, while it presents negligible performance overheads. HPL works on top of OpenCL, the standard for heterogeneous computing, which guarantees that our application can be used in a wide range of devices.

Our heterogeneous implementation prepares in the host the receptor binding site and ligand conformation and then it transfers it to the accelerator (the GPU). This device then creates the first population and starting from this point to the final result, all the GA operations are computed on the GPU. The design is a very important contribution of our work, as the fact that there is no CPU involvement during the execution of the GA, so that everything is locally computed in the GPU, avoids costly data movements between CPU and GPU, which would affect performance negatively. Following this strategy our implementation achieved in our tests average speedups of around 14x with respect to a single core CPU. We have also analyzed the reason for the performance obtained, identifying a several memory optimizations to improve the performance even further.

2 Docking Algorithm

Small molecule docking is the algorithm to predict the preferred binding pose of a small molecule to a target molecule (a protein). Small molecules are known as *ligands* and they may have different *poses,* which are different orientations on 3D space of the same chemical component. These orientations can be stored with the coordinates of an indicator atom and *torsion angles* of ligand molecule. Torsion angles represent the rotation between two imaginary geometric planes of the molecule. The number of torsions is known as the *torsion size* and it is fixed on all poses.

The receptor has many atoms that are not related to the docking of the ligand. This way, only a part of the receptor atoms interact with the ligand, and for the sake of fast execution docking algorithms target directly to a local site on the molecule. This *site* (also called; *binding site, binding pocket, binding cavity*) is decided by the user and defined as a sphere by a center point and a radius. If the binding site covers all receptor (this means that the user does not have information about the specific binding site) the docking is called *blind docking*. This kind of docking algorithm should find the pocket first. As a result blind docking requires more GA runs and uses all the atoms of receptor.

Molecular Docking consists of three basic steps, which are seeding, selection and diversity. In general both MC and GA based molecular docking simulations perform these three basic steps with different styles. We now briefly describe these three steps in turn:

2.1 Seeding

Decoys from the reference coordinates of the ligand populate the first population from generation zero. These decoys are called as seeds and each one of them is generated by means of a random rotation and translation beginning from the reference coordinates [3].

2.2 Selection

After the creation of the first generation, the scoring function is applied to all the individuals. Then, the parents of the next generation and the leaving individuals are identified based on the rankings obtained.

The scoring function is important for the accuracy of the docking algorithm. Unfortunately, its complexity can largely increase run-time. If we are doing the virtual screening of millions of compounds, we can employ a lightweight scoring function for faster turnaround. On the other hand, some docking algorithm employs adaptive scoring techniques such as changing the complexity of the functions in the last iterations of the simulation [3].

2.3 Diversity

In order to generate a child, two parents are randomly chosen according to their rank after the selection step. Although the features of each child come from its parents and yet this step can also create new features by adjusting valence angles and bond lengths [3].

3 GPU Parallelization: Algorithms and Implementation

Figure 2 describes the docking software architecture, which should be executed for each receptor ligand complex. The figure is simplified to represent only the single docking experiment but the actual preparation of the receptor and the ligand requires some more operations according to the docking software.

3.1 Overview of Data Structure

Our docking simulation is based on a genetic algorithm for its *conformational search*. This particular conformational search is the process to find the best possible pose with respect to the receptor. The *pose* is the single individual which includes the 3D conformation of a molecule. There are different properties of each pose like *atom types*, *atom coordinates*, *number of branches*, *branch values* and *calculated scores*. These properties are stored inside multiple arrays that are referred as **property arrays of individuals (PAI)** in this article.

Algorithms and Memory Access Patterns. Our program has two parts. One part runs on CPU and we refer to it as the host program. The other part runs on a GPU and we call it the kernel program, so that our application applies heterogeneous programming.

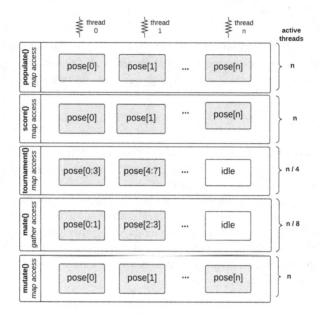

Fig. 2. Threads & memory access patterns

ALGORITHM 1: PSEUDO-CODE FOR THE HOST SIDE:

```
conf = read_conf()
receptor = read_receptor(conf.receptor_path)
ligand = read_ligand(conf.ligand_path)

// trim rest of the receptor,
// only binding site remains
site = prepare_binding_site(conf, receptor)

// move ligand into the center of binding site
initial = move_ligand_into_site(conf, ligand)

// genetic algorithm for conformational search
result = dock(conf, site, initial)  // GPU accel.

// create the pdb file of best results
write_pdbs(result)
```

Algorithm 1 presents the pseudo-code of our host program. The host code first reads all the required files and prepares with them the required data structures for the receptor and ligand. Then the unused parts of receptor are trimmed, the binding site is generated, and the movement of initial pose is carried out in order to make possible the score calculation.

ALGORITHM 2: PSEUDO-CODE FOR THE GA:

```
// get receptor <receptor>,
// ligand <ligand>,
// configuration parameters <conf>
population = populate(ligand)

// loop: with number of generations
foreach generation from conf {

  // energy calculation
  score(population, receptor)

  // selection
  tournament(population)

  // mate the winners and
  // overwrite 3/4 of population
  mate(population)

  // mutate
  mutate(population)
}
```

Algorithm 2 presents the kernel portion of our GPU program which is the most computationally intensive part and which implements the conformational search. This algorithm consists of five different main subroutines, all of which are offloaded to the accelerator device. Our GA iterates and terminates according to a predefined number of generations, which is set by user. Each subroutine of the GA has slightly different memory access patterns such as map and gather. The map access pattern reads and writes data following a regular fashion. On the other hand the gather access reads and writes following an irregular fashion that restricts the memory bandwidth of the GPU. Figure 2 summarizes these patterns. We now briefly describe in turn the five stages of the kernel program of our application.

The number of threads used in the kernel executions is relative to the population size (number of individuals) so the global thread domain size is equal to population size. Since the algorithm has been optimized the algorithm to take advantage of local work groups the local domain size is 1.

Populate: The first step of the conformational search is the generation of a population of a fixed size. Until this point, there is only one pose of the initial ligand, which is specifically moved into the binding site. It is identified as the 0th element of the property array of individuals that represents the whole population.

In this kernel, each thread t reads the 0th element of the array as input, creates a slightly different variation of it and writes the output to the t-th position of the property arrays. Figure 3 describes the map access type data movement of this step, which is performed only once, just before the simulation loop.

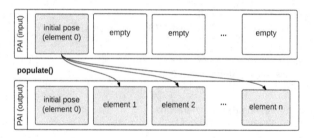

Fig. 3. Data flow for populate

Score: The genetic algorithm relies on the fitness function to take its decisions and find the best fitting individuals. Here the fitness function is the scoring function of the molecular docking.

The score of each pose is calculated a different thread in a map access fashion, as shown in Fig. 4, calculates the score of each pose. The scoring function also uses the binding pocket data, which is the same for all the individuals during the simulation.

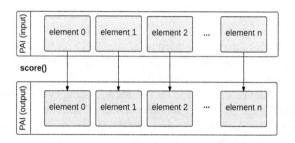

Fig. 4. Data flow for score

Tournament: The tournament selection is designed to terminate the underperforming individuals in order to create space for new generations. The population is divided into groups of four individuals. The selection method has two stages: semi-final and final. The semi-final stage compares the score of two consecutive individuals and eliminates the worst half. Then, the final competition chooses a single winner for each group after comparing the results of the semi-final stage. Figure 5 represents this process. As we can see this method cannot use all the available GPU threads, which is a problem that is typical of reduction processes. Namely, the semi-final stage uses half of the threads, while the final stage only uses a quarter of all threads.

Mate: The tournament process eliminates ¾ of all the individuals. The mating stage should fill the empty spaces because the GA relays on a fixed population size. In this process, illustrated in Fig. 6, each thread randomly picks two parents among the alive individuals, produces new offsprings from the chosen parents, and writes them to the available spaces, marked as empty. The user sets a *crossover ratio*, which is a floating-point number between 0 and 1. Based on this ratio some offsprings will have

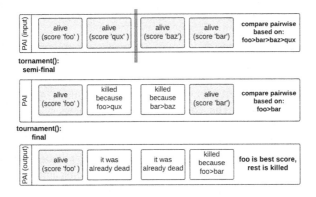

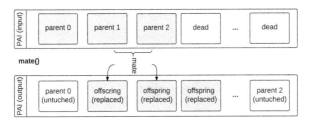

Fig. 5. Data flow for tournament

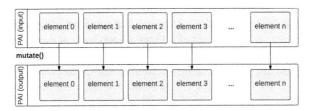

Fig. 6. Data flow for mate

parts from both parents and some may be exact copies of one of their parents. Although mating is the only process that produces newly formed individuals, crossover is applied by chance.

Mutate: Mutation is a key part of GA because it is the main source of biologically inspired variation. In this stage of the algorithm each thread gets an individual and makes variations based on a variation ratio, which is set by user. Figure 7 represents the map access for mutate function.

Fig. 7. Data flow for mutate

4 Performance Results

In this section we show the results of our experiments to compare the performance results of serial and parallel GA. We used a Tesla C2050/C2070 GPU as experimental platform. The device has 448 thread processors with a clock rate of 1.15 GHz and 6 GB of DRAM and it is connected to a host system consisting of 4x Dual-Cores Intel 2.13 GHz Xeon processors. The compiler used for all tests was g ++ 4.7.1 with optimization level O3. This compiler supports C ++11 standards, which is necessary for our code.

Figure 8 represents the execution time over a varying number of GA runs for a single docking performs docking between a ligand of 25 atoms and a large receptor. In our docking algorithm, we choose a binding site from the receptor with 528 atoms.

Users of docking programs (i.e. AutoDock) prefer at least 10 GA runs to make sure the results are satisfying. The reason is that although every GA run includes a full population and iterates for the number of generations specified by the user, a single GA run is not enough for finding a good enough pose because GAs are easy to stuck on local solutions.

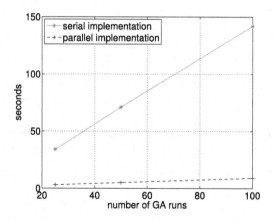

Fig. 8. Execution time over number of GA runs

We have performed our speedup tests using three different compounds (A, B, C). Each one has a different torsion size (7, 5, 8) and a different number of atoms (25, 19, 28). As Table 1 and Fig. 9 represent, the compound size (number of atoms and number of torsions) does not have any important effect on the speedup, but the number of GA runs makes a difference.

The main reason for the increasing speedup as the number of runs grows is the initialization cost of the HPL/OpenCL framework. In addition, our program spends a fixed amount of time for preparing the receptor and the ligand during a single docking experiment. That is, at the beginning of the first simulation there is a one time cost for the whole docking simulation, which is the same no matter how many GA runs will be performed.

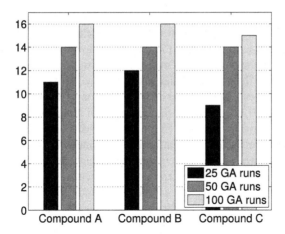

Fig. 9. Speedup on Tesla C2050/C2070 GPU for docking algorithm

Table 1. SPEEDUP

	Speedup		
	25 GA runs	50 GA runs	100 GA runs
Compound A	11	14	16
Compound B	12	14	16
Compound C	9	14	15

We profiled our GPU code for performance bottlenecks. This analysis revealed that our algorithm is not fully using the memory bandwidth of GPU. We concluded that our implementation needs two kind different memory optimizations. The first approach is to use tiling with the shared memory of the GPU streaming multi processors (SM) and in general reconsider the kind of memory used for our data. For example the receptor binding site data is currently accessed from global device memory but this data never changes during the computation, so the memory type for this data should be reconsidered. The second important optimization is to transform the data layout in order to increase the number of coalesced memory access to global memory at the GPU. Namely our code currently stores its data using arrays of structures (AOS). Changing this layout to use structures of arrays (SAO) will allow improving the bandwidth of our accesses to the global memory.

5   Related Work

AutoDock is one of the best-known docking applications. In fact it was the most cited docking program in the ISI Web of Science database in 2005 [8, 9]. In addition to single CPU docking implementations, many research groups have created GPU and FPGA based solutions.

GPU Based Molecular Docking Implementations. Micevski D. et al. [10, 11] profiled the original AutoDock code and identified two different functions (eintcal and trilin-interp) that were suitable to be ported to GPU. Each time these functions are called in their version the corresponding CUDA kernel is executed instead of the original function. In both cases the number of threads within the kernel equals to the number of ligand atoms. This gives place to a very low utilization of the GPU device.

Kannan S. et al. [11, 12] reported the migration of AutoDock to NVIDIA CUDA with the only exception of the local search part, since genetic algorithms are relatively straightforward to be ported to GPU platforms. Their implementation keeps the storage of the ligand coordinates in fast shared memory, which makes the score evaluation faster. For determining the atom-receptor intermolecular energy, each thread first performs a trilinear interpolation for a different ligand atom, and then each thread evaluates the scoring function directly for a different ligand-ligand atom pair. Trilinear interpolation offers a further optimization, since NVIDIA GPUs support the fast access of 3D data by hardware. As a result of this they end up with ~ 50x speedup on fitness function evaluation and 10x to 47x speedup on the core genetic algorithm.

Unfortunately CUDA is a vendor specific model to address parallel execution that is well suited to NVIDIA GPUs, but it is not portable to other architectures. Our implementation however is based on the Heterogeneous Programming Library (HPL) [6, 7], which creates OpenCL code on runtime. Since OpenCL is the portable standard for heterogeneous computing, we are able to run our code on any OpenCL enabled CPU/GPU from different vendors, and other co-processors like Xeon Phi.

Local search is an important factor for the accuracy of AutoDock [8], and performing it on the CPU gives place to underutilization of GPU resources. Pechan I. et al. [11, 13] introduced two different kernels to add local search functionality. Namely one of them creates and evaluates a whole population and the other one performs the local search. They share the same scoring function but they differ in the calculation of the degree of freedom. As a result of that, they achieved speedups of x30 and x64 with respect to CPU executions, when performing 10 and 100 independent runs, for a large set of ligands, respectively.

FPGA Based Molecular Docking Implementations. Pechan I. et al. [14] targeted the parallel execution of distinct docking runs, which is an obvious approach since there is no relation between different docking runs. Also they evaluated different entities of the same population simultaneously, which resulted in remarkably high performance. Their implementation applies pipelines and fine-grained parallelization, and achieves x10–40 speedup over a 3.2 GHz CPU. In addition to this, they manipulated the scoring function in order to fit it better to FPGA devices. Namely instead of computing the scoring function with floating point precision, they used fixed-point arithmetic, as it is likely that the performance of the algorithm does not decrease with this change, while it fits better the capabilities of an FPGA.

VanCourt T. et al. [15] used a 3D correlation method on FPGA devices. Their approach, based on direct summation, allows straightforward combination of multiple forces and enables nonlinear force models. The latter, in particular, are incompatible with the transform-based techniques typically used.

MPI Based Molecular Docking Implementations. Zhang X. et al. [16] created an MPI and multithreading hybrid which is a task parallel solution of the AutoDock Vina [17]. The aim of their research was to develop having a faster virtual screening result, so they did not report the improvement of single docking performance.

6   Conclusions and Future Work

We describe a GPU-accelerated molecular docking code with a genetic algorithm. Our code achieves a speedup of around 14x with respect to a single core. We found that our code is not fully using the memory bandwidth of GPU system, a critical reason for this being that our current data layout is not well suited for the GPU and the special kinds of memories in GPUs, such as local memory, are not fully exploited. Despite this fact, the speedup achieved indicates that the GPU architecture is one of the best possible solutions for GA implementations.

An interesting property of our implementation of molecular docking GA is that it is specifically developed for full GPU-based computation, being the host code just a wrapper for the I/O requirements of our software.

Our work only includes single device executions of independent GA runs for a specified number of iterations. This makes the problem very suitable for multi device implementation, which is part of our future work.

Finally, our GPU implementation of docking makes single docking fast but designing the algorithm for virtual screening of drug candidates, which is another important topic for molecular docking, may lead to better results.

Acknowledgment. Sekan Altuntaş and Zeki Bozkus are funded by the Scientific and Technological Research Council of Turkey (TUBITAK; 112E191). Basilio B. Fraguela is supported by the Ministry of Economy and Competitiveness of Spain and FEDER funds of the EU (ref. TIN2013-42148-P) and by the Galician Government under the Consolidation Program of Competitive Reference Groups (ref. GRC2013-055).

References

1. Taylor, R.D., Jewsbury, P.J., Essex, J.W.: A review of protein-small molecule docking methods. J. Comput. Aided Mol. Des. **16**(3), 151–166 (2002)
2. Huey, R., Morris, G.M., Olson, A.J., Goodsell, D.S.: Software news and update a semiempirical free energy force field with charge-based desolvation (2007)
3. Grosdidier, A.: EADock: design of a new molecular docking algorithm and some of its applications (2007)
4. Manual, R.: Compute unified device architecture. J. Inst. Image Inf. Telev. Eng. ITE **62**, 1–5 (2008)
5. Munshi, A.: OpenCL 1.2 Specification (2012)
6. Bozkus, Z., Fraguela, B.B.: A portable high-productivity approach to program heterogeneous systems. In: Proceedings of the 2012 IEEE 26th International Parallel and Distributed Processing Symposium Workshops, IPDPSW 2012, pp. 163–173 (2012)

7. Viñas, M., Bozkus, Z., Fraguela, B.B.: Exploiting heterogeneous parallelism with the heterogeneous programming library. J. Parallel Distrib. Comput. **73**(12), 1627–1638 (2013)

8. Morris, G.M., Goodsell, D.S., Halliday, R.S., Huey, R., Hart, W.E., Belew, R.K., Olson, A.J.: Automated docking using a Lamarckian genetic algorithm and an empirical binding free energy function. J. Comput. Chem. **19**, 1639–1662 (1998)

9. Huey, R., Morris, G.M., Olson, A.J., Goodsell, D.S.: A semiempirical free energy force field with charge-based desolvation. J. Comput. Chem. **28**, 1145–1152 (2007)

10. Kuiper, M., Micevski, D.: Optimizing Autodock with CUDA. In: VPAC Case Study (2009). http://www.vpac.org/?q=node/290. Accessed 01 January 2012

11. Pechan, I., Fehér, B.: Hardware accelerated molecular docking: a survey (2012)

12. Kannan, S., Ganji, R.: Porting Autodock to CUDA. In: IEEE Congress on Evolutionary Computation (CEC) (2010)

13. Pechan, I., Feher, B.: Molecular docking on FPGA and GPU platforms. Audio Trans. IRE Prof. Gr., 474–477 (2011)

14. Pechan, I., Fehér, B., Bérces, A.: FPGA-based acceleration of the AutoDock molecular docking software, Ph.D. Res. (2010)

15. VanCourt, T., Gu, Y., Mundada, V., Herbordt, M.: Rigid molecule docking: FPGA reconfiguration for alternative force laws. EURASIP J. Adv. Signal Process. **2006**, 1–11 (2006)

16. Zhang, X., Wong, S.E., Lightstone, F.C.: Message passing interface and multithreading hybrid for parallel molecular docking of large databases on petascale high performance computing machines. J. Comput. Chem. **34**(11), 915–927 (2013)

17. Trott, O., Olson, A.J.: AutoDock Vina: improving the speed and accuracy of docking with a new scoring function, efficient optimization, and multithreading. J. Comput. Chem. **31**(2), 455–461 (2010)

EvoRISK

Challenging Anti-virus Through Evolutionary Malware Obfuscation

Marco Gaudesi[1], Andrea Marcelli[1(✉)], Ernesto Sanchez[1], Giovanni Squillero[1], and Alberto Tonda[2]

[1] DAUIN, Politecnico di Torino, Corso Duca Degli Abruzzi 24, 10129 Turin, Italy
{marco.gaudesi,andrea.marcelli,ernesto.sanchez,
giovanni.squillero}@polito.it
[2] INRA, UMR 782 GMPA, 1 Avenue Lucien Brétignières,
78850 Thiverval-Grignon, France
alberto.tonda@grignon.inra.fr

Abstract. The use of anti-virus software has become something of an act of faith. A recent study showed that more than 80 % of all personal computers have anti-virus software installed. However, the protection mechanisms in place are far less effective than users would expect. Malware analysis is a classical example of cat-and-mouse game: as new anti-virus techniques are developed, malware authors respond with new ones to thwart analysis. Every day, anti-virus companies analyze thousands of malware that has been collected through honeypots, hence they restrict the research to only already existing viruses. This article describes a novel method for malware obfuscation based an evolutionary opcode generator and a special ad-hoc packer. The results can be used by the security industry to test the ability of their system to react to malware mutations.

Keywords: Security · Malware · Packer · Computational-intelligence · Evolutionary algorithms

1 Introduction

Malicious software, *malware* for short, plays a part in most computer intrusion and security incidents. The name is used to indicate any software that causes *intentional harm* to a user, computer, or network [1] and may be found in different variants: *worm*, *rootkit*, *trojan*, however *viruses* are the most common.

The term *computer virus* was coined by Fred Cohen in 1983 [2] to indicate those programs that are able to replicate themselves and infect various system files. As many other in computer science, the idea of self-replicating software can be traced back to John Von Neumann in the late 50s [3], yet the first working computer viruses are much more recent.

Creeper, developed in 1971 by Bob Thomas, is generally accepted as the first working self-replicating computer program, but it was not designed with the intent to create damage. A decade later, *Elk Cloner* was the first one known to

© Springer International Publishing Switzerland 2016
G. Squillero and P. Burelli (Eds.): EvoApplications 2016, Part II, LNCS 9598, pp. 149–162, 2016.
DOI: 10.1007/978-3-319-31153-1_11

infect different computers spreading by floppy disks. On the other hand, *Brain*, written by two Pakistani brothers and released in January 1986, is widely considered the first real malware [4]. Since then, malware grows in complexity and dangerousness. As of 2015, malware analysis and detection is still an important field in computer research and computer virus still represents a heavy risk in computer security. Malware analysis is like a cat-and-mouse game. As new analysis techniques are developed, malware authors respond with new ones to thwart analysis.

Fig. 1. Screen shot of malware *Ambulance.com* after infection.

It is interesting to note that from year 2003 the focus has changed from "writing for fun" to "writing for profit". In the early days, viruses were written by hobbyists mainly for joke or for challenge. They usually played with the user or print funny messages or graphics on the screen. For instance, Fig. 1 displays a screen shot of an infection by *Ambulance.com*, a virus written for DOS. When the user got infected, an ambulance crossed the screen and a siren started to sound. Today, when someone is infected by malware does not even know to be infected. The victim does not see anymore funny images on the screen, nor the cd rom trail rom will open and close all the time. The malware runs silently in the background, without crashing the system. In effect, viruses are well tested and debugged in order to not slowing down the system [5]. Indeed, the hiding capability of a malware is a crucial aspect. For this reason the vast majority employs some kind of *obfuscation*, ranging from simple XOR encryption, to more sophisticated anti-analysis tricks. However the most common is *packing* [6,7].

The research presented in this paper aims at developing a new evolutionary obfuscation mechanism. The results could be used by the security industry to evaluate the efficiency and effectiveness of their analysis methodologies, as well as to test the ability of their system to react to new malware mutations. In order to mimic real malware mutations, the research focused on the development of an *evolutionary opcode generator*, which is the foundation of a packer. A mechanism that is able to automatically generate hard to detect programs, can help the security research into developing a new proper countermeasure.

The analysis depicted in this paper represents the third chapter of a work focused on evaluating and testing anti-virus products: in a first paper [8] it was experimented the application of the μGP toolkit to evolve the infection mechanisms of a simple DOS virus. Then, a second work tested commercial products responsiveness to a runtime obfuscated shellcode [9].

The usage of evolutionary computation techniques in the field of malware mutation is not new. Previous works include [10], where an evolutionary framework that exploits genetic algorithms was proposed to evolve a well-known virus family, called Bagle. Furthermore, in [11], the authors proposed an artificial arms race between an automated white-hat attacker and various anomaly detectors for the purpose of identifying intrusion detection system weaknesses. Eventually, in [12], it is provided a theoretical proof behind malware implementation that closely models Darwinian evolution. The malware autonomously incorporates behaviours by including numerous discoverable APIs. The new behaviour profiles would is constantly screened by security software in the same way natural selection acts on biological organisms.

The rest of the paper is organized as follows: Sect. 2 introduces viruses and anti-viruses; Sect. 3 illustrates the goal of the research; Sect. 4 reports experimental evaluation; finally, Sect. 5 concludes the paper, outlining future works.

2   Virus and Anti-virus

Since the first days of malware, the security industry developed anti-virus programs. Their efficiency is evaluated on the base of the *detection ratio* (that should be maximum) and *false-positive ratio* (that should be minimum). However, there is a common misconception about anti-virus scanners: people often get the impression that faster is better; hence all the techniques need to target speed scanning. Commonly used techniques in commercial anti-virus applications include Signature Scanning, Geometric Detection, Disassembler combined with a State Machine and Emulators. *Signature scanning* was the first to be developed and today it is the most common technique of viruses detection. It looks for a specific sequence of bytes in the inspected executables. Instead, *geometric detection* analyses the file structure and identifies a virus if an alteration has occurred. While a *disassembler* is used to separate the byte stream into individual instructions, when it is combined with a state machine, it can be used to record the order in which "interesting instruction are encountered". However, the most advanced and complex solution is the *emulator* which simulates the behaviour of a CPU. The code that runs in an emulator is executed within a controlled environment from which it cannot escape. Moreover, it can be examined periodically or when particular instructions are executed [13].

In the anti-virus field it is often used the term *heuristic* to indicate all those technologies that are employed to detect malware, without requiring a specific detection routine or signature for each known malware. Moreover the usage of heuristic is has the potential to find out new version of malware, by exploiting similarities with known samples. Thus employing heuristics, is a systems that can only reduce the problem against masses of viruses and of course, the perfect detection is more difficult to be done by using pure heuristics without paying some attention to virus-specific details. Heuristic may target both static and dynamic analysis. A proposed system to heuristically detect zero-day malware based on static analysis is PE-Miner, which uses around two hundred features of the Portable Executable file structure [14]. Plenty of research has been done using n-gram distribution or opcode frequency of byte sequences to heuristically detect zero-day malware [15]. However, since these methods depend only on the byte sequence, they can be easily bypassed by obfuscation, code transformation and packing [16]. In the dynamic analysis the program operations are tracked while a heuristic mechanism tries to recognize behavioural patterns typical of malware. In any case it is unknown which is the heuristic technique implemented in anti-virus commercial products, since it represents the value of the software itself.

The usage of anti-virus software has become something of an act of faith [17]. People seem to feel more safe not with a more secure operating system, or with the latest patch, but with some anti-virus software installed in their systems. A recent study [18] showed that over 80 per cent of all personal computers have anti-virus software installed on their computers. Quite clearly, this is a must-have for most users. However, this research outcome showed that viruses are a serious threat and the protection mechanisms in place are less effective than we would expect.

By analysing the evolution of the malware in the last thirty years, it is clear that there has been a growth in the complexity of the hiding mechanisms. Anti-virus software industry historically classifies obfuscating malware in *encrypted*, *olilgomorphic*, *polymorphic* and *metamorphic* viruses. The first ones are the simplest and they are characterised by a constant decryptor that is followed by the encrypted virus body. On the other hand, metamorphics are the most advanced: they are capable of completely disassemble, regenerate the code and re-compile it at run time [19].

The analogies between computer malware and biological viruses are more than obvious. The very idea of an artificial ecosystem where malicious software can evolve and autonomously find new, more effective ways of attacking legitimate programs and damaging sensitive information is both terrifying and fascinating [8]. In the biological world, every time a new species is discovered, it's encountered an adaptation that is not expect. The same is true for malware:

virus analyst needs to train themselves to think like exploratory biologists and be prepared for things they have never seen.

The research, supported by security experts, identifies the *evolutionary malware* as the next possible malware threat. The new category is suggested for those viruses that will exploit the full power of evolutionary algorithms (EA) and computational-intelligence techniques in general. This kind of virus will be able to learn from the surrounding environment, to be trained and to create false positives.

3 Proposed Evolutionary Malware Obfuscation

The section describes the new malware obfuscation mechanism. The developed *ad-hoc packer* is responsible of creating a new working variant of the malicious executable file. The *opcode generator* uses an evolutionary algorithm to create two assembly routines: one is used to *obfuscate* and the other to *de-obfuscate* arbitrary data. Such hiding mechanism is then adopted by the packer to encode the file content, by obfuscating the malicious code and data embedded in the file.

3.1 Evolutionary Opcode Generator

An *Opcode* is the binary representation of an assembly instruction. The opcode generator creates both an encoding and a decoding function starting from randomly-generated, variable-length sequence of IA-32 assembly instructions. Those are directly handled as binary opcodes, so there is no need of a compilation and linking phase.

The generation process exploits a simple evolutionary algorithm with a strategy similar to $(1, \lambda)$, that is, not using any recombination operators. The encoding and the decoding functions are created at the same time in a process that requires to find either reversible assembly instructions (or small blocks of code) and their complementary one. Since even few bytes may represent a signature, it is also necessary to partially shuffle the instructions, although this has the drawback of potentially disrupting the encoding/decoding routines.

In order to assess candidate fitness values both the *reversibility* of the generated routines and the *Jaccard Similarity* is evaluated. In favour of efficiently evaluate a candidate obfuscating routine, the encoding and the decoding routines are applied subsequently to randomly generated sequence of bytes: if the final result is different from the original sequence, the candidate is simply discarded. Then, the *encoder* is used to obfuscate the malicious code and eventually the *Jaccard Similarity*, a statistic coefficient used for comparing the similarity and diversity of sample sets, is appraised. The Jaccard Index has been chosen for keeping the evolution mechanism lightweight while gaining effective results.

Aiming to achieve invisibility through diversity, the process is iterated until a timeout expires, or until the Jaccard coefficient is lower than an experimentally-defined threshold.

The entire process is shown in the following pseudo-code and each phase will be analysed in details in the following paragraphs.

```
Initialise population

while (!stopping condition) {
        create new lambda individuals by choosing one of the
            following methods:
                        1. append instructions to both
                           encoder and decoder
                        2. shuffle decoder instructions
        evaluate new individuals
        keep the best
}

evaluation:
  test reversibility
  if (success)
        evaluate Jaccard
```

IA-32 Instruction Set. The *assembly* language for Intel x86 is a collection of hundreds of instructions. Each of them is translated in several binary representation, called opcodes, according to its *usage* in combination with register or memory and its *version*: 8, 16 or 32 bit.

According to Intel IA-32 manual, instruction can be classified in *Data Transfer, Binary Arithmetic, Decimal Arithmetic, Logic, Shift and Rotate, Bit and Byte, Control Transfer, String, Flag Control, Segment Register, Miscellaneous, MMX Technology, Floating Point* and *System* instructions.

Only a small subset of the IA-32 instruction set have been employed in the realisation of the opcode generator, but more can be added in the future: in the prototype described in this paper no *Floating Point*, nor *MMX Technology* instructions are used. Also *Segment Register, Miscellaneous, Bit and Byte, String* and *System* instructions have been discarded, because their application was not straightforward.

The other categories represent the most used opcodes, hence the most interesting. Some of them have a direct reversible instruction, which is very useful when it comes to create two equal, but reverse sequence of operations. Others, like *mov* or *push and pop*, insert some approximation in the reversibility of the solution. Those who remain are strongly related to *carry management* or to the *evaluation of a condition*.

Among all of them, the most common *twenty* have been chosen, including XOR, INC, DEC, NEG, ROL and ROR. Instructions are stored in a small database of opcode ranges, where each one is associated with a data structure with all the necessary information for the code generation: the opcode length and range, a pointer to the opposite opcode, the need of a parameter and eventually the length and range of this one.

Instruction Generation. When the opcode generator is called, the evolution-ary algorithm starts. An individual is made of a pair of encoding and decoding routines built from randomly chosen sequence of instructions. Since each opcode has a pointer to its directly opposite instruction in the database, both the func-tions are constructed in parallel. Usually instructions are characterised by a base number, that identifies the specific version of the assembly instruction in use and an offset, that identifies which register or memory address is being adopted. If a parameter is required too, the algorithm generates one in the proper range.

In order to assess candidate fitness values both the *reversibility* of the gener-ated routines and the *Jaccard Similarity* of the obfuscated code is evaluated. In order to increase the strength of the obfuscation, the mutation mechanism par-tially shuffle the instruction order, although this has the drawback of potentially disrupting the encoding/decoding routines. Hence, to test the reversibility of the two just created assembly routines, a randomly sequence of byte is generated and copied in the processor registers. If the final register values are different from the original ones, the candidate is simply discarded and the generation restarts from scratch.

The evolutionary process is iterated until a timeout expires, or until the Jaccard coefficient is lower than an experimentally-defined threshold.

Evaluating Similarity. The *Jaccard index*, also known as the *Jaccard simi-larity coefficient*, is evaluated to assess candidate fitness values. It measures the similarity between finite sample sets and it is defined as the size of the intersec-tion divided by the size of the union of the sample sets:

$$J(A, B) = \frac{|A \cap B|}{|A \cup B|}$$

Starting from two variable length byte array of assembly opcodes, the pro-gram cyclically calculates the Jaccard Coefficient for each pair of bytes.

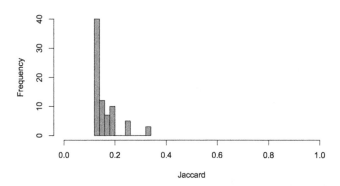

Fig. 2. Histogram of the Jaccard coefficient distribution of a Malware Sample.

The implementation is straightforward: the union is implemented using the binary operation "AND" and the intersection using "OR". Then, the sum of the "1" in each resulting byte is calculated and the result is saved in a temporary array. In the end each coefficient is normalised on a scale of ten values and it could be graphically displayed on an histogram. The algorithm has been tested with different block size and experimental results showed best behaviour considering a block of one byte. Figure 2 shows the *Histogram* result from the calculation between two malware variants that are dissimilar among them. The entire process is repeated untile the Jaccard coefficient is lower than an experimentally defined threshold or when the timeout expires.

3.2   Development of an Ad-Hoc Packer

In the proposed approach, the developed *packer* is responsible of creating new hiding mechanism strong enough to ensure the survival of future generations of malware. The suggested hiding mechanism is based on evolutionary computation and the *Evolutionary Opcode Generator* is embedded directly in the packer. The goal is to develop a packer able to thwart analysis, creating each time a new packing routine exploiting Evolutionary Algorithms. As the decoding routine is embedded in the executable file, once the new malware is run, it will restore each part of the program in memory ready for execution. Differently from a previously published research [9], the packer tackles the entire executable and not only portions of code.

In detail, the developed packer consists of a stand-alone program, the *packer*, that is responsible of creating a new variant of the input program. The research targets the *Portable Executable* file, the format of Windows executable [20]. The file format is logically organised into several sections and the Section Header keeps track of all of them. The packer encode the code and data sections with a custom obfuscating routine and the *unpacker* is directly injected in a new crafted section.

Packing. In order to inject the packing routine, the file size is increased. For this purpose, Windows offers two useful API: *CreateFileMapping* and *MapViewOf-File* which allow to map a file in memory and to work on that by simply using memory addresses. Then the *packing* function is called.

Firstly the *NumberOfSection* is increased by one. Then, the *Characteristic* of each section is set to *read* and *write*. Later, the address of the *RelocationTable* in the *DataDirectory* is set to zero to avoid the Windows Loader to perform "relocation fixups". Finally, a cycle loops on each section encoding it using a previously created obfuscating routine.

A new ".unpack" section is appended at the end of the file, where the unpacker code, directly written in assembly language, is copied and all the necessary informations are written in the *SectionTable*. The *AddressOfEntryPoint* is updated to point to the corresponding decoding function in the new section and the *SizeOfImage* is enlarged to include the new portion of file. Finally the Relative Virtual Address (RVA) of the "Relocation Table" and the *Original Entry Point* are updated in the unpacker code.

Unpacking. The code of the unpacker is the first to be executed when the program is run. It is mainly responsible of restoring the original sections in memory, performing the relocation fixups and loading the required libraries.

ImageBaseAddress: The *Windows Loader* loads the Portable Executable file starting from an arbitrary *ImageBaseAddress*[1]. In the 32bit version of Windows, the FS register points to the *Thread Environment Block*[2], a small data structure for each thread, which contains information about exception handling, thread-local variables and other per-thread state. Among the various fields, one is a pointer to the *Process Environment Block*[3], that contains per process information. Those data structure are not well documented by Microsoft, but over the Internet, it is possible to find a lot of information from independent researchers. An interesting blog post[4] illustrates that the pointer to the *ImageBaseAddress* is constant among all the Windows version, hence the following unpacker is multi version compatible.

PE Parsing and Decoding: The Portable Executable header is parsed, looking for the address of the *SectionTable*, where it is possible to retrieve the starting addresses of the sections to be decoded. Then a loop iterates over each of them and the unpacker restores the original data in memory.

Relocation Fixups: One of the main drawbacks of encoding the code section is that it prevents the Windows Loader to perform the correct relocation adjustment. In summary, the complexity of the unpacker is to mimic the operations typically done by the Windows Loader. The process of relocation is organized in three phases. Firstly it is necessary to calculate the *delta_reloc*: the difference between where the executable is actually loaded, from the one the compiler assumed it would have been loaded. Secondly, it obtains the address of the RelocationTable and thirdly, a loop iterates over each entry of the *RelocationTable* to perform all the necessary fixups. At the end, the program execution is reverted to the *OriginalEntryPoint* and the original program is run.

4   Experimental Evaluation

Experiments have been performed on Windows 7, 32 bit version, with an Intel IA-32 architecture. The choice is dictated by the huge availability of anti-virus

---

[1] Note: test have been executed on Windows 7, by default ASLR is enabled.

[2] https://msdn.microsoft.com/en-us/library/windows/desktop/ms686708(v=vs.85).aspx.

[3] https://msdn.microsoft.com/en-us/library/windows/desktop/aa813706(v=vs.85).aspx.

[4] http://blog.rewolf.pl/blog/.

software for the platform. As a testbed, three well-known online malware scanner have been used: Metascan Online[5], VirusTotal[6] and Jotty[7]. On the whole they make use of tens of different AV products, allowing a direct and simple comparison of the results. Moreover, in order to confirm the outcome precision, several of the most effective AV software have been installed on different hosts.

In order to test the effectiveness of the *packer*, some virus samples have been chosen among a collection of malware from *VX Heaven* (http://vxheaven.org). All of them are characterised by a very high detection rate. Among the ten chosen samples, the packer successfully created a new variant for eight of them. The other two, *Win32.Apparition* and *Win32.Chiton*, make use of anti-dumping and anti-debugging techniques that suspended the packing process. Each generation required an average of 60 individuals to find a working obfuscating routine, making the packing process almost instantaneous. A prudent resource management must be always taken into consideration when dealing with anti-virus software that carefully monitor processes CPU usage.

Table 1 summarises the test results. The first column *Original* shows the detection percentages of the original malware samples, while column *Packed* illustrates the detection percentages of the packed version of each malware sample. Each column is further subdivided into three ones: *Exact* is associated to the exact detection, *Heuristic* is related to the percentage of heuristic detection. The third one, *Total*, is the sum of the previous two values. Finally, column *Worsening* highlights the detection worsening between the original and the packed version of the malware.

Table 1. Comparison of the detection percentages of the Original and Packed version of the Malware samples.

	Original			Packed			Worsening
	Total	Exact	Heuristic	Total	Exact	Heuristic	
Win32.Bee	79 %	74 %	5 %	26 %	18 %	8 %	54 %
Win32.Benny	74 %	64 %	10 %	31 %	23 %	8 %	44 %
Win32.Blackcat	72 %	64 %	8 %	26 %	15 %	10 %	46 %
Win32.Bolzano	64 %	62 %	3 %	28 %	21 %	7 %	36 %
Win32.Crypto	72 %	64 %	8 %	28 %	21 %	7 %	44 %
Win32.Driller	69 %	62 %	8 %	15 %	10 %	5 %	54 %
Win32.Eva	79 %	74 %	5 %	26 %	18 %	8 %	54 %
Win32.Invictus	79 %	67 %	13 %	15 %	0 %	15 %	64 %
Average	74 %	66 %	7 %	24 %	16 %	9 %	49 %

---

[5] https://www.metascan-online.com/.

[6] https://www.virustotal.com.

[7] https://virusscan.jotti.org.

The unencoded version of the malware is characterised by a high detection rate. It ranges from 64 % to 79 % and in most of the cases it is related to *exact* detection. This result is mostly due to virus signature scanning, which is a very effective technique when a malware is not obfuscated. Detecting a packed version of a malware, is much more difficult, indeed the detection rate drops down to an average 24 %, furthermore heuristic is responsible of about half of the uncovering. On average, as illustrated in Table 1, the packer caused a *detection worsening* of the 50 %.

A total of *39* anti-virus have been used during the tests. In the following dissertation they are identified by the acronym *AV1..39*. Figure 3 graphically illustrates the results. *Five* products were not able to detect even a malware sample, whereas *nineteen* detected all the eight original viruses. Among them, *eleven* achieved the perfect detection of the original malware version. As previously stated, heuristic detection plays a small role when the malware is in the original form. Only *AV35* relies more on heuristic than on exact matching.

The most interesting outcomes come from the packed version of the malware. In *fifteen* cases, as shown in Fig. 4, the detection rate drop down to an incredible zero and in other *seven* cases the match worsening is greater than 70 %. Only one, *AV5* achieves the 100 % of malware uncovering, but it is all related to heuristic, then it is less trustworthy than exact detection. *AV9* increased its detection ratio of over 130 % when it analysed the obfuscated variant. However, it is safe to be unconfident of this success: it is likely that the anti-virus detected solely the presence of a packer and not the existence of a real threat. On the other hand, *AV6, AV15 and AV26* reached a great achievement, by worsening their result of only 13 % and other *four* of less than 40 %.

As previously mentioned, when it comes to heuristic detection it is hard to judge the success and it is important to evaluate the credibility of the results. For this purpose a further experiment was performed: a simple "Hello World" program was packed using the same *packer* of the previous test. Of course the original program, that quietly prints to the screen the string *Hello World* was not detected as a threat, when examined and only *four* anti-virus identified the

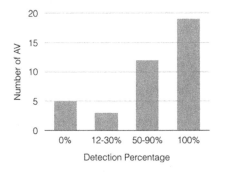 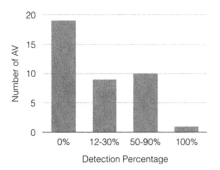

Fig. 3. Comparison of the detection percentage of the original malware samples (on the left) and the packed ones (on the right).

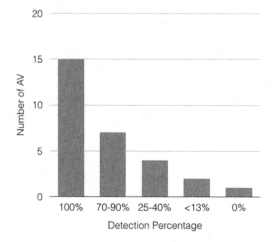

Fig. 4. Detection worsening caused by the packed malware samples.

encoded version as a threat by mean of heuristic detection. Results are quite interesting, as they can be used to further analyse the outcome of the previous test. Simply packing an executable makes *AV9* revealing a threat: this explains the incredible improvement (over 130 %) of the previous experiment. Also *AV6* suffers from false positive detection and its good achievement in detecting packed malware looses value. Finally, despite *AV38* and *AV39* experienced false positive, they were not able to detect none of the packed malware.

5   Conclusion

The basic idea of the research is to use *Evolutionary Computations* techniques to evolve computer viruses to foresee threats of the future. Several proof-of-concept samples have been developed and then tested against most common commercial anti-virus products. Although the relatively simple evolutionary approach, experimental results proved that a packing program that is able to evolve, creating a brand new encoding routine in each infection, may still represent a challenge for the security community. By the way, a mechanism that is able to automatically generate hard to detect programs, can help the security research into developing a new proper countermeasure.

Anti-virus software is one of the most complex application. It has to deal with hundreds of file types and formats: executables, documents, compressed archives and media files. However, it has been showed that the user usually overconfidence anti-virus in their immunity capabilities against all files. Results demonstrated that todays most effective solution to detect known malware is *signature-based* scanning. On average 74 % of the original malware was exactly spotted, in effect by only precisely identifying a threat it is possible to correctly repair the system. However, exact identification is a problem, even in human analysis. How long does it take to be sure if something is really a known variant or a new one? It is believed that when dealing with an *evolutionary malware* it is not feasible to have a custom routine detection for each single sample: the complex infection mechanism coupled with the powerful evolutionary engine make it very difficult to reach full accuracy using only empirical evaluation methods and in depth analysis of the virus code is essential. In the authors' opinion, the anti-virus industry must focus on *heuristic detection*, by increasing the strictness of results and especially targeting the decrease of false positive. Anyway, the need to understand evolutionary code in a quicker way must be the subject of further research.

Malware analysis is like a cat-and-mouse game. As new anti-virus techniques are developed, malware authors respond with new one to thwart analysis. Anyway, as long as old tactics continue to remain effective, we will continue to see them in use: malware needs to propagate, it needs to communicate and it needs to achieve the goals for which it was designed. These are constants that will be seen well into the future.

5.1   Future Developments

While this research is far from being concluded, it has already gained a great deal of insight into the protection provided against viruses. Improvements may concern the developing a more advanced opcode generator and may target analysis of malicious networking interaction.

Future work will focus on investigate *machine learning* techniques for malicious pattern behaviour recognition.

Acknowledgments. A special thank to Peter Ferrie, principal anti-virus researcher at Microsoft, for answering the questions as well as his comments and feedback on latest malware obfuscation technologies.

References

1. Michael, S., Andrew, H.: Practical Malware Analysis - The Hands-On Guide to Dissecting Malicious Software. No Starch Press, San Francisco (2012)
2. Cohen, F.: Computer viruses: theory and experiments. Comput. Secur. **6**(1), 22–35 (1987)
3. Von Neumann, J., Burks, A.W., et al.: Theory of self-reproducing automata. IEEE Trans. Neural Netw. **5**(1), 3–14 (1966)
4. Chen, T.M., Robert, J.-M.: The evolution of viruses, worms. In: Statistical Methods in Computer Security, vol. 1 (2004)
5. Szor, P.: The art of computer virus research and defense. Pearson Education, Indianapolis (2005)
6. Yason, M.V.: The art of unpacking, Chicago (2007). Retrieved 12 February 2008
7. Guo, F., Ferrie, P., Chiueh, T.: A study of the packer problem and its solutions. In: Lippmann, R., Kirda, E., Trachtenberg, A. (eds.) RAID 2008. LNCS, vol. 5230, pp. 98–115. Springer, Heidelberg (2008)

8. Cani, A., Gaudesi, M., Sanchez, E., Squillero, G., Tonda, A.: Towards automated malware creation: code generation and code integration. In Proceedings of the 29th Annual ACM Symposium on Applied Computing pp. 157–160. ACM, March 2014

9. Gaudesi, M., Marcelli, A., Sanchez, E., Squillero, G., Tonda, A.: Malware obfuscation through evolutionary packers. In: Proceedings of the Companion Publication of the 2015 on Genetic and Evolutionary Computation Conference, pp. 757–758. ACM, July 2015

10. Noreen, S., Murtaza, S., Shafiq, M.Z., Farooq, M.: Evolvable malware. In: Proceedings of the 11th Annual Conference on Genetic and Evolutionary Computation, pp. 1569–1576. ACM, July 2009

11. Kayack, H.G., Zincir-Heywood, A.N., Heywood, M.I.: Can a good offense be a good defense? Vulnerability testing of anomaly detectors through an artificial arms race. Appl. Soft Comput. **11**(7), 4366–4383 (2011)

12. Iliopoulos, D., Adami, C., Szor, P.: malware evolution and the consequences for computer security. arXiv preprint arxiv:1111.2503.Chicago

13. Szr, P., Ferrie, P.: Hunting for metamorphic. In: Virus Bulletin Conference, September 2001

14. Nachenberg, C.: Computer virus-coevolution. Commun. ACM **50**(1), 46–51 (1997)

15. Perriot, F., Ferrie, P., Szor, P.: Striking similarities. Virus Bull., 4–6 (2002)

16. Desai, P.: Towards an undetectable computer virus (Doctoral dissertation, San Jose State University), Chicago (2008)

17. Xue, F.: Attacking antivirus. In: Black Hat Europe Conference (2008)

18. Microsoft Security Intelligence Report, vol. 18, December 2014

19. Ferrie, P., Szor, P.: Zmist opportunities. Virus Bull. **3**(2001), 6–7 (2001)

20. Peering Inside the PE: A Tour of the Win32 Portable Executable File Format. https://msdn.microsoft.com/en-us/library/ms809762.aspx

EvoROBOT

Leveraging Online Racing and Population Cloning in Evolutionary Multirobot Systems

Fernando Silva[1,2,4]($\boxtimes$), Luís Correia[4], and Anders Lyhne Christensen[1,2,3]

[1] BioMachines Lab, Lisboa, Portugal
fsilva@di.fc.ul.pt, anders.christensen@iscte.pt
[2] Instituto de Telecomunicações, Lisboa, Portugal
[3] Instituto Universitário de Lisboa (ISCTE-IUL), Lisboa, Portugal
[4] BioISI, Faculdade de Ciências, Universidade de Lisboa, Lisboa, Portugal
luis.correia@ciencias.ulisboa.pt

Abstract. Online evolution of controllers on real robots typically requires a prohibitively long time to synthesise effective solutions. In this paper, we introduce two novel approaches to accelerate online evolution in multirobot systems. We introduce a racing technique to cut short the evaluation of poor controllers based on the task performance of past controllers, and a population cloning technique that enables individual robots to transmit an internal set of high-performing controllers to robots nearby. We implement our approaches over odNEAT, which evolves artificial neural network controllers. We assess the performance of our approaches in three tasks involving groups of e-puck-like robots, and we show that they facilitate: (i) controllers with higher performance, (ii) faster evolution in terms of wall-clock time, (iii) more consistent group-level performance, and (iv) more robust, well-adapted controllers.

Keywords: Online evolution · Multirobot systems · Racing · Population cloning

1 Introduction

Online evolution is one of the most open-ended approaches to adaptation and learning in robotic systems. In online evolution, an evolutionary algorithm is executed onboard robots during task execution to continuously optimise behavioural control. The main components of the evolutionary algorithm, namely evaluation, selection, and reproduction, are performed by the robots without any external supervision. In this way, robots may continuously self-adapt and modify their behaviour in response to changes in the task or in the environmental conditions.

In 1994, Floreano and Mondada [1] conducted the first study on online evolution using a single, mobile robot. In 2002, Watson *et al.* [2] developed *embodied evolution*, an approach in which the evolutionary algorithm is distributed across a collective of robots. The key objective behind the use of multiple robots was to enable a speed-up of evolution due to robots that evolve controllers in parallel and exchange partial solutions to the task. Such approach allows for a form of

© Springer International Publishing Switzerland 2016
G. Squillero and P. Burelli (Eds.): EvoApplications 2016, Part II, LNCS 9598, pp. 165–180, 2016.
DOI: 10.1007/978-3-319-31153-1_12

knowledge transfer that has been shown to speed up the evolutionary process and to facilitate more effective collective problem solving [3, 4].

Over the past years, research in online evolution lead to the development of a number of different approaches. Examples include the $(\mu + 1)$-online evolutionary algorithm by Haasdijk *et al.* [5], r-ASiCo [6] by Prieto *et al.*, mEDEA by Bredeche *et al.* [7], and odNEAT by Silva *et al.* [3], among others. However, there are currently a number of key issues that must be addressed before online evolution becomes a feasible approach to adaptation and learning in real robots. One major impediment to widespread adoption is the long time that online evolution requires to synthesise solutions to any but the simplest of tasks (several hours or days), which currently renders the approach infeasible on real robots [8].

In the vast majority of online evolution algorithms, see [5, 7] for examples, each controller is assessed for a fixed, predefined amount of time. Because the evaluation of inferior controllers amounts to poor task performance, previous studies have focused on how to optimise the evaluation period online via: (i) self-adaptation of the evaluation period [9], (ii) roulette wheel-based selection [10], (iii) a stochastic heuristic rule that monitors the performance of controllers to adjust the evaluation period [10], and (iv) a racing technique that runs independently on every robot in a group [11]. Despite their potential, such techniques have a number of inherent limitations, namely: (i) require the experimenter to decide, for instance, on the maximum evaluation period [9–11], which may be infeasible in practice, (ii) are significantly sensitive to the parameter settings [9–12], and (iii) require a controller to be reevaluated multiple times [10].

In this paper, we propose two novel approaches to speed up online evolution of robotic controllers in multirobot systems. Firstly, we introduce a racing approach for multirobot systems that relies on the task performance of controllers alone, and therefore does not require the definition of a maximum evaluation period. Because the racing approach relies on a modified version of the non-parametric Hoeffding's bounds [13], it can be applied to an unrestricted set of tasks and algorithms. We then extend racing with a population cloning approach, which enables each individual robot to clone and transmit a set of high-performing controllers stored in its internal population to other robots nearby. The underlying motivation is to effectively leverage the genetic information accumulated by multiple robots that evolve together.

We implement our approaches over odNEAT [3], which optimises artificial neural network (ANN) controllers online in a distributed and decentralised manner. One of the main advantages of odNEAT is that it evolves both the weights and the topology of ANNs, thereby bypassing the inherent limitations of fixed-topology algorithms [3]. odNEAT is used here as a representative efficient algorithm that has been successfully used in a number of simulation-based studies related to adaptation and learning in robot systems, see [3, 4, 14, 15] for examples. We assess the performance of our proposed approaches in three tasks involving groups of e-puck-like robots [16], namely in two foraging tasks with differing complexity and in a dynamic phototaxis task. Our results show that racing and population cloning facilitate: (i) synthesis of controllers with higher performance,

(ii) faster evolution in terms of wall-clock time, (iii) more consistent group-level performance, and (iv) more robust, well-adapted controllers.

2   Related Work

In this section, we review the background on racing approaches in machine learning and in evolutionary computation, we discuss current approaches to the exchange of genetic information between robots and the principles behind population cloning, and we describe odNEAT.

2.1   Racing

The general racing framework was originally introduced by Maron and Moore [17] as a technique for model selection in machine learning. The key principle behind racing is to iteratively test multiple models in parallel, use the error values of each model to discard those that are statistically inferior as soon as there is enough evidence, and then concentrate the computational effort on the remaining models. In this way, models race against each other.

Given the similarity between model selection in machine learning and parameter tuning in meta-heuristics, such as evolutionary algorithms, previous contributions have assessed how evolutionary techniques could benefit from racing procedures [18–20]. In [11], Haasdijk *et al.* studied how racing could be used to cut short the evaluation of poor controllers in online evolution. The authors used $(\mu + 1)$-online [5], an encapsulated algorithm in which there is no exchange of genetic information among robots. In the native $(\mu + 1)$-online algorithm, a new controller is produced at regular time intervals, and operates for a fixed amount of time called the *evaluation period*. When the evaluation period elapses, a new controller is synthesised and its evaluation starts. In the racing version of $(\mu + 1)$-online, the current controller is compared with those previously evaluated as it operates. If the fitness score of the controller is below a lower bound, the evaluation is aborted. The lower bound is computed based on a modified version of Hoeffding's bounds [13] taking into account the fitness score of the worst controller in the population.

Haasdijk *et al.* [11]'s study was the first demonstration of how racing techniques could be used to speed up online evolution. There are, however, a number of disadvantages regarding the authors' approach. Firstly, algorithms such as $(\mu + 1)$-online can lead to incongruous group behaviour and poor performance in collective tasks due to the periodic substitution of controllers [3]. Secondly, $(\mu + 1)$-online is an encapsulated algorithm, meaning that an isolated instance runs independently on each robot. In this way, $(\mu + 1)$-online does not benefit from the parallelism in multirobot systems or from exchange of genetic information between robots, which can effectively speed up online evolution [3]. Thirdly, Haasdijk *et al.*'s approach was tailored to the elitist dynamics of $(\mu+1)$-online as the lower bound of performance considers that the worst fitness score in the population does not decrease. Thus, the approach may be subject to backtracking when applied to non-elitist algorithms. Finally, even with the racing technique, the combined approach still requires the experimenter to decide on the maximum evaluation period, and is significantly sensitive to such parameter settings [12].

2.2   Exchange of Genetic Information in Online Evolution

The exchange of genetic information between robots is a crucial feature in distributed, online evolutionary algorithms. This process can be viewed as a set of *inter-robot reproduction events*, in which reproduction is implemented using other robots in the same group [2]. In traditional evolutionary algorithms, selection precedes reproduction and is accomplished by having [2]: (i) more-fit individuals becoming parents and supplying genes, (ii) less-fit individuals being replaced by the offspring, or (iii) by a combination of the two. In online evolution, this amounts to individual robots transmitting to neighbouring robots either part of a genome [2] or a complete genome [3,7]. That is, the genome is the unit in the selection process, and the population of robots is a distributed substrate which genetic information can spread across.

In our population cloning approach, see Sect. 3.2 for a description, we take on a novel approach to the exchange of genetic information between robots. We place the selection and reproductive processes at a higher level. That is, we consider the elements in the selection process to be the internal population of each robot. A robot can therefore transmit to neighbouring robots *a copy* of any part of its population (e.g. a single genome or a set of genomes representing high-performing controllers) or of the complete population. In this way, robots have the potential to leverage the genetic information they have accumulated, and to enable a more effective knowledge transfer to solve the current task.

2.3   Online Evolution with odNEAT

This section provides an overview of odNEAT; a comprehensive description of odNEAT can be found in [3]. odNEAT is distributed across multiple robots that exchange candidate solutions to the task. Specifically, the evolutionary process is implemented according to a physically distributed island model. Each robot optimises an internal population of genomes (directly encoded artificial neural networks) through intra-island variation, and genetic information between two or more robots is exchanged through inter-island migration. In this way, each robot is potentially self-sufficient and the evolutionary process opportunistically capitalises on the exchange of genetic information between multiple robots for collective problem solving [3,4].

One of the key features of odNEAT it that it starts with minimal artificial neural networks (ANNs) with no hidden neurons, that is, with each input neuron connected to every output neuron. Throughout evolution, topologies are gradually complexified by adding new neurons and new connections through mutation. In addition, the internal population of each robot implements a niching scheme comprising speciation and fitness sharing, which allows each robot to maintain a healthy diversity of candidate solutions with differing topologies. In this way, odNEAT is able to evolve a suitable degree of complexity for the current task, and an appropriate ANN topology is the product of the evolutionary process [3].

During task execution, each robot is controlled by an ANN that represents a potential solution to the task. Each controller maintains a virtual energy level reflecting its individual task performance. The fitness score is defined as the mean energy level, sampled at regular time intervals. When the virtual energy level reaches a minimum threshold, the current controller is considered unfit for the task. A new controller is then synthesised via selection of a parent species and two genomes from that species (the parents), crossover of the parents' genomes, and mutation of the offspring. Mutation is both structural and parametric, as it adds new neurons and new connections, and optimises parameters such as connection weights and neuron bias values. A new controller is guaranteed a maturation period during which the controller is not replaced.

odNEAT has been successfully used in a number of simulation-based studies related to long-term adaptation and learning in robot systems. Previous studies have shown key features of odNEAT, including: (i) adaptivity, as odNEAT effectively evolves controllers for robots that operate in dynamic environments [15], (ii) scalability, in the sense that odNEAT allows groups of different size to leverage their multiplicity [4], (iii) robustness, as the controllers evolved can often adapt to changes in environmental conditions without further evolution [3,14], and (iii) fault tolerance: robots executing odNEAT are able to adapt and learn new behaviours in the presence of sensor faults [3], and (v) how to incorporate and optimise behavioural building blocks prespecified by the human experimenter [21]. Given previous results and the ability to efficiently optimise ANN weights and topology, odNEAT is used in our study as a representative distributed online evolutionary algorithm.

3 Racing and Cloning in Multirobot Systems

We propose a combined racing and cloning approach to speed up online evolution of robotic controllers in multirobot systems. The objective is to leverage racing to cut short the evolution of poor controllers, and the genetic information accumulated by individual robots evolving in parallel.

3.1 Racing

We extended odNEAT with a racing approach based on a modified version of the non-parametric Hoeffding's bounds [13]. The major advantage of Hoeffding's bounds is that they do not require any assumption regarding the underlying fitness distribution. In our racing approach, an evaluation is aborted if the controller's performance $F_{current}$ is below a lower bound L_b given by:

$$L_b = M_c(t) - 2 \cdot \xi(t) \tag{1}$$

$$\xi(t) = \sqrt{\frac{(F_{best} - F_{worst})^2 \cdot \log(2/\alpha)}{2 \cdot t}} \tag{2}$$

where $M_c(t)$ is a dynamic fitness threshold henceforth referred to as *minimal criterion* (see below), t is the current control cycle since the controller started executing, F_{worst} and F_{best} are respectively the fitness scores of the worst and best controllers in the internal population, and α is the significance level of the comparison. The minimal criterion is computed based on the fitness of the internal population. Whenever there is a change to the fitness scores of a given controller in the population (e.g. a robot receives a new controller or the fitness score of a controller is updated), $M_c(t)$ is computed based on the value v_n, which corresponds to the P-th percentile of the fitness scores in the population:

$$M_c(t) = M_c(t-1) + max(0, (v_n - M_c(t-1)) \cdot W) \tag{3}$$

where W is a weighting parameter that enables fine-grained control over the magnitude of the changes to the minimal criterion. Because racing approaches require a certain number of measurement points to produce reliable results [20], we take advantage of the maturation period of odNEAT to put a lower boundary on the sample size for the racing approach. That is, racing can only abort the evaluation of a controller after the maturation period has expired.

3.2   Population Cloning

To implement our population cloning technique according to the principles described in Sect. 2.2, we adopt an approach in which internal populations compete when robots meet. Specifically, when two robots are in communication range, a connection link between the robots is created if none of them has been involved in a population cloning process within a predefined period of time P_c. Winner and loser are determined by comparing the $M_c(t)$ value of each robot, as defined in Eq. 3, which is indicative of the performance of each population. The robot with the highest $M_c(t)$ value is considered the winner. The genomes injected in the losing robot are those from the population of the winning robot that have a fitness score above $M_c(t)$.

We consider two variants of the population cloning approach. In one variant, genomes are injected from one robot to another as described above. In a second variant, the internal population of the losing robot is subject to an extinction event. The genomes in the receiving population that yield a fitness score below the $M_c(t)$ of the winner robot are removed before the injection of new genomes, thus potentially pushing evolution towards higher quality solutions.

4   Methods

In this section, we define our experimental methodology[1], including the simulation platform and robot model, and we describe the three tasks used in the study: two foraging tasks with differing complexity and a dynamic phototaxis.

---

[1] The source code of the experiments can be found at: http://fgsilva.com/?page_id=302.

Table 1. Controller details. Light sensors have a range of 50 cm (phototaxis task). Other sensors have a range of 25 cm.

Foraging tasks – controller details	
Input neurons: 25	
	4 for IR robot detection
	4 for IR wall detection
	1 for energy level reading
	8 for resource A detection
	8 for resource B detection
Output neurons: 3	
	2 for left and right motor speeds
	1 for controlling the gripper
Phototaxis task – controller details	
Input neurons: 25	
	8 for IR robot detection
	8 for IR wall detection
	8 for light source detection
	1 for energy level reading
Output neurons: 2	
	Left and right motor speeds

4.1   Experimental Setup

We use the JBotEvolver platform [22] to conduct our simulation-based experiments. JBotEvolver is an open-source, multirobot simulation platform and neuroevolution framework. The robots are modelled after the e-puck [16]. The e-puck is a small circular (7.5 cm in diameter) differential drive robot that can move at speeds of up to 13 cm/s. Similarly to the e-puck, each simulated robot is equipped with infrared sensors that multiplex obstacle sensing and communication between robots at a range of up to 25 cm. The controller details, namely input and output configurations for the tasks, are listed in Table 1. Each sensor and each actuator are subject to noise, which is simulated by adding a random Gaussian component within $\pm\,5\,\%$ of the sensor saturation value or of the current actuation value. The controllers are discrete-time ANNs with connection weights in the range [-10,10]. The inputs of the neural network are the readings from the sensors, normalised to the interval [0,1]. The output layer is composed of two neurons. The values of the output neurons are linearly scaled from [0,1] to [-1,1] to set the signed speed of each wheel. In the two foraging tasks, each robot is also equipped with a gripper that enables the robot to collect the closest resource within a range of 2 cm, if there is any. In these two tasks, a third output neuron is used to set the state of the gripper. The gripper is activated if the output value of the neuron is higher than 0.5, otherwise it is deactivated.

Foraging Tasks. In the foraging tasks, robots have to search for and collect objects spread across the environment. Foraging is a canonical task in cooperative robotics, and is evocative of tasks such as search and rescue, harvesting, and toxic waste clean-up [23].

Similarly to [4], we setup a foraging task with different types of resources that have to be collected. Robots spend virtual energy at a constant rate, and gain energy when they collect resources. When a resource is collected by a robot, a new resource of the same type is placed randomly in the environment in order to keep the number of resources constant. We conduct experiments with two variants of a foraging task: (i) in one variant there are only type A resources, henceforth called *standard foraging task*, and (ii) in the other variant there are two types of resources, namely type A and type B resources, henceforth called *concurrent foraging task*. In the concurrent foraging task, type A and type B have to be consumed alternately. In this way, besides having to learn the foraging aspects of the task, robots need to actively decide which type of resource to collect. The energy level of each controller is initially set to 100 units, and limited to the range [0,1000]. At each control cycle, E is updated as follows:

$$\frac{\Delta E}{\Delta t} = \begin{cases} reward\_item & \text{if right type of resource is collected} \\ penalty\_item & \text{if wrong type of resource is collected} \\ \text{-0.02} & \text{if no resource is consumed} \end{cases} \tag{4}$$

where $reward\_item = 10$ and $penalty_item = $ -10. The constant decrement of 0.02 means that each controller will execute for a period of 500 s if no resource is collected since it started operating. Note that the $penalty\_item$ component applies only to the concurrent foraging task. The number of resources of each type is set to the number of robots multiplied by 10.

Phototaxis Task. In the classic phototaxis task, a widely used benchmark in evolutionary robotics experiments, robots have to find and move towards a light source. Following previous studies [3,4], we setup a dynamic phototaxis task. In this task, the light source is periodically moved to a new random location. The robots thus have to continuously search for and reach the light source, which eliminates controllers that find the light source by chance. The virtual energy level is limited to the range [0,1000] units. Each controller is assigned an initial value of 100 units. At each control cycle, E is updated as follows:

$$\frac{\Delta E}{\Delta t} = \begin{cases} S_r & \text{if } S_r > 0.5 \\ 0 & \text{if } 0 < S_r \leq 0.5 \\ penalty & \text{if } S_r = 0 \end{cases} \tag{5}$$

where $penalty = $ -0.01, S_r is the maximum value of the readings from light sensors, between 0 (no light) and 1 (brightest light). Light sensors have a range of 50 cm, meaning that robots are only rewarded if they are close to the light source. The remaining sensors have a range of 25 cm.

4.2   Experimental Parameters and Treatments

We assess the performance of four approaches: (i) standard odNEAT, henceforth called **odNEAT**, (ii) odNEAT with racing alone, which we simply refer to as **racing**, and (iii, iv) racing plus population cloning, with and without extinction events (**racing-ppc-rem** and **racing-ppc-norem**, respectively). For each task and each algorithm considered, we conduct 30 independent runs. Each run lasts 100 hours of simulated time. Robots operate in a square arena surrounded by walls. The size of the arena is chosen to be 3×3 meters. odNEAT parameters are set as in previous studies [3], including a population size of 40 genomes per robot. Each robot executes a control cycle every 100 ms. Regarding the minimal criterion for racing, we set P to the 50th percentile of the fitness scores found in the population and $W = 1$, meaning that $M_c(t)$ amounts to the median fitness score, and $\alpha = 0.95$. In the population cloning technique, we set P_c to 100 s of simulated time. These parameter settings are robust to moderate variation, and were found to perform effectively in preliminary experiments.

5   Experimental Results

In this section, we present and discuss the experimental results. We use the two-tailed Mann-Whitney test to compute statistical significance of differences between sets of results because it is a non-parametric test, and therefore no strong assumptions need to be made about the underlying distributions.

5.1   Comparison of Performance

Figure 1 shows the mean fitness score of controllers throughout the simulation trials. In the two foraging tasks, **racing-ppc-rem** and **racing-ppc-norem** typically produce high-performing solutions in the early stages of evolution, which contributes to their superior performance. These approaches consistently outperform **racing** and **odNEAT**. In addition, the solution-synthesis process of racing with and without population cloning contrasts with that of **odNEAT**, which synthesises increasingly higher-performing controllers in a more progressive manner. In the dynamic phototaxis task, differences in performance between the most effective approaches (**racing-ppc-rem** and **racing-ppc-norem**) and the less effective ones further accentuate, which provides additional evidence regarding the benefits of racing plus population cloning.

Regarding the fitness score of the final controllers, see Fig. 2, **racing-ppc-rem** and **racing-ppc-norem** lead to superior collective performance across the three tasks, which is validated by the distribution of the mean fitness of each group of robots ($\rho < 0.001$ and $\rho < 0.05$ in the standard foraging task, $\rho < 0.01$ and $\rho < 0.001$ in the concurrent foraging task, $\rho < 0.0001$ and $\rho < 0.001$ in the dynamic phototaxis task, respectively). Differences between **racing-ppc-rem** and **racing-ppc-norem** are not statistically significant across all comparisons.

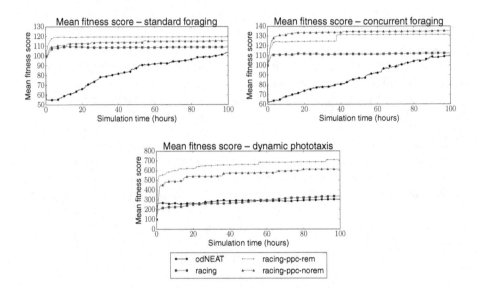

Fig. 1. Mean fitness score of controllers throughout the simulation trials. Top: standard foraging and concurrent foraging. Bottom: dynamic phototaxis.

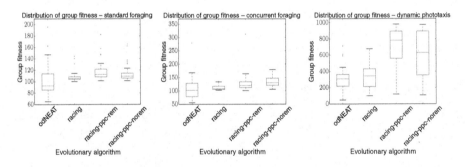

Fig. 2. Distribution of the mean group fitness of the final controllers. From left to right: standard foraging, concurrent foraging, and dynamic phototaxis.

Our results show distinct features of racing and population cloning techniques. Comparing with **odNEAT** alone, the main benefit of **racing** is a potential speed up of evolution. The benefits of combining racing with population cloning, on the other hand, are twofold: **racing-ppc-rem** and **racing-ppc-norem** significantly speed up the evolutionary process *and* lead to the synthesis of superior controllers. This result is particularly significant to online evolution because **racing-ppc-rem** and **racing-ppc-norem** effectively minimise the time spent assessing the quality of poor controllers, which is time spent not performing adequately at the task to which solutions are sought.

5.2   Analysis of the Evolutionary Dynamics

We first analysed how the fitness score of the final controllers vary within each group to better understand the evolutionary dynamics of the multiple approaches. To measure the *intra-group fitness variation*, we computed the relative standard deviation (RSD) of the fitness scores of each group of robots. Values close to zero indicate similar fitness values within the group. Higher values, on the other hand, indicate increasingly larger variation of fitness scores.

The distribution of RSD values is shown in Fig. 3. Across the three tasks, **racing-ppc-rem** typically displays the smaller intra-group fitness variation. **odNEAT** displays significantly larger variation than **racing** ($\rho < 0.001$). In turn, the variation of **racing** is also larger than that displayed by both **racing-ppc-rem** and **racing-ppc-norem** ($\rho < 0.0001$ in the standard foraging and in the dynamic phototaxis tasks, $\rho < 0.01$ in the concurrent foraging task, respectively). Population cloning thus leads not only to more capable controllers, but also to more consistent groups fitness-wise. Overall, the results suggest a *boosting effect* on the evolutionary process and an interplay between racing and cloning towards stable collective performance (hypothesis 1). In addition, the high RSD values of **odNEAT** across the three tasks and the relatively higher RSD values of **racing** in the dynamic phototaxis task suggest that the performance of such approaches may be more sensitive to the task requirements and environmental pressure than the performance of **racing-ppc-rem** and **racing-ppc-norem** (hypothesis 2).

We conducted two sets of complementary experiments using the dynamic phototaxis task in order to verify the two hypotheses. In the first set of experiments, we removed the racing component of both **racing-ppc-rem** and **racing-ppc-norem**, and we assessed the performance of population cloning in isolation, henceforth **ppc-rem** and **ppc-norem**. In the second set of experiments, we studied the relation between evolutionary pressure and evolutionary dynamics by varying the value of the *penalty* component defined in Eq. 5 when the light source is not in a robot's line of sight. The *penalty* was increased by a factor

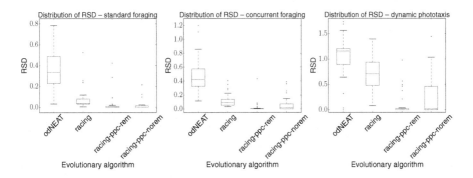

Fig. 3. Distribution of the RSD of the final controllers. From left to right: standard foraging, concurrent foraging, and dynamic phototaxis.

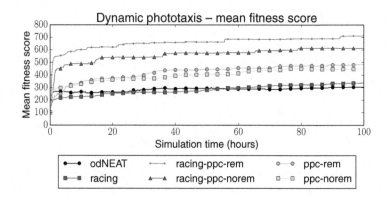

Fig. 4. Mean fitness score of controllers throughout the simulation trials for the first set of complementary experiments (see text for details). The **ppc-rem** and **ppc-norem** approaches refer to population cloning with and without extinction events, but without the racing component.

of 2, 5, 8, and 10, that is, to a value of -0.02, -0.05, -0.08, and -0.10 per control cycle. In this way, each controller unable to find the light source respectively executes for a period of 500, 200, 125, and 100 s since it started operating. These experimental setups are henceforth referred to as **p2**, **p5**, **p8**, and **p10** setups, respectively. For each configuration in each set of complementary experiments, we conducted 30 independent runs.

Isolating the Effects of Population Cloning. Figure 4 shows the mean fitness score throughout the simulation trials for the first set of complementary experiments. Overall, the results confirm that adding population cloning can effectively boost the evolutionary process and push towards higher-quality solutions. In addition, the median RSD of the final controllers is of 1.158 for **odNEAT**, 0.708 for **racing**, 0.375 for **ppc-rem**, 0.829 for **ppc-norem**, $4.17 \cdot 10^{-4}$ for **racing-ppc-rem**, and $1.89 \cdot 10^{-2}$ for **racing-ppc-norem**, which confirms the interplay between racing and cloning in the evolution of controllers with less disparate performance levels. The key reason for such interplay is that population cloning typically increases the fitness scores of the receiving robot's internal population and therefore the $M_c(t)$ value, which in turn contributes to further increasing the lower bound of performance of racing.

Modifying the Evolutionary Pressure. In terms of the second set of complementary experiments, as shown in Fig. 5, an analysis of the mean fitness score throughout evolution shows that both **racing-ppc-rem** and **racing-ppc-norem** achieve qualitatively similar performance levels across the setups with varying evolutionary pressures. The remaining two approaches, **odNEAT** and **racing**, yield different performance levels as the evolutionary pressure varies. From **p5** onwards, **racing** is the approach that evolves, on average,

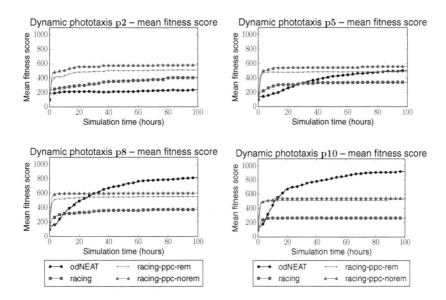

Fig. 5. Mean fitness score of controllers throughout the simulation trials for the second set of complementary experiments (see text for details).

the controllers with lowest performance. The mean fitness score of the final controllers is 410.08, 342.16, 371.75, and 265.98, respectively. **odNEAT**, on the other hand, synthesises controllers with superior performance levels as the evolutionary pressure is increased. The mean fitness score of the final controllers varies from 241.64 in the **p2** setup to 924.24 in the **p10** setup, outperforming both **racing-ppc-rem** and **racing-ppc-norem** in the two most demanding configurations of the dynamic phototaxis task.

One way to understand the responses of **racing** and of **odNEAT** to the evolutionary pressure is to study: (i) the operation time (age) of the controllers used by the robots during the experiments, and (ii) the number of controllers produced. The operation time of controllers relates to the robustness of solutions evolved and ability to adapt to changes in the position of the light source. Complementarily, the number of controllers produced is an indicator of the difficulty of the evolutionary process to adapt the behaviour of the robots.

Figure 6 shows the mean operation time of controllers during the experiments. For **racing-ppc-rem** and **racing-ppc-norem**, the operation time typically increases linearly, with a gentle slope, proportionally to the simulation time, which indicates that new controllers are increasingly rarely synthesised after the early stages of evolution. In effect, the final solutions synthesised by **racing-ppc-rem** and **racing-ppc-norem** operate, on average, up to 98 consecutive hours, which indicates that the controllers are robust and well-adapted to the task. This result, combined with the number of controllers produced, showcases the ability of the algorithms to quickly assess and abort the evaluation of

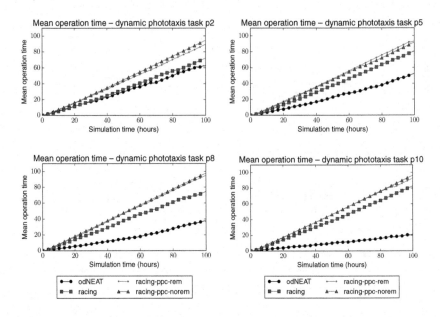

Fig. 6. Mean operation time of controllers produced throughout the simulation trials for the second set of complementary experiments (see text for details)

inefficient controllers. Regarding **racing**, the algorithm displays a trend to that of **racing-ppc-rem** and **racing-ppc-norem** but yields a relatively lower mean operation time and higher number of controllers produced, which indicates that it typically requires more evaluations and therefore more time to evolve stable solutions to the task. Complementarily, **odNEAT** substitutes the controller of each robot more frequently as the evolutionary pressure increases. Specifically, each robot executing **odNEAT** produces on average 24 controllers in the **p2** setup, 120 controllers in the **p5** setup, 420 controllers in the **p8** setup, and 920 controllers in the **p10** setup. This result confirms that, in this particular case, **odNEAT**'s dynamics are more sensitive than the dynamics of the racing approaches to the magnitude of the evolutionary pressure.

6   Concluding Discussion and Future Work

In this paper, we proposed two novel approaches to speed up online evolution of controllers in multirobot systems: (i) a racing technique, and (ii) a population cloning technique. To implement our approaches, we used odNEAT, a decentralised online evolution algorithm in which robots optimise controllers in parallel and exchange candidate solutions to the task. We conducted experiments with four approaches (**odNEAT**, **racing**, **racing-ppc-rem**, and **racing-ppc-norem**) in three tasks: (i, ii) two foraging tasks with differing complexity, and (iii) dynamic phototaxis.

We showed the benefits of our racing approach, and of a population cloning technique that allows evolution to effectively leverage the genetic information accumulated by each individual robot. The combined racing plus population cloning approaches typically yielded: (i) the highest task performance in terms of the fitness score, (ii) the fastest evolution of effective solutions to the task, (iii) the most consistent and stable group-level performance, and (iv) the highest degree of robustness as the evolutionary pressure to solve the task increases. However, if the evolutionary pressure is set above a certain limit, algorithms such as odNEAT can, in certain conditions, display superior performance in the long-term. One key research question is therefore how to enable robots to find the best evolutionary algorithm to solve a given task during the actual task execution.

The immediate follow-up work is to investigate the performance of our proposed approaches in real multirobot systems. In this respect, we also intend to investigate: (i) the effects of *heterogeneous* racing at the *algorithm level*, that is, of allowing multiple robots to race with different configurations of the online evolutionary algorithm in order to find the most effective one, and (ii) cloning techniques and their limits, including potential robustness vs. stagnation trade-offs, and how sensitive is the performance of population cloning to the frequency of interactions between robots.

Acknowledgements. This work was partly supported by FCT under grants SFRH/BD/89573/2012, UID/EEA/50008/2013, and UID/Multi/04046/2013.

References

1. Floreano, D., Mondada, F.: Automatic creation of an autonomous agent: genetic evolution of a neural-network driven robot. In: 3rd International Conference on Simulation of Adaptive Behavior, pp. 421–430. MIT Press, Cambridge (1994)
2. Watson, R., Ficici, S., Pollack, J.: Embodied evolution: distributing an evolutionary algorithm in a population of robots. Rob. Auton. Syst. **39**(1), 1–18 (2002)
3. Silva, F., Urbano, P., Correia, L., Christensen, A.L.: odNEAT: an algorithm for decentralised online evolution of robotic controllers. Evol. Comput. **23**(3), 421–449 (2015)
4. Silva, F., Correia, L., Christensen, A.L.: A case study on the scalability of online evolution of robotic controllers. In: Pereira, F., Machado, P., Costa, E., Cardoso, A. (eds.) EPIA 2015. LNCS, vol. 9273, pp. 189–200. Springer, Heidelberg (2015)
5. Haasdijk, E., Eiben, A., Karafotias, G.: On-line evolution of robot controllers by an encapsulated evolution strategy. In: IEEE Congress on Evolutionary Computation, pp. 1–7. IEEE Press, Piscataway (2010)
6. Prieto, A., Becerra, J., Bellas, F., Duro, R.J.: Open-ended evolution as a means to self-organize heterogeneous multi-robot systems in real time. Rob. Auton. Syst. **58**(12), 1282–1291 (2010)
7. Bredeche, N., Montanier, J.M., Liu, W., Winfield, A.: Environment-driven distributed evolutionary adaptation in a population of autonomous robotic agents. Math. Comput. Model. Dyn. Syst. **18**(1), 101–129 (2012)

8. Silva, F., Duarte, M., Correia, L., Oliveira, S.M., Christensen, A.L.: Open issues in evolutionary robotics. Evol. Comput. In press (2016). http://www.mitpressjournals.org/doi/pdf/10.1162/EVCO_a_00172

9. Dinu, C.M., Dimitrov, P., Weel, B., Eiben, A.: Self-adapting fitness evaluation times for on-line evolution of simulated robots. In: 15th Genetic and Evolutionary Computation Conference, pp. 191–198. ACM, New York (2013)

10. Arif, A., Nedev, D., Haasdijk, E.: Controlling maximum evaluation duration in on-line and on-board evolutionary robotics. Evolving Syst. 5(4), 275–286 (2014)

11. Haasdijk, E., Atta-ul Qayyum, A., Eiben, A.: Racing to improve on-line, on-board evolutionary robotics. In: 13th Genetic and Evolutionary Computation Conference, pp. 187–194. ACM, New York (2011)

12. Haasdijk, E., Smit, S.K., Eiben, A.E.: Exploratory analysis of an on-line evolutionary algorithm in simulated robots. Evol. Intell. 5(4), 213–230 (2012)

13. Hoeffding, W.: Probability inequalities for sums of bounded random variables. J. Am. Stat. Assoc. 58(301), 13–30 (1963)

14. Silva, F., Correia, L., Christensen, A.L.: Dynamics of neuronal models in online neuroevolution of robotic controllers. In: Correia, L., Reis, L.P., Cascalho, J. (eds.) EPIA 2013. LNCS, vol. 8154, pp. 90–101. Springer, Heidelberg (2013)

15. Silva, F., Urbano, P., Christensen, A.L.: Online evolution of adaptive robot behaviour. Int. J. Nat. Comput. Res. 4(2), 59–77 (2014)

16. Mondada, F., Bonani, M., Raemy, X., Pugh, J., Cianci, C., Klaptocz, A., Magnenat, S., Zufferey, J., Floreano, D., Martinoli, A.: The e-puck, a robot designed for education in engineering. In: 9th Conference on Autonomous Robot Systems and Competitions, IPCB, Castelo Branco, Portugal, pp. 59–65 (2009)

17. Maron, O., Moore, A.W.: The racing algorithm: model selection for lazy learners. Artif. Intell. Rev. 11(1), 193–225 (1997)

18. Lobo, F.G.: The Parameter-Less Genetic Algorithm: Rational and Automated Parameter Selection for Simplified Genetic Algorithm Operation. Ph.D. thesis, Universidade Nova de Lisboa, Lisbon, Portugal (2000)

19. Birattari, M., Stützle, T., Paquete, L., Varrentrapp, K.: A racing algorithm for configuring metaheuristics. In: 4th Genetic and Evolutionary Computation Conference, pp. 11–18. Morgan Kauffmann, San Francisco (2002)

20. Yuan, B., Gallagher, M.: Combining meta-EAs and racing for difficult EA parameter tuning tasks. In: Lob, F.G., Lima, F.C., Michalewicz, Z. (eds.) Parameter Setting in Evolutionary Algorithms. Studies in Computational Intelligence, vol. 54, pp. 121–142. Springer, Heidelberg (2007)

21. Silva, F., Correia, L., Christensen, A.L.: Speeding up online evolution of roboticcontrollers with macro-neurons. In: Esparcia-Alcázar, A.I., Mora, A.M. (eds.) EvoApplications 2014. LNCS, vol. 8602, pp. 765–776. Springer, Heidelberg (2014)

22. Duarte, M., Silva, F., Rodrigues, T., Oliveira, S.M., Christensen, A.L.: JBotEvolver: a versatile simulation platform for evolutionary robotics. In: 14th International Conference on the Synthesis and Simulation of Living Systems, pp. 210–211. MIT Press, Cambridge (2014)

23. Cao, Y., Fukunaga, A., Kahng, A.: Cooperative mobile robotics: antecedents and directions. Auton. Rob. 4(1), 1–23 (1997)

Multi-agent Behavior-Based Policy Transfer

Sabre Didi$^{(\boxtimes)}$ and Geoff Nitschke

Department of Computer Science, University of Cape Town,
Rondebosch, Cape Town 7700, South Africa
sabredd0@gmail.com, gnitschke@cs.uct.ac.za

Abstract. A key objective of transfer learning is to improve and speed-up learning on a target task after training on a different, but related, source task. This study presents a neuro-evolution method that transfers evolved policies within multi-agent tasks of varying degrees of complexity. The method incorporates behavioral diversity (novelty) search as a means to boost the task performance of transferred policies (multi-agent behaviors). Results indicate that transferred evolved multi-agent behaviors are significantly improved in more complex tasks when adapted using behavioral diversity. Comparatively, behaviors that do not use behavioral diversity to further adapt transferred behaviors, perform relatively poorly in terms of adaptation times and quality of solutions in target tasks. Also, in support of previous work, both policy transfer methods (with and without behavioral diversity adaptation), out-perform behaviors evolved in target tasks without transfer learning.

Keywords: Multi-agent learning · Evolutionary algorithms · Transfer learning · Behavioural diversity adaptation

1 Introduction

Transfer learning[1] is a technique that attempts to improve learning a task by leveraging knowledge from learning a related but simpler task [1]. Specifically, transfer learning is the process of reusing learned information across tasks, where information is shared between a source and target task. Transferring knowledge that is learned on a source task accelerates learning and increases solution quality in target tasks by exploiting relevant prior knowledge.

Transfer learning has been widely studied in the context of *Reinforcement Learning* (RL) [2], for various single-agent tasks including pole-balancing [3], game-playing [4], robot navigation as well as multi-agent tasks including predator-prey [5]. For such single and multi-agent tasks, policy (behavior) transfer is typically done within the same task domain for varying task complexity [2]. To facilitate the learning of generalized problem solving behavior various agent (controller) representations have been used including *Decision Trees* [4], *Artificial Neural Networks* (ANNs), *Cerebellar Model Arithmetic Computer*, and *Radial Basis Functions* [6].

---

[1] *Transfer learning* and *policy transfer* are used interchangeably in this paper.

© Springer International Publishing Switzerland 2016
G. Squillero and P. Burelli (Eds.): EvoApplications 2016, Part II, LNCS 9598, pp. 181–197, 2016.
DOI: 10.1007/978-3-319-31153-1_13

Such representations are typically selected as they are amenable to decomposition for transfer of partial policies between source and target tasks as well as further adaptation in target tasks [4]. In multi-agent transfer learning, policies learned in source tasks are often shared between agents and used as a starting point for learning new policies in target tasks [5]. A popular multi-agent test-bed is *RoboCup Keep-Away Soccer* [7], which has received significant attention in multi-agent transfer learning research [6].

Recently, in addition to RL, there has been an increasing amount of work on transfer learning using evolutionary algorithms to adapt policies with various representations. For example, Doncieux [8] used *neuro-evolution* [9] to search for effective ANN controllers in a simulated robot ball collecting task, and investigated methods for extracting behavioral features shared between versions of the task. These extracted features were then used as stepping stones to shape rewards in the evolution of controllers transferred to more complex versions of the ball collecting task.

In related research, Moshaiov et al. [10] used *Multi-Objective Evolutionary Algorithms* [11] to devise a *Family Bootstrapping* method that evolved groups of complementary ANN controllers to robot navigation tasks. These controllers were then used as an evolutionary starting point for controller evolution in robot navigation tasks with different objectives. Taylor et al. [12] used the NEAT neuro-evolution method [13] to further evolve a population of ANN controllers already evolved for a source keep-away soccer task. The authors demonstrated that biasing and further evolving a fittest population of controllers for more complex versions of keep-away significantly decreased evolution time.

Verbancsics et al. [14] used an indirect encoding neuro-evolution method (HyperNEAT [15]) to facilitate evolved solutions encoding the geometry of the keep-away soccer task. HyperNEAT facilitated the transfer of evolved multi-agent behaviors between source and target tasks with varying numbers of agents and soccer field sizes, without the need for any further adaptation. The authors also used HyperNEAT to demonstrate successful transfer of multi-agent behaviors between the *Knight's Joust* (a multi-agent predator-prey task variant) [6] and keep-away soccer tasks. The efficacy of this approach was further supported by improved task performance on target tasks after further neuro-evolution and evolved behaviors that were comparable to RL derived policies [16,17]. In support of this approach, related work [14,18] has also highlighted the effectiveness of indirectly coded representations for facilitating transfer learning between multi-agent task variants as well as between different multi-agent tasks.

A key challenge in transfer learning is to ensure that a policy, learned in a source task can be meaningfully transferred to a policy in a target task, with a typically more complex representation [6]. Hence a mapping function is required in order that learned policies are transferable between tasks with different numbers of state and action variables. For example, tasks of increasing complexity or different but related tasks such as keep-away soccer [7] and Knight's Joust [6]. To address this, Taylor et al. [6], devised a *inter-task mappings for policy search* method to transfer a population of control policies (ANN controllers) between keep-away soccer, knight's joust and *Server Job Scheduling* tasks [17]. This method was successfully applied with full (hand-coded) inter-task mapping functions, where inter-task mapping functions were only partially available or where inter-task mapping functions had to be learned prior to policy transfer.

This study combines and extends previous work on *inter-task mappings for policy search* [6] and facilitating transfer learning with HyperNEAT [14]. Specifically, we investigate the adaptation of multi-agent behaviors in the keep-away soccer task domain with the *Novelty Search* [19] behavioral diversity mechanism. Whilst many studies support the efficacy of objective-based (fitness function) search approaches in transfer learning [6, 14, 17, 18], the impact of behavioral diversity maintenance on transfer learning remains unexplored. This study also investigates the benefits of behavioral diversity and objective based search with policy transfer using direct (NEAT) and indirect encoding (HyperNEAT) methods to evolve behaviors in target keep-away tasks.

First, we hypothesize that if behavioral diversity maintenance is used in multi-agent behavior evolution, this will yield a higher *task performance* than objective-based evolution in all tasks tested. Second, we hypothesize that NEAT and HyperNEAT are appropriate policy (multi-agent behavior) search methods for enabling policy transfer where transferred behaviors yield a *higher task performance* and *efficiency* compared to those without policy transfer. Efficiency refers to the average number of generations until the *average maximum fitness* (for the given method) was attained.

These hypotheses were devised given related research results [20–22], and were tested by a comparison of keep-away behaviors evolved with NEAT and HyperNEAT (using either objective-based search or behavioral diversity maintenance) in keep-away target tasks with and without policy transfer.

2 Methods

2.1 NEAT: Neuro-Evolution of Augmenting Topologies

This research uses *Neuro-Evolution of Augmenting Topologies* (NEAT) [13] as the direct encoding policy search method. NEAT evolves both connection weights and ANN topologies, and applies three key techniques to maintain a balance between performance and diversity of solutions. First, it assigns a unique historical marking to every new gene so as crossover can only be performed between pairs of matching genes. Second, NEAT speciates the population so as ANNs (genotypes) compete primarily within their own niches (identified by historical markings) instead of competing with the population at large. Third, NEAT begins evolution with a population of simple ANNs with no hidden nodes but gradually adds new topological structure (nodes and connections) using two special mutation operators called *add hidden node* and *add link*.

NEAT was selected as this study's direct encoding method as it has been successfully used for a broad range of multi-agent control tasks [2, 10, 12, 23, 24]. However, there has been relatively little research as to efficacy of NEAT as a policy search method for multi-agent transfer learning [12].

2.2   HyperNEAT: Hypercube-Based NEAT

Hypercube-based NEAT (HyperNEAT) [15] is an indirect (generative) encoding neuro-evolution method that extends NEAT and uses two networks, a *Composite Pattern Producing Network* (CPPN) [25] and a *substrate* (ANN) (Fig. 1). The CPPN is the generative encoding mechanism that indirectly maps evolved genotypes to ANNs and encodes pattern regularities, symmetries and smoothness of the geometry of a given task in the form of the substrate. This mapping functions via having coordinates of each pair of nodes connected in the substrate fed to the CPPN as inputs. The CPPN outputs a value assigned as the synaptic weight of that connection and a value indicating whether that connection can be expressed or not. HyperNEAT uses the evolutionary process of NEAT to evolve the CPPN and determine ANN fitness values. The main benefit of HyperNEAT is scalability as it exploits task geometry and thus effectively represents complex solutions with minimal genotype structure [15]. This makes HyperNEAT an appropriate choice for evolving complex multi-agent solutions [14,26].

HyperNEAT was selected as this study's indirect encoding neuro-evolution method since previous research indicated that transferring the *connectivity patterns* [27] of evolved behaviors is an effective way for facilitating transfer learning in multi-agent tasks [14,18]. HyperNEAT's capability to evolve controllers that account for task geometry also makes HyperNEAT appropriate for deriving controllers that elicit behaviors robust to variations in state and action spaces [28] as well as noisy, partially observable environments of multi-agent tasks. Also, it has been demonstrated that HyperNEAT evolved multi-agent policies can be effectively transferred to increasingly complex versions of keep-away soccer [7] without further adaptation [14] and that transferred behaviors often yield comparable task performance to specially designed learning algorithms [16].

2.3   Behavioral Diversity

Encouraging behavioral diversity is a well studied concept in neuro-evolution and has been used to discover novel solutions, increase solution performance in a wide range of tasks as well as out-perform controller evolution approaches that encourage genotypic diversity [20,29–31].

One such approach is *Novelty search* (NS) [19], that is not driven by a fitness (objective) function but rather rewards evolved phenotypes (behaviors) based on their novelty. Thus, a genotype is more likely to be selected for reproduction if its encoded behavior is sufficiently different from all other behaviors produced thus far in an evolutionary run. Recent results indicate that controllers evolved with a NS metric attained some degree of generality. For example, in a maze solving task, controllers evolved to solve one maze were successfully transferred to solve different mazes [32]. Also, NS has been demonstrated as yielding solutions that out-perform objective based search in various tasks [20,22] including complex multi-agent tasks with large numbers of agents [21]. Hence, NS was selected as the behavioral diversity mechanism to be applied to our selected policy search methods (NEAT and HyperNEAT).

In this study, the function of NS is to consistently generate novel team (keep-away) behaviors. Hence, we define team behavior in terms of properties that potentially influence team behavior but are not directly used for task performance evaluation. For the keep-away task, the behavioral properties we use are the *average number of passes, average dispersion of team members*, and *average distance of the ball to the center of the field*.

In line with previous research on hybrid NS and fitness metrics supporting performance gains in various tasks [33], including multi-agent tasks [34], we use a behavioral diversity metric that linearly combines NS with objective-based search (NEAT and HyperNEAT), in order to improve keep-away policy search.

Several hybrid metrics have been proposed including fitness sharing and linear combination [21], restarting converged evolutionary runs using NS [33], a minimal criteria NS (for genotype survival and reproduction) [35], and a progressive minimal criteria (incrementing reproduction requirements throughout evolution) [34]. Here we use a linear combination [21] (Eq. 1):

$$score_i = \rho \cdot \overline{fit_i} + (1 - \rho) \cdot \overline{nov_i} \tag{1}$$

Where, $\overline{fit_i}$ and $\overline{nov_i}$ are normalized fitness and novelty of ith genotype respectively. Then $\rho \in [0,1]$ is a parameter selected by the experimenter ($\rho = 0.4$, in this study) to control the relative contribution of each metric to the selection pressure. To measure novelty we use normalized task specific behavioral vectors: *Average number of passes, Mean team mates dispersion*, and *Average distance of ball to the center of the field*.

This team level behavioral characterization has been used previously [21] and out-performs individual behavioral characterizations and fitness based search. Behavioral distance is computed as a Euclidean distance (Eq. 2):

$$\delta(x,y) = \|x_i - y_i\| \tag{2}$$

Where, x_i and y_i are normalized behavioral characterization vectors of genotype x and y. The novelty is then quantified by Eq. 3:

$$nov_x = \frac{1}{3k} \sum_{i=1}^{k} \sum_{j=1}^{3} \delta(x_j, y_{ij}) \tag{3}$$

Where, x_j is the jth behavioral property of genotype x, y_{ij} is the jth behavioral property of the ith nearest neighbour of genotype x and δ is the behavioral distance between two genotypes x and y computed in Eq. 2 which is based on the behavioral characterization vector. The nov_x is then derived from the mean of behavioral distance of an individual with k nearest neighbors. The parameter k is specified by the experimenter to represent the number of nearest neighbors, where $k = 15$ has been widely used in NS experiments [22]. A few researchers have used $k = 20$ [22,36] and k in the range of $[3,10]$ though it is unclear if such k values were derived experimentally. Gomes et al. [22] discovered that the choice of k value heavily depended on the type of novelty archive used and that

$k = 15$ yielded relatively good performance across all tested archive types. Hence in this study we use $k = 15$.

2.4   Policy Transfer Method

For both NEAT and HyperNEAT, and their non-objective (novelty) and objective based search variants, we tested three policy transfer approaches. First, the entire evolved population was transferred from the source task (at the final generation of neuro-evolution) and set as the initial population for neuro-evolution in the target task. Second, target population was seeded with the fittest genotype in the source task and used as a bias for initialising the remainder of the target population. Third, the fittest 50 % of the population evolved for the source task was selected to seed and bias initialization of the rest of the starting population in the target task. The first approach was found to be the most effective for all methods and tasks tested in this case study and was thus used in company with the selected mapping function for policy transfer (Algorithm 1). Algorithm 1 is a transfer mapping function that is an extension of that proposed by Taylor et al. [6] and used is used in this study's keep-away policy transfer experiments.

Algorithm 1. Transfer Mapping Function

Generate a network with same number of inputs and outputs as in the Π_{source}
Add the same number of hidden nodes to Π_{target} as in Π_{source}
Repeat
For each pair of nodes (n_i, n_j) in Π_{target} **do**
    If $\exists$ link $L_{i,j} \in \Pi_{source}$ **then**
        add link $L_{i,j}$ to Π_{target} with $w^t{}_{i,j} = w^s{}_{i,j}$ in Π_{source}
    Else
        If $\nexists$ nodes$(n_i, n_j) \in \Pi_{source}$
            add link $L_{i,j}$ to Π_{target} with $w^t{}_{i,j} = $ random weights
Until all pairs of nodes are visited

3   Experiments

Experiments test this study's research objectives (Sect. 1). First, to test the impact of using a non-objective (behavioral diversity) versus objective (fitness) based search approach for two given policy search methods (NEAT and Hyper-NEAT). Second, to test the efficacy of NEAT and HyperNEAT as appropriate methods for yielding task performance and efficiency boosts after policy transfer.

Experiments are run in a source keep-away task (using NEAT or Hyper-NEAT to evolve multi-agent keep-away behavior), where populations evolved after 20 generations, are transferred to a target task, and evolved for a further 50 generations (Table 2). Results are compared to those where no policy transfer takes place, that is where NEAT and HyperNEAT are used to evolve keep-away

Table 1. Sensory inputs (13 input nodes) and motor outputs (three outputs) for a team's ANN controller in the *3vs2 keep-away task*. Keeper 1 is the agent with the ball.

Sensory Inputs	Description
$\text{dist}(K_b, C)$, $\text{dist}(K_{t1}, C)$, $\text{dist}(K_{t2}, C)$	Distance of each keeper to field center
$\text{dist}(T1, C)$, $\text{dist}(T2, C)$	Distance of each taker to field center
$\text{dist}(K_b, K_{t1})$, $\text{dist}(K_b, K_{t2})$	Distance of each taker to keeper 1
$\text{dist}(K_b, T1)$, $\text{dist}(K_b, T2)$	Distance of each taker to keeper 1
$min_{j \in 1,2} dist(K_{t1}, T_j)$, $min_{j \in 1,2} dist(K_{t2}, T_j)$	Distance of closest taker to keeper 1
$min_{j \in 1,2} angle(K_{t1}, T_j)$, $min_{j \in 1,2} angle(K_{t1}, T_j)$	Angle of closest keeper, taker, keeper 1
Motor Outputs	
Hold	Do not pass ball
Pass to K_{t1}, Pass to K_{t2}	Pass to keeper 2, keeper 3

behaviors from *scratch* in the target tasks. For both NEAT and HyperNEAT experiments, each genotype (agent team) is evaluated over 30 task trials per generation, where each task trial tests different (random) agent positions. The ball always starts in the possession of a (randomly selected) keeper. Average fitness (task performance) per genotype is computed over these 30 task trials. Table 2 specifies the neuro-evolution and simulation parameters for these experiments.

The efficacy of policy transfer was evaluated in terms of time (genotype evaluations) taken to attain a policy transfer threshold, with and without policy transfer. The threshold was the *average maximum fitness* attained after applying NEAT and HyperNEAT to evolve behaviors *from scratch* in each target task. Policy transfer occurs between source and incrementally complex target tasks. That is, first we evolve keep-away behavior for three keepers versus two takers (denoted as *3vs2*) in a 20×20 virtual field[2] (Table 2). Evolved behaviors (policies) are then transferred (and neuro-evolution continued) in one of three keep-away target tasks, four keepers versus three takers (*4vs3*), five keepers versus three takers (*5vs3*) or six keepers versus four takers (*6vs4*).

3.1   NEAT Experiments

Table 1 describes the 13 sensory input nodes in a team's ANN controller for the *3vs2 keep-away* task. The output nodes represent an agent's decision to *hold* the

---

[2] All experiments were run in *RoboCup Keep-Away version 6* [6]. Source code and executables can be found at: http://people.cs.uct.ac.za/~gnitschke/EvoStar2016/.

Table 2. Left: *Neuro-Evolution* (NE), *Novelty Search* (NS) parameters (final three rows). **Right:** CPPN (HyperNEAT) activation Functions and simulation parameters.

NE / NS Parameters	Setting
Population Size	150
Generations (Source task)	20
Generations (Target task)	50
Maximum number of species	10
Maximum species population	30
Weight mutation	±0.01
NEAT Weight value range	[-5.0, 5.0]
HyperNEAT Weight value range	[-5.0, 5.0]
Mutation rate	0.05
Survival threshold	0.2
NS nearest neighbor k	15
Maximum archive size	1000
Compatibility threshold	3
Behavioral threshold	0.03

HyperNEAT CPPN	Functions
Identity	x
Gaussian	$e^{-2.5x^2}$
Bipolar Sigmoid	$\frac{2}{1+e^{-4.9x}} - 1$
Absolute value	$|x|$
Sine	$sine(x)$

Simulation Parameters	Setting
Number of Runs	20
Iterations per task trial	4500
Trials per generation	30
Agent positions	Random
Environment size	20x20 grid
Agent speed (per iteration)	1 grid cell
Ball speed (per iteration)	2 grid cells

ball, *pass to keeper 2* or *pass to keeper 3*, where *keeper 1* has the ball. At any task trial iteration, the output with the highest activation is the action selected.

NEAT is direct encoding method, so the genotype representation (encoding sensory-motor elements of a keep-away team's controller) needs to change as task complexity and the number of agents changes. For example, as task complexity increases, from *3vs2* to *4vs3 keep-away*, an ANN topology with 19 input nodes and 4 output nodes is required. The additional output node represents the decision of *keeper 1* to *pass to keeper 4*. The extra six input nodes represent: (1) distance of *keeper 4* from the field's center, (2) distance of *taker 3* from the field's center, (3) distance of *keeper 1* from *taker 3*, (4) distance between *keeper 4* and the closest taker, (5) angle formed between keeper 1 and the closest keeper and taker, and (6) distance of *keeper 1* to *keeper 4*. Similarly, for the *5vs3 task* an ANN with 27 inputs and six outputs is needed.

However, for all keep-away tasks tested (*3vs2*, *4vs3*, and *5vs4*) the ANN sensory-motor layer topology was kept static (13 sensory inputs and three motor outputs) in order to facilitate transfer across tasks of increasing complexity. Thus, as the number of agents increased with task complexity, a heuristic selected which agents in the environment would be processed by the ANN's 13 sensory input nodes. At each sensory-motor cycle (task trial iteration), the heuristic selected the closest two keeper and taker agents to be processed by the ANN, but had the potential to process any agent as sensory input. In keep-away task simulation this was tantamount to noise preventing the keeper with the ball from processing agents too far away and thus accounting for them in action selection.

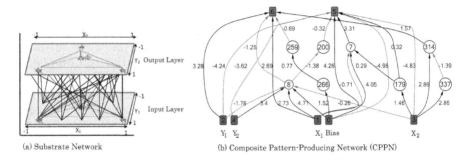

(a) Substrate Network   (b) Composite Pattern-Producing Network (CPPN)

Fig. 1. Left: Substrate encoding the virtual field (20×20 grid of inputs and outputs). Connection values ($[-1.0, 1.0]$) between these input-output nodes represent positions of agents relative to the keeper with the ball. **Right:** Connections from pairs of nodes in the substrate are sampled and the coordinates passed as inputs to the CPPN, which then outputs the synaptic weight of each sampled connection.

Table 3. Average normalized maximum fitness (over 20 runs) for the three experimental setups. Values are portions of the maximum possible hold time (possession of the ball) for the team of keepers. Standard deviations are shown in parentheses.

Experiment	4vs3	5vs3	6vs4
	Keep-Away	Keep-Away	Keep-Away
No Policy Transfer			
NEAT	0.438 (0.037)	0.473 (0.052)	0.419 (0.057)
HyperNEAT	0.587 (0.059)	0.765 (0.050)	0.533 (0.044)
Fitness-Based Policy Transfer			
NEAT	0.482 (0.059)	0.580 (0.069)	0.464 (0.033)
HyperNEAT	0.729 (0.089)	0.873 (0.089)	0.632 (0.038)
Fitness + NS Policy Transfer			
NEAT	0.545 (0.047)	0.638 (0.0048)	0.520 (0.036)
HyperNEAT	0.752 (0.054)	0.943 (0.029)	0.697 (0.032)

3.2   HyperNEAT Experiments

HyperNEAT uses indirect encoding and can thus represent changes in task complexity without changing genotype representation [14]. In this experiment *Bird's Eye View* (BEV) representation [14] is used to encode keep-away's physical state (layout of the field and locality of agents) and actions onto a substrate network. The virtual keep-away soccer field is divided into a 20×20 grid world, where each agent can occupy one grid cell per task trial iteration. The input and output layers of the substrate network are two dimensional, with coordinates in the x, y plane in the range of $[-1.0, 1.0]$. Each grid cell in the virtual space is represented by a node in the substrate network layer, so the 20×20 grid world is represented by 400 nodes in the substrate network. Hence, the layout of nodes in the substrate network (network geometry) directly maps to the tasks geometry and this enables HyperNEAT to exploit the task's geometric regularities and relationships. The position of each agent is marked on the substrate input layer, where each position of the keeper is marked by a value 1.0, and takers by −1.0. Physical paths between agents are drawn. Each direct path from a keeper with a ball to another keeper is marked by a value 0.3 and to a taker by a value −0.3. The region to pass the ball to is highlighted on the substrate output by activating the node with the highest output.

The CPPN queries each connection between input and output layers of the two dimensional substrate network taking coordinates $(x1, y1)$ and $(x2, y2)$ as input. The CPPN output represents the weight of that connection and the connection expression value. The connection weights are then produced as a function of their endpoints. The functions used are listed in Table 2 (right).

4   Results and Discussion

Policy transfer was applied between the source *3vs2* keep-away task and incrementally complex *4vs3*, *5vs3* and *6vs4* keep-away tasks (Sect. 3). Keep-away behaviors were evolved for 20 generations with NEAT or HyperNEAT (using either the novelty-objective hybrid or objective-based search) in the source task, transferred to the target task and then further evolved for 50 generations. For policy transfer, three population initialization methods in the target task were tested (Sect. 2.4). However, the transfer of the *entire population* (from generation 20 in the source task) to target tasks best facilitated policy transfer. Hence only results for this population initialization method are presented here.

Table 3 presents the average normalized maximum fitness (attained during each run and averaged over 20 runs) for the three experimental setups. Experiment 1 (*No Policy Transfer*) presents results from evolving keep-away behaviors in each of the target tasks from *scratch* (without policy transfer). Experiment 2 (*Fitness-Based Policy Transfer*) presents results from evolving keep-away behaviors using objective (fitness) based NEAT and HyperNEAT in the source and then in target tasks (after policy transfer). Experiment 3 (*Fitness + NS Policy Transfer*) presents results from evolving keep-away behaviors with NEAT and HyperNEAT using the novelty-objective hybrid based search. In experiments 2 and 3, NEAT or HyperNEAT is applied in the source task for 20 generations and thereafter for 50 generations in the target task.

Results data was found to be non-parametric using the *Kolmogorov-Smirnov* normality test with *Lilliefors* correction [37]. The *Mann-Whitney U* test [38] was then applied in a series of pair-wise comparisons to gauge if there was a statistically significant difference between corresponding result sets of the three experiments (Table 3). Pair-wise comparisons were conducted between average results data for NEAT or HyperNEAT (for a given experiment). The null hypothesis stated that two comparative data sets were not significantly different, and $\alpha = 0.05$ was selected as the significance threshold.

4.1   Policy Versus No-Policy Transfer: Performance Comparisons

First, statistical tests indicated that for all policy transfers (from *3vs2* to *4vs3*, *5vs3* and *6vs4*), there was a statistically significant difference (p-value < 0.05) between the *novelty-objective hybrid* and *objective-based NEAT*. That is, NEAT with behavioral diversity maintenance yielded a significantly higher average maximum task performance for all policy transfer cases (Table 3).

Second, statistical tests indicated that for all policy transfers transfer cases, both NEAT and HyperNEAT using behavioral diversity maintenance yielded a higher average maximum task performance compared to objective-based NEAT and HyperNEAT (Table 3).

This result supports this study's first hypothesis (Sect. 1), that encouraging behavioral diversity facilitates the evolution of higher performance keep-away behaviors in all tasks tested, compared to keep-away behavior adaptation without behavioral diversity maintenance.

Statistical tests also indicated that NEAT (using either the novelty-objective or objective-based search) yielded a higher average task performance (with statistical significance) for all policy transfers, compared to objective-based NEAT without policy transfer. That is, where NEAT was applied to evolve keep-away behavior (*from scratch*) in each of the target tasks (*4vs3*, *5vs3* and *6vs4*). Similarly, statistical tests indicated that HyperNEAT (*with* and *without* behavioral diversity maintenance), yielded significantly higher task performances in all target tasks compared to HyperNEAT without policy transfer (Table 3).

These results partially support this study's second hypothesis, that NEAT and HyperNEAT are appropriate as policy search methods where policy transfer enables the evolution of significantly higher performance keep-away behaviors in all target tasks tested (compared to keep-away behaviors evolved *from scratch*). These results are further supported by previous work demonstrating that transfer learning enables multi-agent behavior adaptation with significantly higher task performances compared to adaptation without transfer learning [5,12,14].

4.2   Policy Versus No-Policy Transfer: Efficiency Comparisons

To further support this study's second hypothesis, the efficiency of NEAT and HyperNEAT (*with* and *without* behavioral diversity maintenance) is compared in the target tasks where policy transfer was applied versus where keep-away behaviors were evolved in the target tasks *from scratch*.

Results (*performance threshold*) of applying NEAT and HyperNEAT to adapt keep-away behaviors *from scratch* (without policy transfer) in the target tasks (*4vs3*, *5vs3* and *6vs4*) were used as a benchmark for comparisons with the same methods applied with policy transfer. This threshold was the *average maximum task performance* of NEAT and HyperNEAT in the target tasks, where keep-away behavior was evolved without policy transfer (Table 3).

First, for *objective-based NEAT without policy transfer*, an average maximum task performance of 0.443 (as a portion of maximum task performance) for all three target tasks was attained after approximately 40 generations[3]. After 40 generations negligible task performance increases were observed. Additional experiments that used relatively few runs, but 100 generations of evolution indicated that objective-based NEAT, without policy transfer, gets stuck in a local optima. However, this is not the case when policy transfer is used (for both objective and hybrid objective-novelty based variants of NEAT).

Comparatively, results from *objective-based NEAT with policy transfer* indicated efficiency gains for all target tasks. Objective-based NEAT with policy transfer yielded an average maximum task performance of 0.482, 0.580 and 0.464 for the *4vs3*, *5vs3* and *6vs4* keep-away tasks, respectively. These tasks performances were attained after approximately 48 generations. However, additional experiments using relatively few runs but 100 generations indicated that the task performances yielded by objective-based NEAT with policy transfer continued to increase. Also, all task performances yielded by objective-based NEAT *with policy transfer* were significantly higher (Mann-Whitney test, p-value < 0.05) compared to those yielded by NEAT without policy transfer in the same tasks.

Objective-based HyperNEAT with policy transfer also yielded greater efficiency for all target tasks. That is, objective-based HyperNEAT with policy transfer resulted in an average maximum task performance of 0.632 for *6vs4* keep-away after 38 generations, *with policy transfer*. This was compared to the significantly lower 0.533 (Mann-Whitney test, p-value < 0.05) average maximum performance after 49 generations, *without policy transfer* in the same *5vs3* keep-away task.

Task performance in *5vs3* keep-away steadily reached an average maximum of 0.873 after 50 generations, *with policy transfer*. This was compared to the significantly lower 0.765 (Mann-Whitney test, p-value < 0.05) average maximum performance after 48 generations, *without policy transfer* in the same task.

Task performance in *4vs3* keep-away steadily reached an average maximum of 0.729 after 47 generations, *with policy transfer*. This was compared to the significantly lower 0.587 (Mann-Whitney test, p-value < 0.05) average maximum performance after 45 generations, *without policy transfer* in the same task. Also, additional experiments using relatively few runs but 100 generations indicated that the task performances yielded by objective-based HyperNEAT with policy transfer continued to increase.

Second, for *novelty-objective based NEAT with policy transfer*, average maximum task performances of 0.545, 0.638 and 0.520 were attained for tasks *4vs3*, *5vs3* and *6vs4*, after 48, 49 and 48 generations, respectively. This compared to the significantly lower task performances (Mann-Whitney test, p-value < 0.05) of *novelty-objective based NEAT without policy transfer* for the same tasks. That is, 0.443, 0.473, and 0.473 for the *4vs3*, *5vs3* and *6vs4* tasks, attained after 40, 37 and 43 generations respectively.

Also, *novelty-objective based HyperNEAT with policy transfer* yielded average maximum task performances of 0.752, 0.943 and 0.697 for tasks *4vs3*, *5vs3*

---

[3] NEAT and HyperNEAT average maximum task performance progression graphs can be found at: http://people.cs.uct.ac.za/~gnitschke/EvoStar2016/.

and *6vs4*, after 45, 49 and 48 generations, respectively. This compared to the significantly lower task performances (Mann-Whitney test, p-value < 0.05) of *objective based HyperNEAT without policy transfer* for the same tasks (Table 3), yielded in a comparable number of generations (45, 48 and 49 generations, for the *4vs3*, *5vs3* and *6vs4* tasks, respectively).

Hence, these results further support this study's second hypothesis, that NEAT and HyperNEAT are appropriate policy search methods, where policy transfer enables a higher efficiency in the target tasks tested. Thus, NEAT and HyperNEAT with policy transfer (using objective-based search or behavioral diversity maintenance) converge to a higher task performance faster compared to the same methods without policy transfer.

4.3   Behavioral Diversity Maintenance and Policy Transfer

The results of this study have important implications for current policy transfer research, specifically multi-agent policy transfer where neuro-evolution is used for policy search (agent behavior adaptation).

First, the results indicated significant *task performance* and *efficiency* (speed-up of evolution) benefits of policy transfer in a multi-agent task (*Keep-away RoboCup Soccer*) where team behavior was evolved with NEAT or HyperNEAT in a source task and then further evolved in more complex target tasks. This was compared to the same methods for evolving keep-away behavior *from scratch* in the target tasks. These results are also supported by related policy transfer research that used neuro-evolution for policy search [8,10,12,14].

Second, results indicated that HyperNEAT with behavioral diversity maintenance yielded the greatest benefits for policy transfer overall. These transferred behaviors leveraged the most benefits of behaviors evolved in the source task such that further evolution in target tasks yielded the highest overall task performances. This was compared to NEAT with behavioral diversity maintenance, objective-based NEAT and HyperNEAT and the same methods without policy transfer. Such benefits of behavioral diversity maintenance coupled with objective-based search is supported by related work [33–35]. Also, advantages of indirect encoding neuro-evolution methods such as HyperNEAT have been highlighted in a broad range of task domains [14,15,26,28].

However, a key contribution of this study is that this is the first time (to the authors' knowledge), the benefits of behavioral diversity maintenance coupled with neuro-evolution, have been demonstrated in multi-agent policy transfer.

The significantly higher performance of HyperNEAT with behavioral diversity maintenance across all tasks is theorized to be a result of beneficial interactions between a more effective search for high performance behaviors (aided by behavioral diversity maintenance) and HyperNEAT's indirect encoding of agent behaviors. Consider that in the source task the novelty-objective hybrid based search employed by HyperNEAT facilitated an effective exploration versus exploitation trade-off in the search for high-performance keep-away behaviors. This is supported by results from previous work [21,33–35] that similarly report the benefits of hybrid novelty-objective based search approaches (including task performance advantages over objective based search approaches).

Also, when effective high-performance behaviors are discovered as solutions to the source task (*3vs2 keep-away*), HyperNEAT's indirect encoding of such behaviors and the spatial geometry of the keep-away task facilitates more effective policy transfer to incrementally complex target tasks. That is, HyperNEAT evolves CPPNs that are able to represent complex ANN controllers with their own symmetries and regularities and exploit the sensory-motor geometry of multi-agent tasks [15,26]. This controller representation significantly impacted the efficacy of evolved keep-away behavior across all tested target tasks.

Thus, we hypothesize that adapting controllers with HyperNEAT in company with the aid of behavioral diversity maintenance allows first, for the discovery of novel robust and effective multi-agent behaviors (that might not have otherwise been discovered with pure objective-based search). Second, HyperNEAT encodes team behaviors that do not rely upon specific sensory-motor mappings in the agent team controller and thus set task environment configurations (such as specific agent and ball positions and numbers of agents). That is, HyperNEAT evolves *connectivity patterns* [27] that are broadly applicable to tasks of varying complexity (in keep-away, numbers of agents). This is supported by related research that similarly demonstrates the robustness of HyperNEAT evolved controllers in tasks of varying complexity [28].

The performance of HyperNEAT evolved teams was contrasted to the significantly lower task performance of NEAT (with and without behavioral diversity maintenance). In NEAT, team behaviors were directly encoded with an ANN with fixed sensory-motor layers (13 sensory inputs and three motor outputs), where the number of hidden nodes and connections were evolved. This static sensory-motor layer ANN topology prevented a smooth and effective transfer from the source task to the more complex target tasks. However, behavioral diversity maintenance did boost the task performance and efficiency of NEAT evolved behaviors in all target tasks after policy transfer (Table 3).

The lower performance of both NEAT and HyperNEAT (with and without behavioral diversity maintenance) in the *6vs4* target task (Table 3) remains the subject of current research. Though this is hypothesized to be a result of the increased complexity of four takers on the same sized virtual field, making taker interception of ball passes more likely. Also this increases the required complexity of evolved keep-away behaviors, meaning evolved behaviors must effectively scale to coordinate larger numbers of keepers while accounting for more takers, but with the same spatial constraints on the virtual field as the *4vs3* and *5vs3* tasks.

5   Conclusion

This study investigated methods for improving the current state of the art in multi-agent transfer learning. That is, improving task performance and efficiency (speed of adaptation) of *Keep-away RoboCup Soccer* behaviors evolved in a source task but then further evolved on more complex versions of the same task.

Experiments compared two neuro-evolution methods, NEAT and HyperNEAT, applying them to evolve keep-away behaviors. This study's main contribution was elucidating that behavioral diversity maintenance coupled with these methods yielded increased task performance in increasingly complex keep-away tasks.

Results indicated that behavioral diversity maintenance used in company with NEAT and HyperNEAT is an appropriate approach for increasing task performance and efficiency in keep-away tasks of increasing complexity. Using behavioral diversity maintenance enabled NEAT and HyperNEAT to out-perform objective-based NEAT and HyperNEAT with and without policy transfer in all tested target keep-away tasks. Also, results indicated that HyperNEAT using behavioral diversity maintenance yielded the highest overall task performance and efficiency. This was theorized to be a result of HyperNEAT's indirect encoding of keep-away behaviors, facilitating effective transfer of evolved behaviors between tasks of varying complexity.

Future work will further investigate the efficacy of indirect encoding methods for facilitating effective policy transfer between similar but related multi-agent tasks (for example, keep-away to multi-agent predator-prey [26]), thus addressing the larger goal of devising controller design methods capable of producing generalized problem solving behaviors.

References

1. Pan, S., Yang, Q.: A survey on transfer learning. IEEE Trans. Knowl. Data Eng. **22**(10), 1345–1359 (2010)
2. Torrey, L., Shavlik, J.: Transfer learning. In: Olivas, E.S. (ed.) Handbook of Research on Machine Learning Applications, pp. 17–23. IGI Global, Hershey (2009)
3. Ammar, H., Tuyls, K., Taylor, M., Driessens, K., Weiss, G.: Reinforcement learning transfer via sparse coding. In: Proceedings of the Eleventh International Conference on Autonomous Agents and Multiagent Systems, Valencia, Spain, pp. 4–8. AAAI (2012)
4. Ramon, J., Driessens, K., Croonenborghs, T.: Transfer learning in reinforcement learning problems through partial policy recycling. In: Kok, J.N., Koronacki, J., Lopez de Mantaras, R., Matwin, S., Mladenič, D., Skowron, A. (eds.) ECML 2007. LNCS (LNAI), vol. 4701, pp. 699–707. Springer, Heidelberg (2007)
5. Boutsioukis, G., Partalas, I., Vlahavas, I.: Transfer learning in multi-agent reinforcement learning domains. In: Sanner, S., Hutter, M. (eds.) EWRL 2011. LNCS, vol. 7188, pp. 249–260. Springer, Heidelberg (2012)
6. Taylor, M., Stone, P., Liu, Y.: Transfer learning via inter-task mappings for temporal difference learning. J. Mach. Learn. **8**(1), 2125–2167 (2010)
7. Stone, P., Kuhlmann, G., Taylor, M.E., Liu, Y.: Keepaway soccer: from machine learning testbed to benchmark. In: Bredenfeld, A., Jacoff, A., Noda, I., Takahashi, Y. (eds.) RoboCup 2005. LNCS (LNAI), vol. 4020, pp. 93–105. Springer, Heidelberg (2006)
8. Doncleux, S.: Knowledge extraction from learning traces in continuous domains. In: AAAI 2014 Fall Symposium on Knowledge, Skill, and Behavior Transfer in Autonomous Robots, Arlington, USA, pp. 1–8. AAAI Press (2014)
9. Floreano, D., Dürr, P., Mattiussi, C.: Neuroevolution: from architectures to learning. Evol. Intel. **1**(1), 47–62 (2008)

10. Moshaiov, A., Tal, A.: Family bootstrapping: a genetic transfer learning approach for onsetting the evolution for a set of realated robotic tasks. In: Proceedings of the Congress on Evolutionary Computation, pp. 2801–2808. IEEE Press (2014)

11. Deb, K.: Pareto Based Multi-objectives Optimization Using Evolutionary Algorithms. Wiley, New York (2001)

12. Taylor, M., Whiteson, S., Stone, P.: Transfer learning for policy search methods. In: ICML 2006: Proceedings of the Twenty-Third International Conference on Machine Learning Transfer Learning Workshop, Pittsburgh, USA, pp. 1–4. ACM (2006)

13. Stanley, K., Miikkulainen, R.: Evolving neural networks through augmenting topologies. Evol. Comput. 10(2), 99–127 (2002)

14. Verbancsics, P., Stanley, K.: Evolving static representations for task transfer. J. Mach. Learn. Res. 11(1), 1737–1763 (2010)

15. Stanley, K., D'Ambrosio, D., Gauci, J.: A hypercube-based indirect encoding for evolving large-scale neural networks. Artif. Life 15(2), 185–212 (2009)

16. Stone, P., Sutton, R., Kuhlmann, G.: Reinforcement learning for robocup-soccer keepaway. Adapt. Behav. 13(3), 165–188 (2006)

17. Whiteson, S., Stone, P.: Evolutionary function approximation for reinforcement learning. J. Mach. Learn. Res. 7(1), 877–917 (2006)

18. Bahceci, E., Miikkulainen, R.: Transfer of evolved pattern-based heuristics in games. In: Proceedings of the IEEE Symposium on Computational Intelligence and Games, Perth, Australia, pp. 220–227. Morgan Kaufmann (2008)

19. Lehman, J., Stanley, K.: Abandoning objectives: evolution through the search for novelty alone. Evol. Comput. 19(2), 189–223 (2011)

20. Mouret, J., Doncieux, S.: Encouraging behavioral diversity in evolutionary robotics: an empirical study. Evol. Comput. 20(1), 91–133 (2012)

21. Gomes, J., Mariano, P., Christensen, A.: Avoiding convergence in cooperative coevolution with novelty search. In: Proceedings of the International Conference on Autonomous Agents and Multi-agent Systems, pp. 1149–1156. ACM (2014)

22. Gomes, J., Mariano, P., Christensen, A.: Devising effective novelty search algorithms: a comprehensive empirical study. In: Proceedings of the Genetic Evolutionary Computation Conference, Madrid, Spain, pp. 943–950. ACM (2015)

23. Degrave, J., Burm, M., Kindermans, P., Dambre, J., Wyffels, F.: Transfer learning of gaits on a quadrupedal robot. Adapt. Behav. 23, 9–19 (2015)

24. Knudson, M., Tumer, K.: Policy transfer in mobile robots using neuro-evolutionary navigation. In: Proceedings of the Genetic and Evolutionary Computation Conference, Philadelphia, USA, pp. 1411–1412. ACM Press (2012)

25. Stanley, K.: Compositional pattern producing networks: a novel abstraction of development. Genet. Program Evolvable Mach. 8(2), 131–162 (2007)

26. D'Ambrosio, D., Stanley, K.: Scalable multiagent learning through indirect encoding of policy geometry. Evol. Intell. J. 6(1), 1–26 (2013)

27. Gauci, J., Stanley, K.: A case study on the critical role of geometric regularity in machine learning. In: Proceedings of the AAAI Conference on Artificial Intelligence, Menlo Park, USA, pp. 628–633. AAAI Press (2008)

28. Risi, S., Stanley, K.: Confronting the challenge of learning a flexible neural controller for a diversity of morphologies. In: Proceedings of the Genetic and Evolutionary Computation Conference, pp. 255–261. ACM (2013)

29. Gomes, J., Christensen, A.: Generic behavior similarity measures for evolutionary swarm robotics. In: Proceedings of the Genetic and Evolutionary Computation Conference, Amsterdam, The Netherlands, pp. 199–206. ACM Press (2013)

30. Urbano, P., Georgiou, L.: Improving grammatical evolution in santa fe trail using novelty search. In: Proceedings of the 12th European Conference on Artificial Life, Taormina, Italy, pp. 917–924. MIT Press (2013)
31. Lehman, J., Stanley, K.: Efficiently evolving programs through the search for novelty. In: Proceedings of the 12th Annual Conference on Genetic and Evolutionary Computation, Portland, USA, pp. 837–844. ACM (2010)
32. Velez, R., Clune, J.: Novelty search creates robots with general skills for exploration. In: Proceedings of the Genetic and Evolutionary Computation Conference, Vancouver, Canada, pp. 737–744. ACM (2014)
33. Cuccu, G., Gomez, F., Glasmachers, T.: Novelty-based restarts for evolution strategies. In: Proceedings of the Congress on Evolutionary Computation, New Orleans, USA, pp. 158–163. IEEE Press (2011)
34. Gomes, J., Urbano, P., Christensen, A.L.: Progressive minimal criteria novelty search. In: Pavón, J., Duque-Méndez, N.D., Fuentes-Fernández, R. (eds.) IBERAMIA 2012. LNCS, vol. 7637, pp. 281–290. Springer, Heidelberg (2012)
35. Lehman, J., Stanley, K.: Revising the evolutionary computation abstraction: minimal criteria novelty search. In: Proceedings of the 12th Annual Conference on Genetic and Evolutionary Computation, pp. 103–110. ACM (2010)
36. Liapis, A., Yannakakis, G., Togelius, J.: Constrained novelty search: a study on game content generation. Evol. Comput. **23**(1), 101–129 (2015)
37. Ghasemi, A., Zahediasl, S.: Normality tests for statistical analysis: a guide for non-statisticians. Int. J. Endocrinol. Metab. **10**(2), 486–489 (2012)
38. Flannery, B., Teukolsky, S., Vetterling, W.: Numerical Recipes. Cambridge University Press, Cambridge (1986)

On-line Evolution of Foraging Behaviour in a Population of Real Robots

Jacqueline Heinerman[1]([⊠]), Alessandro Zonta[2], Evert Haasdijk[1],
and A.E. Eiben[1]

[1] Vrije Universiteit Amsterdam, Amsterdam, The Netherlands
{j.v.heinerman,e.haasdijk,a.e.eiben}@vu.nl
[2] University of Padua, Padua, Italy
alessandro.zonta@outlook.com

Abstract. This paper describes a study in evolutionary robotics conducted completely in hardware without using simulations. The experiments employ on-line evolution, where robot controllers evolve on-the-fly in the robots' environment as the robots perform their tasks. The main issue we consider is the feasibility of tackling a non-trivial task in a realistic timeframe. In particular, we investigate whether a population of six robots can evolve foraging behaviour in one hour. The experiments demonstrate that this is possible and they also shed light on some of the important features of our evolutionary system. Further to the specific results we also advocate the system itself. It provides an example of a replicable and affordable experimental set-up for other researches to engage in research into on-line evolution in a population of real robots.

Keywords: Evolutionary robotics · Neural networks · Distributed on-line learning · Embodied evolution · Foraging

1 Introduction

Evolutionary robotics is a research area that "applies the selection, variation, and heredity principles of natural evolution to the design of robots with embodied intelligence" [5]. In particular, evolutionary robotics aims to evolve the controllers, the morphologies, or both, for real and simulated autonomous robots [18]. Considering the complexity of interactions between environment, morphology and controller, employing evolution is a very promising approach to designing intelligent robots for a range of circumstances [1,13]. However, forced by technical constraints the usual modus operandi in evolutionary robotics is quite limited: evolve robot controllers in simulation and transfer the outcome to real hardware afterwards. Thus, even though the final goal is to obtain physical robots with intelligent behaviour, the evolutionary process is usually digital, which leads to the notorious reality gap problem [9]. Furthermore, the evolutionary algorithm is usually (ab)used as an off-line optimizer. It is only employed during the design stage that ends with finding a good controller that is then deployed on the real robot and kept fixed during the operational stage.

© Springer International Publishing Switzerland 2016
G. Squillero and P. Burelli (Eds.): EvoApplications 2016, Part II, LNCS 9598, pp. 198–212, 2016.
DOI: 10.1007/978-3-319-31153-1_14

This paper describes a study in evolutionary robotics conducted completely in hardware, without recourse to any simulations. The experiments use Thymio II robots as hardware platform [14]. These robots – recently on the market– are quite small, cheap, and easily available. Therefore, setting up a physical robot population is much less demanding than a few years ago. The experiments consider on-line evolution, where a collective of robots evolve controllers in the robots' task environment as the robots perform their tasks. This results in a distributed on-line setup where robots learn individually and socially, meaning the exchange of individually evolved controllers with others. The experiments are motivated by a long-term vision of an ongoing evolutionary process that enables adaptation to the environment and the task. This research differs from most evolutionary robotics that employs evolution as an optimisation procedure to obtain well-performing controllers before deploying the robots.

There are earlier experiments in distributed on-line evolutionary robotics that take place exclusively in hardware, e.g., the work of Watson et al. and Simões et al., both in 1999 [15,19]. Watson et al. coined the phrase "embodied evolution" for such systems. These experiments consider very straightforward robot tasks such as phototaxis and obstacle avoidance, and subsequent research into embodied evolution-like settings has not yet ventured beyond that.

The research in this paper serves as a further stepping stone towards on-line evolutionary adaptation in complex tasks and complex environments. An important goal of this paper is, therefore, to provide a replicable, accessible and affordable experimental set-up for further research into the on-line evolution of non-trivial behaviour in real robots. This paper investigates the feasibility of tackling more complex tasks in a realistic timeframe. In particular, it considers the on-line evolution of foraging behaviour in a population of real robots, with a fitness function that rewards appropriate behaviours for a number of subgoals. We try to answer the following research questions:

1. Can a group of robots evolve appropriate controllers for a non-trivial task in one hour?
2. How important is the feature that the robots share the controllers they evolve individually (social learning)?
3. How important is the most task-specific part of the compound objective function?

An evolutionary process implemented in real hardware has the obvious benefit of avoiding the reality gap, and when considering multiple robots with sophisticated sensors such as cameras simulations may actually run slower than real time, even for groups as small as six robots. An additional benefit is that experimenting with real robots encourages the researcher to review the robots' actual behaviour during the experiments rather than allowing only post-facto analysis of the metrics gathered by unattended simulation runs. Thus, experimenting with real robots enhances the understanding of robot behaviour.

2   Related Work

In principle, the scope of the related work includes all on-line distributed evolution in populations of physical robots as well as research where foraging behaviour evolves in simulation and is subsequently transferred to hardware. For the latter category, we restrict the scope of related work to research applying neuro-evolutionary methods to control differential drive robots.

There are few papers that consider on-line distributed evolution in populations of physical robots. Pioneering works in this field are those by Watson *et al.* and Simões *et al.*, both published in 1999. Watson *et al.* implemented "embodied evolution" in a population of six robots [19]. The robot controllers evolved to tackle a phototaxis task in an on-line fashion by broadcasting parts of their genome at a rate proportional to their task performance. Simões *et al.* evolved morphological features as well as the controllers in the same genome for collision-free behaviour [15]. These are the first examples of evolutionary algorithms distributed in a population of physical robots where the robots are not only evaluated, but are also performing real tasks. Since then, four more studies show on-line distributed evolution in populations of physical robots where the task to learn is either only obstacle avoidance [7,17], obstacle avoidance, phototaxis and robot seeking [10] or survival without a specified fitness function [3]. These studies focus on showing that on-line evolution of the controllers for different tasks is possible and that communication between the robots is always beneficial. To our knowledge, the current study is the first to consider evolving foraging behaviour in hardware and in an on-line fashion.

Foraging is often considered as a task in off-line evolution. Here, we consider research where the controllers evolve in simulation and the best controllers are transferred onto robotic hardware where they remain fixed. We found no examples of off-line evolution of foraging behaviour where evaluations were performed on robotic hardware. The first example is [12], where controllers evolved in simulation to locate, recognise and grasp "garbage" objects and transport them outside the arena. The platform of choice was a Khepera robot with a gripper module. With a population size of 100 it took 1,000 generations to generate the desired, complex behaviour. One controller evaluation took 8 s in simulation (and 300 s in reality). The number of components in the fitness function (penalties and rewards for different observed behaviours) was found to have substantial impact. Transferring the best evolved controller onto hardware resulted in only a small performance decline.

In [8], the author incorporated dynamic rearrangement of neural networks with the use of neuromodulators to push a "peg" to a light source. Both peg and target were arranged in a straight line in front of the robot. A population of 100 controllers evolved for 200 generations, with controllers evaluated by simulating them for 500 time steps. When transferred to a Khepera robot, the best controller showed appropriate behaviour.

In [16], the task was limited to box pushing, and this is the only example where the required behaviour was simple enough that the neural networks did not require hidden nodes. The authors investigated the difference in performance with fitness functions that use internal/external and global/local information.

They showed that a global external fitness function (diametrically opposed to the notion of distributed on-line evolution) performed best for their task. The best controller was transferred onto a Khepera robot and showed adequate behaviour. The controller resulted from a (30+1) evolutionary strategy running for 250 generations; controllers were evaluated by simulating them for 100 time steps, starting from different positions in the arena.

The most recent work that we know of is [4]. Here, multi-objective evolutionary algorithms were applied for two conflicting tasks: protecting another robot by following it closely and collecting objects in the arena. Controllers evolved in an unknown population size over 100 generations, with each controller evaluated in simulation for 70 s. Even though the input sensor values for the simulated and real robot (CR robot) did not overlap perfectly, the robot showed the desired behaviour.

Thus, evolving foraging behaviours in simulation and subsequently transferring the results to the appropriate hardware platform has been shown to work in a number of settings. However, once the final controller is transferred, it does not adapt further and so cannot cope with unforeseen circumstances or changes in the environment. Our ambition is to develop foraging behaviour through on-line evolution, with some speed-up afforded by distributing the evolutionary process across a small number of robots. Given the population sizes and number of generations used in simulation-based research, this is not a trivial task to accomplish within a realistic timeframe (i.e., within the robot battery's operational period).

3   System Description

3.1   Robot

The Thymio II robot includes seven Infra-Red (IR) proximity sensors for obstacle detection; five are arranged along the front and two along the back of the robot. The sensors return values between 0 and circa 4,500, with high values corresponding to close obstacles. The robot has differential drive with the maximum wheel actuators set between −500 and 500. The operating speed is set to 30 % of the maximum speed, which is between −150 and 150, to prevent overheating and to adapt to the camera's image processing speed. We extend the standard Thymio set-up with a more powerful logic board, a camera, wireless com-

Fig. 1. Thymio II robot, developed by The École Polytechnique Fédérale de Lausanne (EPFL) and École Cantonal d'Arts de Lausanne (ÉCAL), with Raspberry Pi 2, Raspberry Pi NoIR camera, WiFi dongle, external battery and a LEGO gripper.

munication, and a high capacity battery. We use a Raspberry Pi 2 that connects to the Thymio's sensors and actuators an processes the data from the Raspberry

Pi NoIR Camera. A WiFi dongle (Edimax 150 Mbps Wireless 802.11b/g/n nano USB WiFi Adapter model EW-7811Un) attached to the Raspberry Pi enables inter-robot communication. Battery life is etended through a Verbatim Dual USB 12,000 mAh battery, allowing for a total experimental time of 10 h. The extended Thymio II is shown in Fig. 1. The hand made LEGO gripper helps the robot maintain control of the pucks when manoeuvring.

3.2 Environment

Two robots operate together in a 1×1 meter arena with six pucks and one target area. This set-up is duplicated three times, facilitating a total population of six robots, where communication across arena instances is possible. A set-up with three arenas with two robots each (as opposed to having all six robots in a single arena) reduces the likelihood that the robots' grippers become entangled.

The arena size allows the robot to see the target area from across the whole arena. The arenas have a white floor and white walls for improved obstacle detection. The pucks are red and the target area, located in a corner, is blue, with white stripes added for better colour recognition when the robot is close to the target.

A WiFi router placed close to te arenas facilitates reliable wireless communication between all robots in the three arenas. A motion tracking system monitors the robots and pucks in one of the three arenas for post-hoc qualitative analysis of robot traces.

Fig. 2. The 1×1 meter arena with two Thymio II robots searching for the red pucks. The target location is indicated by the blue corner. The arena is duplicated three times to facilitate six robots while minimising robot collisions (Color figure online).

3.3 Task and Objective Function

In the foraging task, the robots must collect items (pucks) and deliver them to a target location. Although this task may not seem difficult, there is no consensus on how to define an objective function for this task, and in particular there is no prior experience with suitable functions for on-line evolution. Our experiments use an objective function that rewards appropriate behaviour for a number of subgoals to provide a relatively smooth fitness gradient; only rewarding the successful delivery of a puck results in a largely featureless fitness landscape. The objective function assesses robot behaviour over a period of T timesteps as follows:

$$f_{total} = \sum_{t=0}^{T} f_{obs} + f_{puck} + f_{target} + f_{bonus}, \tag{1}$$

where:

$$f_{obs} = v_{trans} \cdot (1 - v_{sens}), \tag{2a}$$

$$f_{puck} = b_{puck} + 2 \cdot b_{push}, \tag{2b}$$

$$f_{target} = b_{tar}, \tag{2c}$$

$$f_{bonus} = b_{push} \times b_{tar}, \tag{2d}$$

and:

- v_{trans} is the translational speed (normalised between 0 and 1), calculated as the sum of the absolute speeds assigned to the left and right motor;
- v_{sens} is the value of the proximity sensor closest to an obstacle and normalised between 0 and 1;
- b_{puck} is a boolean value indicating whether the camera detects a puck in sight;
- b_{push} is a boolean value indicating whether the robot is pushing a puck;
- b_{tar} is a boolean value indicating whether the camera detects the target area.

Note, that the analyses in Sect. 5 report overall system performance as the number of pucks collected over a period of time, not as the fitness function as described above.

3.4   Controller

The controller is a feed forward neural network with 13 input nodes, 5 hidden nodes and 2 output nodes as depicted in Fig. 3. The nodes have *tanh* activation functions. Five input nodes, denoted PS_1 to PS_5, connect to the proximity sensors (three in the front and two in the back of the robot). The remaining seven inputs relate to camera information. To extract the salient information from the camera image, the image is divided into four parts (left, middle, right and bottom, or l, r, m and b) as shown in the top left in Fig. 3.

Masking red values to detect pucks in the four areas yields the (P_l, P_m, P_r, P_b) inputs that denote the size of the largest red area in each sub-image. A further three inputs denote the total percentage of blue in the three top sub-images (T_l, T_m, T_r). The input and hidden layer each have an additional bias node. The output nodes drive the left and right motor actuators. All input nodes are connected to all hidden nodes (except the bias node) and all hidden nodes are connected to the output nodes, resulting in a neural network with 62 weights. The sensor input values are normalised between -1 and 1 and the weights are between -4 and 4.

3.5   Evolutionary or Learning Mechanisms

The evolutionary process employs the common direct encoding scheme of an array of neural network weights. In this case, the length of that array is 62. Controllers evolve on-line and on-board; the robots encapsulate a self-contained evolutionary algorithm and augment this with a distributed evolutionary system,

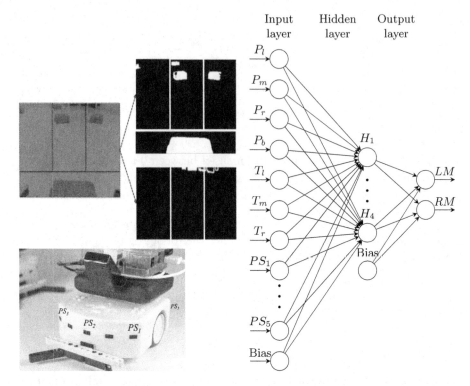

Fig. 3. Visualisation of salient information extracted from the camera image (left) and the neural network controller with 13 input nodes, 5 hidden nodes and 2 output nodes (right). Input nodes P_x detect pucks as the largest area of red pixels in each sub-image (l, m, r, b denoting the left, middle, right and bottom sun-image), T_x detect the total amount of blue pixels in each of the three upper sub-images and PS_x detect obstacles (Color figure online).

amounting to what was labelled as a hybrid scheme in [6]. Such a combination of evolutionary processes can also be cast as a combination of individual learning (the encapsulated evolutionary process) and social learning (the distributed evolutionary process). We adopt the individual and social learning terminology to align with the DREAM project[1] of which this research is part.

Individual Learning. The robots implement individual learning through a $(1+1)$ evolutionary strategy, similar to the approach in [2,7], using the objective function defined in Eq. 1. The neural network weights are mutated using Gaussian perturbation with $N(0, \sigma)$, where the value of σ doubles when the offspring controller (the challenger) does not improve on the current controller (the champion). When a challenger outperforms the current champion, it replaces the champion. Each controller is evaluated over a testing epoch of 1,000 time steps, equating to a period of circa 3.6 s.

[1] http://robotsthatdream.eu.

This is a very short evaluation time, certainly too short to complete one round of detecting and approaching a puck and then transporting it to the target area. One could argue that this evaluation time should be longer, but the difficulty of a longer evaluation time is the need for more controller evaluations because it will take the robot longer to discover all the subtasks required to perform the overall task. The short evaluation time results in a quick response to objects of interest — the pucks and target. To enable robust assessment of controllers, even though the robot cannot experience all relevant situations during a single evaluation, the robots re-evaluate their champion controllers with a 20 % chance. The champion's fitness value is updated upon re-evaluation by taking the weighted sum of the current and the re-evaluated fitness with a 20–80 weight distribution (20 % of the re-evaluated and 80 % of the current fitness value).

Social Learning. The robots broadcast their champion genome after every evaluation, provided that its fitness exceeds 50 % of the maximum fitness. Robots cache received genomes in their *social learning storage*. When a robot embarks on a round of social learning, it takes the most recently received genome from the storage and either evaluates it directly as a challenger (in 75 % of the cases) or creates a challenger through averaging crossover with the current champion. Directly re-evaluating received controllers rather than copying the sender's fitness value makes sense because the robot may be in a very different situation than the sending robot when it evaluated the controller. Just as with individual learning, the challenger replaces the champion if it outperforms it.

The robots randomly alternate between individual and social learning with a rate of 70:30.

4   Experimental Set-up

The duration of each experiment is one hour, allowing for 1,000 controller evaluations of 1,000 time steps (ca 3.6 s) each. Unless indicated otherwise, each set-up was repeated 20 times. The table on the right lists the most important system parameter[2].

The robots start every experiment in the same position, facing the middle of one of the walls that is not adjacent to the target area. The robots are not repositioned during an experiment unless two robots' grippers or wiring become entangled or when a robot is pushing multiple pucks against the wall (in that case, the pucks prevent the robot from moving close enough to the wall for obstacle

System parameters	
Total controller evaluations	1000
Evaluation duration (sec)	3.6
Re-evaluation chance	20 %
Re-evaluation champion weight	80 %
Individual learning chance	70 %
Social learning chance	30 %
Individual learning	
Maximum fitness	6000
Weight range	8
σ initial	1
σ maximum	4
σ minimum	0.01
Social learning	
Broadcast threshold	50 %
Averaging crossover	25 %
Import	75 %
Social learning storage size	20

[2] The code for implementation is available on https://github.com/ci-group/Thymio_swarm.

sensors to detect it). In case the robot loses its gripper, it is reattached and the experiment continues. Furthermore, the temperature of the robots is constantly tracked to prevent the robots from overheating. As a result, no robots broke during the experiments and all collected data can be used.

The robots are preprogrammed with a threshold for the amount of blue that needs to be exceeded before we can say that a puck is collected. When a puck is pushed to the target area, the robot internally registers the time and emits a sound to alert the experimenter. The puck is then manually replaced in the centre of the arena and the experimenter records the number of collected pucks.

5   Experimental Results

To quantify the performance of the robots, we consider the number of pucks collected in ten-minute intervals. The objective function defined in Eq. 1 is not suitable for this purpose: controller evaluations are so short that a perfectly good controller scores poorly, e.g., because the robot spends the entire evaluation period looking for pucks.

Figure 4 compares a baseline experiment with the performance when learning is enabled. For the baseline experiment, the evolutionary mechanisms are disabled and the robots run with randomly generated weights for every evaluation.

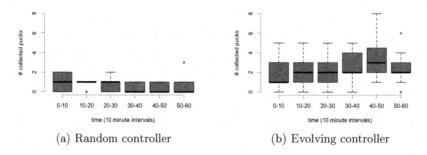

(a) Random controller   (b) Evolving controller

Fig. 4. Number of collected pucks per 10 min intervals by the robot population for random and evolving controllers. The box plots show the median, interquartile range, min and max values. The evolutionary process was repeated 20 times, the random baseline was repeated 10 times. Although the increase in pucks collected when learning increases only marginally, the increase is significant (Mood's Median test with Fisher's Exact Test, $p = 0.02$).

There is a clear difference in performance between the random and evolving controllers, so learning definitely improves task performance. If the robots do learn the foraging task successfully, the number of pucks collected per interval should increase over the course of an experiment. The performance does indeed increase over time in Fig. 4(b), but the increase is slight. To ascertain whether the increase in performance is statistically significant, we use Mood's Median test on the median number of pucks collected the first and the second half hour.

Mood's Median test is calculated as follows: the median number of pucks collected in the first and second half of each experiment is calculated. The data for the 20 repeats is then summarised in a 2 × 2 table: each cell in the table contains the count of repeats in its class. Classes are assigned to the cells according to whether the performance metric was calculated over the first or the second half hour (first or second column, respectively) and whether the metric is lower or equal or higher than the overall median (first and second row, respectively). The standard Mood's Median test then calls for a χ^2 test, but in this comparison requires a one-tailed test to establish whether the robots collect significantly more pucks in the seconds half hour. Therefore, we substitute it with Fisher's Exact test. For these data, this yielded $p = 0.02$ – a significant increase at $\alpha = 5\,\%$.

Performance is in fact somewhat disappointing and also worse than at least some behaviour that was observed (one of the benefits of running experiments in real robots is that the experimenters monitor robot behaviour as a matter of course and do not rely only on quantitative post-hoc analysis as is often the case with simulation-based experiments). From observations of the robots during the experiment, it appeared that the robots do learn appropriate behaviour, but subsequently revert to much less efficient behaviour.

To investigate whether the robots do indeed learn (and subsequently forget) efficient foraging behaviour, we ran ten repeats of an experiment with a high-fitness controller without further development. This controller was selected manually from controllers that consistently showed high fitness values, also over multiple re-evaluations. All six robots were programmed with the neural network weights set according to this individual genome, positioned in the standard starting position and subsequently ran for one hour, with all other settings as described for the runs with the learning mechanisms enabled.

The box plots in Fig. 5 show the performance of ten repeats of this experiment in ten-minute intervals. The first ten minutes show exceptional performance due to the convenient starting position of the pucks in the middle of the arena. But also in the other intervals, performance beats the on-line performance with an active learning process by a factor of three (note, that the y-axis in Fig. 5 had a larger range than that in Fig. 4).

Thus, it appears that the evolutionary process does discover very good controllers, but subsequently discards these individuals because they perform poorly at some point. We hypothesise that this is a result of the brief evaluation periods that these experi-

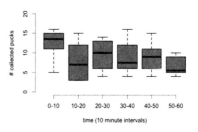

Fig. 5. Number of collected pucks over ten runs per 10 min interval of one of the better controllers developed during an evolutionary run. The box plot shows the median, interquartile range, min and max values.

ments require: if, for instance, no pucks are picked up during a re-evaluation of a controller, its assessment is revised, even if it is actually very relevant. Combating this 'forgetting' behaviour is the focus of further research that is now underway.

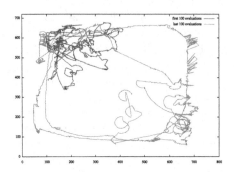

Fig. 6. Traces of one robot showing the first (red) and last (blue) 100 controller evaluations. The target location is at position (800,0). The traces show the mid-point of each robot, meaning that the robots learn wall-following behaviours (Color figure online).

For a qualitative analysis of robot behaviour and the effect of evolution, Fig. 6 shows traces of one robots at the beginning and towards the end of an experiment (the first and last 100 evaluations in red and blue, respectively). The traces indicate typical behaviour that was observed in most runs. The robots learn wall-following behaviour – the traces show the mid-point of each robot, its sides are actually very close to the wall for large parts of the trace.

Because the robots often push pucks towards the wall in the early stages of learning, this is quite efficient behaviour to move pucks to the target area. When the robots deviate from this behaviour, they often end up pushing more pucks to the wall to resume wall-following.

The Need for Social Learning. To investigate the influence of the social learning mechanisms, another set of 20 experiments was conducted where communication between robots was disabled, so that they only learn using the encapsulated $1+1$ evolution strategy. Other than that, all settings are as in the original runs presented above. Figure 7 compares the results from the experiments with both social and individual learning with experiments where social learning is disabled.

The plot shows that the social learning mechanisms yields better performance and is in fact necessary to prevent a decline in performance. Social learning implements a shared repository of controllers that goes some way to exacerbate the 'forgetting' behaviour.

Objective Function Design. This is known to be a difficult problem in general [11]. In our case, the observation of robot trajectories (cf. Fig. 6) suggests that evolution solves the task differently than we, the experimenters, have expected. In particular it seems that the behaviours are optimsed for grabbing a puck and carrying it around along the walls.

This raises the question whether rewards f_{target} and f_{bonus} that correspond to the subtask of reaching the target area are important or not. To find out we ran a series of experiments with a reduced objective function that contains only the first two terms f_{obs} and f_{puck} of formula 1. Based on this function the optimal behavior is continuously pushing a puck without hitting the walls.

The results of these experiments are shown in red in Fig. 8, with the original results in blue for reference. Using these data we can ask two questions regarding the role of the target related rewards: (1) Whether the robots are still learning

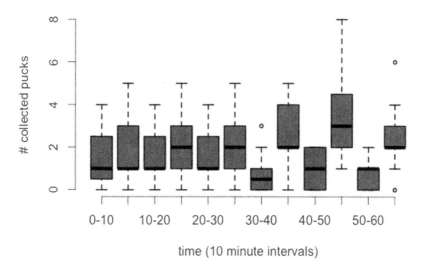

Fig. 7. Number of collected pucks per 10 min interval by the robot population for evolving controllers without (red) and with communication (blue). Box plots show median, interquartile range, min and max values. Both experiments were repeated 20 times. Communication is shown to be necessary to ensure a performance increase for the foraging task (Color figure online).

the puck collecting task without these rewards? and (2) Whether the omission of these rewards decreases performance?

Calculating the Mood's Median test for the increase in performance for first and second half an hour, gives a value of $p = 0.57$. In other words, the performance in the second half of the experiment is not significantly higher than in the first half. Thus, with the reduced fitness function the population does not seem to learn over time.

To answer the second question, Mood's Median tests are performed for every ten minutes between the two algorithms. Values of $p = 0.62, 0.63, 0.9, 0.5, 0.37, 0.09$ indicate that there is no significant decrease in performance when removing the rewards related to the target area – using a significance level $\alpha = 5\%$.

These results are seemingly contradictory. On the one hand, the system with the full fitness function (blue) is different from the one with the reduced fitness function (red), because it does learn over time, while the other one does not. On the other hand, the system with the full fitness function is not different from the one with the reduced fitness function, because their performance differences over the whole experiment are not significant. This "contradiction" is caused by the statistics and indeed the last 10 min when the red system exhibits poor performance.

A more interesting conclusion we can draw from these data concerns the role of the target related fitness rewards. The fact that red and blue do not differ significantly regarding the number of pucks collected can be explained by

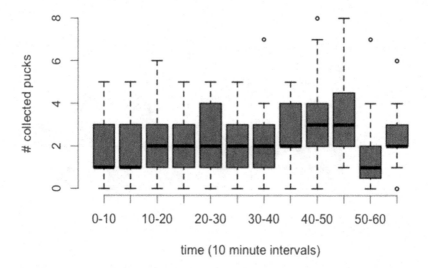

Fig. 8. Number of collected pucks per 10 min interval by the robot population for evolving controllers without (red) and with the bonus component of the fitness function (blue). Box plots show median, interquartile range, min and max values. Both experiments were repeated 20 times. Although there seems to be an increase in pucks collected when there is no bonus component, the increase is not significant (Mood's Median test with Fisher's Exact Test, $p = 0.57$) (Color figure online).

taking a closer look at the environment. The target area is positioned in a corner and can be reached by wall following. This feature implies that a robot that is continuously pushing around a puck without hitting the walls will eventually deliver the puck to the target area. This solution was not intended by us, but it nicely illustrates how evolution can exploit environmental features to the surprise of the experimenters.

6   Conclusions and Future Work

In this paper we described an evolutionary robotics study completely conducted in hardware, without using software simulations up front. An important result is the experimental evidence that a population of six robots can evolve foraging behaviour in one hour. The significance of this finding is apparent from a comparison with existing work; on-line evolution embodied in real robots has previously been only applied to relatively simple tasks, such as obstacle avoidance and phototaxis.

Our results show that the robots do learn foraging behaviour and that performance rises steadily if marginally. The robots learn appropriate control as shown by an analysis of one of the better controllers considered, but these controllers are lost, presumably when the robots re-evaluate them in inauspicious circumstances. This is probably linked to the short evaluation times necessary for our experiments. Further research is required to investigate possibilities to mitigate this 'forgetting' of good controllers.

Experiments disabling certain components of the evolutionary system revealed the importance of these components. In particular, we observed that allowing the robots to communicate and cross-fertilise brings significant improvements with respect to a system where each robot is running an isolated internal evolutionary learning process whose results are not shared with the others. The robots' performance increase was found to be only significant when the robots are able to share information about their controller. Without the sharing of information, performance declines, probably because of the short evaluation time where the robot is not able to experience all subtasks in the arena within an evaluation. Thus, individual learning augmented with social learning outperforms individual learning alone. In the terminology of [2] this shows the advantage of a hybrid (encapsulated plus distributed) system over encapsulated evolution.

We also looked into the composition of the fitness function and learned that rewarding the subtask of reaching the target area is not necessary to obtain overall good performance. This is a consequence of a particular feature of the environment that can be exploited by evolution.

The main ambition of the research in this paper was to push for more complex tasks to be considered in on-line evolutionary robotics outside of simulations. Although the performance achieved by these experiments can be improved upon, this paper provides an important stepping stone towards this goal by providing the research community with a well-documented and affordable example of an experimental set-up to investigate an evolving collective of robots with rich sensory inputs in a non-trivial task. We hope that this will provide inspiration and opportunity for other researchers to research physical, not simulated, robot collectives that evolve to tackle tasks beyond the complexity of obstacle avoidance.

Acknowledgements. This work was made possible by the European Union FET Proactive Initiative *Knowing, Doing, Being: Cognition Beyond Problem Solving,* funding the Deferred Restructuring of Experience in Autonomous Machines (DREAM) project under grant agreement 640891.

References

1. Bongard, J.: Evolutionary robotics. Comm. ACM **56**(8), 74–85 (2013)
2. Bredeche, N., Haasdijk, E., Eiben, A.E.: On-line, on-board evolution of robot controllers. In: Collet, P., Monmarché, N., Legrand, P., Schoenauer, M., Lutton, E. (eds.) Proceedings of the 9th International Conference on Artificial Evolution, pp. 110–121. Springer, Berlin (2009)
3. Bredeche, N., Montanier, J.M., Liu, W., Winfield, A.F.: Environment-driven distributed evolutionary adaptation in a population of autonomous robotic agents. Math. Comput. Model. Dyn. Syst. **18**(1), 101–129 (2012)
4. Capi, G.: Multiobjective evolution of neural controllers and task complexity. IEEE Trans. Robot. **23**(6), 1225–1234 (2007)

5. Doncieux, S., Bredeche, N., Mouret, J.B., Eiben, A.: Evolutionary robotics: what, why, and where to. Front. Robot. AI **2**(4) (2015)
6. Eiben, A.E., Haasdijk, E., Bredeche, N.: Embodied, on-line, on-board evolution for autonomous robotics. In: Levi, P., Kernbach, S. (eds.) Symbiotic Multi-Robot Organisms: Reliability, Adaptability, Evolution, pp. 361–382. Springer, Heidelberg (2010). http://www.springer.com/engineering/mathematical/book/978-3-642-11691-9
7. Heinerman, J., Rango, M., Eiben, A.: Evolution, individual learning, and social learning in a swarm of real robots. In: Proceedings of the 2015 IEEE International Conference on Evolvable Systems (ICES). IEEE (2015, to appear)
8. Ishiguro, A., Tokura, S., Kondo, T., Uchikawa, Y., Eggenberger, P.: Reduction of the gap between simulated and real environments in evolutionary robotics: a dynamically-rearranging neural network approach. In: IEEE SMC 1999 Conference Proceedings, IEEE International Conference on Systems, Man, and Cybernetics, vol. 3, pp. 239–244. IEEE (1999)
9. Morán, F., Merelo, J.J., Moreno, A., Chacon, P. (eds.): ECAL 1995. LNCS, vol. 929. Springer, Heidelberg (1995)
10. Nehmzow, U.: Physically embedded genetic algorithm learning in multi-robot scenarios: the pega algorithm. In: Prince, C., Demiris, Y., Marom, Y., Kozima, H., Balkenius, C. (eds.) Proceedings of The Second International Workshop on Epigenetic Robotics: Modeling Cognitive Development in Robotic Systems, No. 94 in Lund University Cognitive Studies, LUCS, Edinburgh, UK, August 2002
11. Nelson, A.L., Barlow, G.J., Tsidis, L.: Fitness functions in evolutionary robotics: a survey and analysis. Robot. Autonom. Syst. **57**(4), 345–370 (2009). http://www.sciencedirect.com/science/article/B6V16-4TTMJV3-1/2/2549524d8e0f3982730659e49ad3fa75
12. Nolfi, S.: Evolving non-trivial behaviors on real robots: a garbage collecting robot. Robot. Autonom. Syst. **22**(3), 187–198 (1997)
13. Nolfi, S., Floreano, D.: Evolutionary Robotics: The Biology, Intelligence, and Technology of Self-Organizing Machines. MIT Press, Cambridge (2000)
14. Riedo, F., Chevalier, M.S.D., Magnenat, S., Mondada, F.: Thymio II, a robot that grows wiser with children. In: 2013 IEEE Workshop on Advanced Robotics and its Social Impacts (ARSO), pp. 187–193. IEEE (2013)
15. Simoes, E., Dimond, K.R.: An evolutionary controller for autonomous multi-robot systems. In: 1999 IEEE International Conference on Systems, Man, and Cybernetics, IEEE SMC 1999 Conference Proceedings, vol. 6, pp. 596–601. IEEE (1999)
16. Sprinkhuizen-Kuyper, I.G., Kortmann, R., Postma, E.O.: Fitness functions for evolving box-pushing behaviour. In: Proceedings of the Twelfth Belgium-Netherlands Artificial Intelligence Conference, pp. 275–282 (2000)
17. Usui, Y., Arita, T.: Situated and embodied evolution in collective evolutionary robotics. In: Proceedings of the 8th International Symposium on Artificial Life and Robotics, pp. 212–215 (2003)
18. Vargas, P., Paolo, E.D., Harvey, I., Husbands, P. (eds.): The Horizons of Evolutionary Robotics. MIT Press, Cambridge (2014)
19. Watson, R., Ficiei, S., Pollack, J.B., et al.: Embodied evolution: Embodying an evolutionary algorithm in a population of robots. In: Proceedings of the 1999 Congress on Evolutionary Computation, CEC 99, vol. 1, pp. 335–342. IEEE (1999)

Hybrid Control for a Real Swarm Robotics System in an Intruder Detection Task

Miguel Duarte[1,2,3]([✉]), Jorge Gomes[1,2,4], Vasco Costa[1,2,3],
Sancho Moura Oliveira[1,2,3], and Anders Lyhne Christensen[1,2,3]

[1] BioMachines Lab, Lisbon, Portugal
[2] Instituto de Telecomunicações, Lisbon, Portugal
[3] Instituto Universitário de Lisboa (ISCTE-IUL), Lisbon, Portugal
{miguel_duarte,vasco_craveiro_costa,sancho.oliveira,
anders.christensen}@iscte.pt
[4] Faculdade de Ciências, BioISI, Lisbon, Portugal
jgomes@di.fc.ul.pt

Abstract. Control design is one of the prominent challenges in the field of swarm robotics. Evolutionary robotics is a promising approach to the synthesis of self-organized behaviors for robotic swarms but it has, so far, only produced been shown in relatively simple collective behaviors. In this paper, we explore the use of a hybrid control synthesis approach to produce control for a swarm of aquatic surface robots that must perform an intruder detection task. The robots have to go to a predefined area, monitor it, detect and follow intruders, and manage their energy levels by regularly recharging at a base station. The hybrid controllers used in our experiments rely on evolved behavior primitives that are combined through a manually programmed high-level behavior arbitrator. In simulation, we show how simple modifications to the behavior arbitrator can result in different swarm behaviors that use the same underlying behavior primitives, and we show that the composed behaviors are scalable with respect to the swarm size. Finally, we demonstrate the synthesized controller in a real swarm of robots, and show that the behavior successfully transfers from simulation to reality.

Keywords: Swarm robotics · Evolutionary robotics · Hybrid control · Self-organization · Aquatic robots

1 Introduction

Swarm robotics [1] is an approach to multirobot systems that relies on large numbers of relatively simple robots with decentralized control. While swarm robotics systems tend to display a number of desirable properties, such as scalability, flexibility, and fault tolerance [1–3], synthesizing controllers for such systems is still an open challenge, as there is no general way to derive individual rules that result in the desired swarm behavior [3]. One popular approach used

© Springer International Publishing Switzerland 2016
G. Squillero and P. Burelli (Eds.): EvoApplications 2016, Part II, LNCS 9598, pp. 213–230, 2016.
DOI: 10.1007/978-3-319-31153-1_15

to address this challenge is evolutionary robotics (ER) [4], which applies evolutionary techniques to the synthesis of robot controllers. ER has been shown capable of automatically synthesizing behaviors for robots without the need of manually programming the individual rules that govern the robots' behavior, and of generating self-organized behavior [5].

While ER has the potential to synthesize behaviors for robotic swarms [5], a number of issues [6] have prevented it from being applied to real-world tasks, and most successful demonstrations have thus been confined to simple tasks and either simulation-based studies or real-robot studies in strictly controlled laboratory conditions [2,3]. An alternative to standard, monolithic evolutionary techniques is the hierarchical control synthesis approach [7], which consists of recursively dividing a task into sub-tasks, synthesizing control for each sub-task individually. The final controller is then obtained by composing the different sub-controllers hierarchically. The hierarchical control synthesis approach uses engineering-oriented [8], pragmatic techniques to enable the use of evolution-based control in scenarios for which it might be difficult to apply standard evolution.

In this paper, we explore how the hierarchical control synthesis approach can be used in a real-world task to produce swarm behaviors that: (i) are complex, in the sense that they require multiple sub-behaviors; (ii) are modular and easily reconfigurable; (iii) can operate under realistic constraints; (iv) scale with the number of robots in the swarm; and (v) can be successfully transferred from simulation to reality. Our study focuses on an intruder detection task, where robots have to first navigate from a base station to an area of interest, monitor the area, detect and follow intruders that cross the area, and simultaneously avoid running out of energy by recharging periodically at the base station. Tasks that take several realistic constraints into account typically require the robots to exhibit multiple different behaviors, which contrasts with the vast majority of previous ER studies [6,9].

We demonstrate for the first time that the hierarchical control synthesis approach [7] can be successfully applied to swarms of real robots in a task with realistic constraints. First, we demonstrate the flexibility of the approach by showing how variations of the top-level behavior arbitrator result in different collective behaviors. We then assess the scalability of the composed controllers in terms of swarm size and task configuration. Finally, we evaluate the composed controller in a real swarm of aquatic surface robots, and assess how the hierarchical control transfers from simulation to reality.

2 Related Work

2.1 Swarm Robotics

The concept of swarm robotics systems (SRS) is inspired by the observation of social insects, in which large numbers of autonomous and simple units self-organize to display collectively intelligent behavior [1]. Due to the decentralized control and self-organization, SRS typically display a number of advantageous properties such as scalability, versatility, parallelism of operation, and robustness to individual faults [1,3]. These properties make SRS especially suited for tasks that require distributed sensing and/or action. So far, however, SRS have only been demonstrated in relatively simple collective tasks such as aggregation, flocking, navigation, path formation, and foraging, among others [2]. Furthermore, demonstrations of self-organized control are typically only carried out in simulation, and when they are conducted in real robotic hardware, studies are only performed in strictly controlled laboratory environments [3].

In a recent work, Duarte et al. [10] showed for the first time a swarm robotics system operating in realistic conditions, using evolved controllers exclusively. This work used a swarm of aquatic surface robots with up to ten robots, and demonstrated homing, clustering, dispersion, and monitoring behaviors. It was shown that the evolved behaviors transferred successfully from simulation to reality, and were able to display key swarm properties, such as scalability, robustness, and flexibility.

2.2   Challenges in Evolutionary Robotics

While it has been shown that evolutionary techniques can produce successful swarm behaviors, the complexity of such behaviors is still relatively limited [2,3]. Evolving controllers for tasks where the robots have to display a number of different behaviors and fulfill multiple objectives poses a number of challenges for the evolutionary process [6,11]. First, if the task objective is too difficult to achieve, the initial solutions will tend to drift to uninteresting regions of the search space and evolution will fail to bootstrap — the *bootstrap problem* [12]. Second, to bootstrap the evolutionary process and to express the multiple objectives, the fitness function typically becomes convoluted [13]. The fitness landscape therefore becomes more rugged, which can lead to *deception* and *premature convergence* [14], where the evolutionary process is lead towards local optima, instead of (near-)optimal solutions.

Several approaches have been proposed in order to address the bootstrap problem and deception [15]. Incremental evolution [12], for instance, allows the experimenter to decompose and incrementally evolve behavior. It can be applied by progressively complexifying the task [12,16], or progressively making the objectives more challenging [15]. An alternative approach is the use of novelty/diversity driven evolution [17,18], in which individuals are rewarded for displaying novel behaviors instead of their performance with respect to a traditional fitness function. Novelty search has been demonstrated to find controllers that can solve deceptive tasks [19,20].

Another limitation commonly associated with ER techniques is that the evolved behaviors often cannot be modified a posteriori without having to run the evolutionary process again with the new requirements. This is especially true if neural network controllers are used (which is the case in the majority of ER works), as the controller is essentially a black box that receives sensory inputs and outputs the actuator values.

2.3   Behavioral Composition Schemes

An alternative to attain more complex and reusable behaviors is to divide the task into sub-tasks that are easier to solve, and then synthesize control for each sub-task independently. In the past, several approaches have been proposed for the synthesis of controllers by hierarchical task decomposition, applying different techniques such as genetic programming [21], neuroevolution [22, 23], or fuzzy logic control [24]. In the recently proposed *hierarchical control synthesis approach* [7], a complex task is first decomposed in several simpler sub-tasks. Control is then synthesized for the different sub-tasks, resulting in a number of *behavior primitives*, i.e., sub-controllers that have direct control over the robot's actuators. These behavior primitives can then be combined using a *behavior arbitrator*, which delegates control to one of it's sub-controllers. The approach allows for controller hierarchies of arbitrary depth, where behavior arbitrators can delegate control to lower-level arbitrators until a behavior primitive is reached.

One of the advantages of the hierarchical control synthesis approach is that it allows for hybrid controllers, that is, different types of control synthesis techniques can be used. In this way, it is possible to: (i) apply evolutionary techniques in order to automatically synthesize control for particular sub-tasks, (ii) compose increasingly complex controllers that leverage previously synthesized behaviors, and (iii) enable the transfer of control from simulation to reality, by maintaining the complexity of each behavior low enough to allow for the evolution of general and robust behaviors. The hierarchical control synthesis approach has been demonstrated to solve tasks beyond the complexity limits of traditional ER techniques in real robotic hardware for single-robot systems [7], and it has also been shown in a proof-of-concept simulation-based study how such scheme could also be applied to swarms of robots [25].

In [10], it was shown how evolved swarm behaviors can be combined by executing them in a fixed sequence to produce a controller for an environmental monitoring task. Such a simple composition scheme cannot, however, accommodate stochastic events in the environment. In this paper, we extend this work by employing hierarchical control synthesis. Using such an approach allows robots to switch behaviors based on their current state and sensory inputs, which enables a broader range of tasks to be addressed.

3   Experimental Setup

In our maritime intruder detection task, the robots are initially located at a base station, to which they must periodically return in order to recharge their batteries. The swarm must monitor a previously designated area, delimited by a geo-fence, and pursue intruders that try to cross the area. Below, we describe the robotic platform used for our real experiments, and the simulation setup used to synthesize control.

Fig. 1. A photo of a group of robots in the area where the experiments were performed.

3.1 Robotic Platform

For the experiments, we use an aquatic swarm robotics system, which is composed of small (60 cm in length) and inexpensive ($\approx$300 EUR/unit) differential drive mono-hull robots [10] (see Fig. 1). The robots can move at speeds of up to 1.7 m/s (3.3 kts) and are equipped with a GPS and a compass for navigation. The robots broadcast information (such as their position) to their neighbors using Wi-Fi up to a range of 40 m. The robots are completely autonomous and each robot is controlled by an onboard single-board computer (Raspberry Pi 2). The swarm is homogeneous, meaning that all robots are identical both in terms of morphology and of control. Each behavior primitive receives the sensor values and has two outputs that control the linear speed and the angular velocity of the robot. These values are then converted to left and right motor speeds.

We implemented four sensors for the detection of points and objects of interest in the task environment. The sensor values are obtained by calculating the relative distances and orientations of entities in the environment of which the robot is currently aware. The following sensors were implemented (see [10] for details):

Waypoint sensor: Detects waypoints in the environment and returns two values: (i) the relative angle from the robot to the waypoint, and (ii) the distance from the robot to the waypoint.
Robot sensor: Measures the distance to nearby robots using four equally-sized slices around the perimeter of the robot.
Intruder sensor: Measures the distance to nearby intruders using four equally-sized slices around the perimeter of the robot.
Collective intruder sensor: Measures the distance to nearby intruders based on information shared by neighboring robots.
Geo-fence sensor: Detects geo-fences in the environment and returns two values: (i) the distance to the closest edge of a geo-fence using four equally-sized slices around the perimeter of the robot, and (ii) whether or not the robot is inside a geo-fence.

The role of the intruder is played by a robot executing a path-following controller. The robots are aware of the intruder's position up to the range of the onboard intruder sensors. When a robot detects an intruder using its onboard sensors, it shares the intruder's position with neighboring robots. Neighboring robots, in turn, use the received positions to compute the reading for their collective sensors, effectively extending their sensing capabilities [26]. The readings for the collective sensors are calculated by taking into account the robot's own location and orientation, as well as the received location. For the robotic controller, the collective sensor is indistinguishable from onboard sensors, since the communication and integration of received positions is part of the low-level firmware and not of the controller itself. We used a range of 40 m (equivalent to the communication range of the robots) for the robot sensors, the collective intruder sensors, and the geo-fence sensors, and a range of 20 m for the onboard intruder sensors.

3.2   Simulation Model

All controllers were synthesized offline, in simulation, using JBotEvolver, a simulation platform and neuroevolution framework [27]. We used a two-dimensional simulation environment, where the robots are abstracted as circular objects with a certain heading and position. In order to maintain the computational feasibility of the evolutionary process, we simplified the robot model by not including complex physics simulation and fluid dynamics. Instead, the robot model was based on simple measurements of real robots, both in terms of movement, and in terms of sensing.

The real robots' sensors and actuators introduce a high degree of stochasticity, such as GPS and compass errors/drifts, or varying motor performance. In order to facilitate the transfer from simulation to reality, noise was applied to sensors, actuators, and environmental features during the evolutionary process (as advocated in [28,29]), based on measurements obtained from real robots. We used the following different types of noise: (i) water currents that affected the robots' position with a random magnitude in the range $[0, 1]$ m/s and a random direction, (ii) GPS errors up to 1.8 m and compass errors up to 10°, and (iii) motor speed offsets up to 10 % of the maximum speed, motor output noise of up to 5 %, and heading offset of up to 5 %. A complete description of the robot model, the evolutionary parameters, and the different types of noise used can be found in [10].

4   Evolution of Behavior Primitives

Each behavior primitive is an artificial neural network, which is synthesized using the NEAT [30] evolutionary algorithm. We decomposed the complete intruder detection task into four different sub-tasks: (i) go to area, (ii) recharge, (iii) area monitoring, and (iv) pursue intruder. A homing behavior primitive is used as a general navigation behavior both for the go to area and recharge sub-tasks. Controllers for the homing and area monitoring behavior primitives were reused from experiments presented in a previous study [10], and therefore were not re-evolved for these experiments. The pursue intruder behavior primitive was evolved specifically for the experiments presented in this paper. Each sub-task was evolved for a total of 100 generations, with a population size of 150 individuals. The behavior primitives were then combined in a hybrid controller using a manually programmed behavior arbitrator (see Sect. 5). The evolutionary setups for the individual behavior primitives are summarized below.

Homing: In the homing task, the swarm of robots had to navigate to a given waypoint while avoiding collisions among the robots. During evolution, the candidate solutions were evaluated with different numbers of robots, and waypoints in different locations. The robots were positioned up to 50 m away from the waypoint, and were rewarded for minimizing their distance to the waypoint, according to the following equation:

$$f_{homing} = \left(\frac{1}{T} \sum_{t=1}^{T} \frac{1}{R} \sum_{r=1}^{R} \frac{startingDist_r - dist_{r,t}}{startingDist_r} \right) \times S, \tag{1}$$

where T is the maximum number of time steps, R is the number of robots, $startingDist_r$ is the initial distance from robot r to the waypoint, and $dist_{r_t}$ is the distance of robot r to the waypoint at time t.

Collisions between robots are undesirable in any task, as they can damage the real robots. To avoid collisions, we introduced a *safety coefficient* (S) when assessing the fitness in all task setups, which penalized solutions where robots get too close to one another (less than 3 m). The safety coefficient S is in the range $[0.1,1]$, and is inversely proportional to the minimum distance between any two robots in the current simulation trial ($minDist$):

$$S = 0.1 + \max(0, \min(3, minDist))/3 \times 0.9. \tag{2}$$

Area Monitoring: In the area monitoring task, a geo-fence delimits an area of interest, and the robots should coordinate to continuously cover as much of the area as possible. During evolution, the controllers were assessed in a variety of randomly generated areas with different shapes. For the purpose of this task, we considered that each robot covers a circular area around it with a fixed radius. The challenge in this task is to find a general movement pattern that takes into account many different shapes of the monitoring area and varying numbers of robots.

In order to assess the performance of candidate solutions, we divided the monitoring area into a grid with a fixed cell size of $1\,m^2$. Each robot could visit cells within the coverage radius V, setting its value to 1. The value of previously visited cells decayed linearly over a time frame of 100 s, down to 0. Controllers were scored based on how much of the grid was covered over time, according to the following equation:

$$f_{monitor} = \left(\frac{1}{T} \sum_{t=1}^{T} \frac{1}{C} \sum_{c=1}^{C} val(c_t) \right) \times S, \tag{3}$$

where T is the maximum number of time steps, C is the number of cells, and $val(c_t)$ is the value of cell c at time t.

Pursue Intruder: To evolve the pursue intruder behavior primitive, we used an onboard intruder sensor with a range of 20 m, and a collective [26] intruder sensor with a range of 40 m. During evolution, robots were randomly spread in a given area, and one intruder traversed the area with a randomly generated trajectory. When a robot detected an intruder, the robot should attempt to remain within range of it. In simulation, we rewarded controllers according to the following equation:

$$f_{intruder} = \left(\frac{1}{T} \sum_{t=1}^{T} \frac{1}{R} \sum_{r=1}^{R} found(r_t) \right) \times S, \quad found(r_t) = \begin{cases} 1, & \textit{within range} \\ 0, & \textit{else} \end{cases}$$

(4)

where T is the maximum number of time steps, R is the number of robots, and $found(r_t)$ indicates whether or not robot r is detecting the intruder at timestep t.

Evolutionary Results: The results of the evolutionary algorithm for all tasks can be found in Fig. 2. As it can be seen in the figure, each individual task was relatively simple to solve and effective solutions for all tasks were found within 100 generations.

5   Top-Level Behavior Arbitrator

In this section, we detail the task setup for the complete intruder detection task and conduct a series of experiments to: (i) study how changes to the high-level

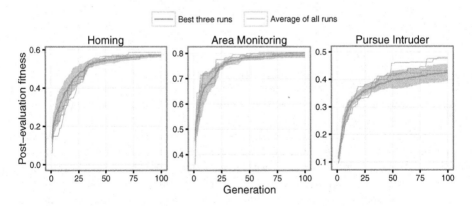

Fig. 2. Fitness plots for the three different behavior primitives. The plots show the highest fitness scores found so far. The red lines depict the three highest-scoring evolutionary runs, while the blue line depicts the average of the ten runs, with the respective standard deviation shown in gray (Color figure online).

behavior arbitrator impact the overall swarm behavior, (ii) study how the swarm size and environment's size affect the performance of system, and (iii) validate the transferability of the controllers by testing them in a real robotic swarm.

5.1 Task Setup

The intruder detection task is carried out in square-shaped monitoring areas with different sizes. We evaluated the controllers in six experimental setups with different numbers of robots and monitoring area sizes, maintaining the number of robots proportional to the size of the area, see Table 1. The base station, where the robots start the task, is located 20 m from the monitoring area (see Fig. 3). In each trial, an intruder makes a total of four crossings of the area, and the robots should detect and remain within range of their onboard intruder sensors (20 m). The task terminates after the intruder's fourth crossing. The onboard battery allows a robot to operate for 8 min. The battery limitation was imposed to facilitate real-robot experiments of relatively short duration while still requiring the use of the recharging behavior. The robots need to return to the base station in order to recharge the battery when it reaches 10 %. In the beginning of the task, the battery level of each robot is set to between 50 % and 100 %, and the robot receives and stores a waypoint and a geo-fence that correspond to the position of the base station and to the monitoring area, respectively.

5.2 Testing Different Arbitrators

After we evolved the pursue intruder behavior primitive, we combined it with the two previously evolved [10] homing and area monitoring behavior primitives using

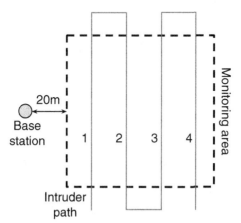

Fig. 3. Representation of the experimental environment. The robots are deployed from a base station, to which they must regularly return in order to recharge their batteries. An intruder makes a total of four crossings (1–4, in red) through the monitoring area (dashed lines), and the experiment ends after the last crossing (Color figure online).

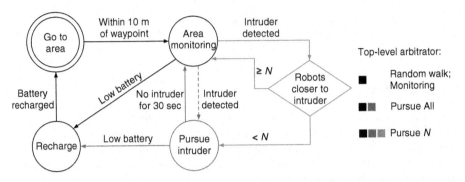

Fig. 4. An FSM representing the preprogrammed top-level arbitrator. The differently colored states and transitions represent incremental extensions to the arbitrator. **Random walk:** the controller will monitor the area individually and manage the battery level. **Monitor:** the controller will monitor the area cooperatively and manage the battery level. **Pursue All:** the controller will actively pursue a detected intruder. **Pursue N:** we limit the number of robots that can actively pursue the intruder. In this last case, the dashed transition from the "Area Monitoring" state to the "Pursue Intruder" state is not used (Color figure online).

a simple finite state-machine (FSM) pre-programmed arbitrator. The arbitrator chooses which behavior should be active at any given time, depending on the current state of the robot and the robot's sensory inputs. We tested different FSMs to incrementally complexify the collective swarm behavior in order to achieve the task goal. Four different FSMs were evaluated in simulation, see Fig. 4:

Random walk: Monitor individually and recharge behavior, no intruder pursuit.
Monitor: Monitor cooperatively and recharge behavior, no intruder pursuit.
Pursue All: Pursue intruder with no restrictions ($N = \infty$).
Pursue 1: Pursue intruder only if there is no one else pursuing ($N = 1$).
Pursue 3: Pursue intruder if there are less than 3 robots closer to the intruder ($N = 3$).

To analyze the collective behaviors obtained with the different FSMs, we relied on three metrics: (i) *observation time*: the proportion of time where the intruder is within the range of at least one robot's sensors when inside the area, (ii) *near robots*: the average number of robots within range of the intruder, and (iii) *coverage*: the average distance from each robot to the closest neighbor.

As the results in Fig. 5 show, the *Monitor* arbitrator performs worse than the remaining arbitrators regarding *observation time* and *near robots* metrics ($p < 0.001$)[1], since the robots do not actively pursue the intruder. The *Random walk* arbitrator, used as a baseline, performed significantly worse than

---

[1] Unless indicated otherwise, the statistical tests were performed using two-sided Mann-Whitney U tests, with the p values adjusted using the Holm-Bonferroni method when multiple comparisons were made.

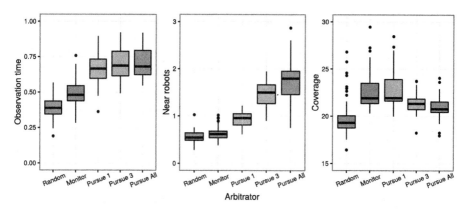

Fig. 5. Three different metrics (*observation time*, *near robots*, and *coverage*) for five variations of the behavior arbitrator. Each boxplot corresponds to 60 data points (10 samples per experimental setup, see Table 1).

all other arbitrators in these two metrics. The arbitrator with no restrictions in the number of robots pursuing (*Pursue All*) achieves the highest number of *near robots*. Adding restrictions to the number of pursuing robots (*Pursue 1* and *Pursue 3*) significantly decreases the number of *near robots* ($p < 0.001$), confirming that the value of N has a significant impact in the behavior of the swarm. Regarding the *observation time*, all the arbitrators that use the pursue behavior show no statistically significant differences ($p = 0.09$, Kruskal-Wallis H test), indicating that as long as at least a single robot actively pursues an intruder, the performance of the swarm remained identical.

The *coverage* metric reveals that there is a trade-off between the number of pursuing robots and the coverage of the monitoring area. Having only one robot pursuing the intruder does not negatively impact the coverage of the area, when compared to the Monitor arbitrator ($p = 0.56$). Allowing more robots to pursue the intruder, however, negatively impacts the coverage of the monitoring area (the *coverage* differences between *Pursue 1*, *Pursue 3*, and *Pursue All* are statistically significant, $p < 0.05$). This trade-off is explained by the higher concentration of robots near the intruder. The lower coverage of the area did not decrease the performance of the swarm in our setups, since there was only

Table 1. Parameters of each experimental setup.

Experimental setups	S_1	S_2	S_3	S_4	S_5	S_6
Monitoring area side-length (m)	100	141	200	245	283	316
Number of robots	5	10	20	30	40	50
Experiment duration (minutes)	8	11	15	19	22	24

one intruder. A higher coverage, however, is advantageous for scenarios where multiple intruders might cross the monitoring area simultaneously.

These results show how different swarm behaviors can be achieved by making simple modifications to the behavior arbitrator, reusing the same behavior primitives. For the remaining experiments in this paper, we chose the *Pursue 3* behavior arbitrator for its higher behavioral complexity, and because of its potential fault-tolerant characteristics in the case of a failure in one of the pursuing robots.

5.3 Scalability

We assessed the scalability of the *Pursue 3* controller using different combinations of swarm size and monitoring area size. We computed the *observation time* metric in a total of 36 setups, each tested in 100 simulation samples. The results show that the chosen controller was scalable both across the swarm size, as well as the monitoring area size (see Fig. 6). There is always a performance increase as the swarm size increases, or as the monitoring area size decreases. Although we found no evidence of interference between the robots as the swarm density increases [31], such a result could be related with the idiosyncrasies of the task, which benefits from more robots covering the monitoring area. In previous work [10], we have shown that the area monitoring and the homing behaviors were scalable with respect to the swarm size. The results presented here suggest that the combination of multiple scalable swarm behaviors results in a scalable composed behavior.

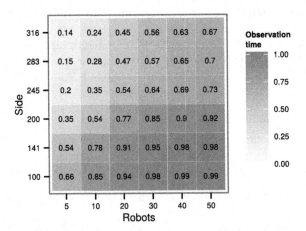

Fig. 6. Proportion of time the intruder was detected when inside the area, in different task setups with varying number of robots and area size. Each setup was repeated in 100 simulations.

6   Transferring Control to Real Robots

We performed real-robot experiments at Parque das Nações, Lisbon, Portugal, in a semi-enclosed area in the margin of the Tagus river. We used experimental setup S_1, a square-shaped monitoring area with a size of 100 m by 100 m (see Fig. 3) and a swarm composed of five robots. We conducted three repetitions of the task with the *Pursue 3* controller. In order to compare the controller's performance on the real robots with its performance in simulation, we ran 100 simulation samples using the same experimental setup. The results are shown in Fig. 7.

In the experiments with the real robots, the swarm was able to find the intruder in all crossings, as shown in Fig. 7 (top). The plot shows that up to four robots followed the intruder simultaneously, and once an intruder was detected, it was always followed continuously until it left the monitoring area. In five of the 12 crossings, the intruder was seen almost as soon as it entered the monitoring area, and then followed during the entirety of its crossing (crossings A-3, B-1, B-3, C-1 and C-2). Figure 8 shows an example of the swarm's behavior in the real-robot experiments. The performance of the controller on the real robots was very similar to the performance in simulation, see Fig. 7 (bottom, *observation*

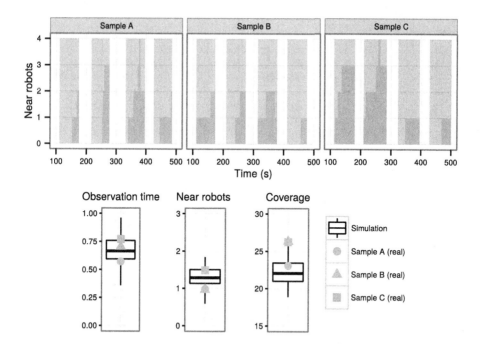

Fig. 7. Top: number of robots pursuing the intruder over time for the three real-robot experiments. The period in which the intruder is traversing the monitoring area is shown in gray, and the number of robots pursuing at any given instant is shown in blue. Note that the intruder was detected in all crossings. Bottom: comparison of the results in simulation and on the real robots, using the metrics presented in Sect. 5.

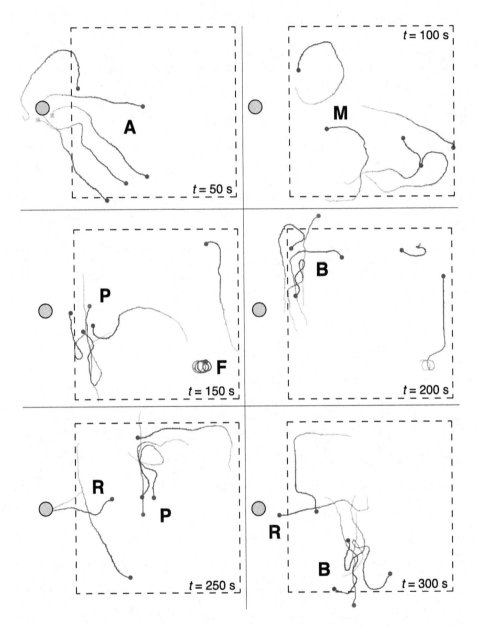

Fig. 8. Traces of the first 300s from sample C of the real-robot experiments, taken at 50s intervals. The swarm is shown in blue, and the intruder is shown in red. The labels identify specific events that highlight the behavioral capabilities of the swarm. **A**: swarm going to the monitoring area. **M**: swarm monitoring the area. **P**: robots pursuing an intruder. **F**: temporary motor failure in one of the robots. **B**: robots abandoning the pursuit when the intruder leaves the monitoring area. **R**: robot returning to the base station to recharge (Color figure online).

time): on average, the intruder was followed during 68 % of the time in the real scenario, and 66 % in simulation. The average number of pursuing robots is also close to what was observed in simulation. The coverage was slightly higher in reality than in simulation, which is consistent with the results reported in [10] for the evolved monitoring controller. Overall, these results show that the hybrid controller successfully crossed the reality gap, achieving a similar behavior and a similar performance on the real robots as in simulation.

7 Conclusions

In this paper, we explored how hybrid control can produce modular and flexible swarm behaviors, and enable the control of swarm robotics systems in complex tasks with realistic constraints. We applied the approach to an intruder detection task, where a swarm of robots had to navigate to a predefined area, monitor the area, and follow any intruders that crossed it. At the same time, the robots had to manage their battery level by returning periodically to a base station in order to recharge.

We reused two evolved behavior primitives from a previous study (homing and area monitoring), and evolved a behavior primitive for intruder pursuit. These behaviors were then combined with a behavior arbitrator consisting of a manually programmed finite state-machine. We first tested multiple variants of the behavior arbitrator, and showed how simple modifications resulted in different swarm behaviors, using the same evolved behavior primitives. We then tested the scalability of the composed behavior, showing that it scales predictably with the number of robots and the size of the monitoring area. Finally, we tested the controller in real robots, using a swarm of five aquatic surface robots and one intruder. The controllers successfully crossed the reality gap by achieving a similar performance on the real robots as in simulation.

The combination of multiple evolved sub-controllers using a finite state-machine allowed us to leverage evolution's automatic synthesis of self-organized behavior, while at the same time provided high-level control of the swarm behavior through simple manually programmed rules. In summary, our results show that hybrid control introduces a series of advantages for the synthesis of control for robotic swarms: (i) the decomposition of the task into simpler sub-tasks facilitates the use of evolutionary techniques for the synthesis of behaviors; (ii) multiple evolved behaviors can be combined to produce control for complex swarm robotics tasks; (iii) previously evolved control can be reused in different applications, simplifying the synthesis of control for new tasks, (iv) the flexibility of manually programmed behavior arbitrators can enable realistic task-specific constraints to be addressed quickly, without the need to re-evolve control, and (v) the reality gap can be effectively overcome in complex tasks by relying on general and robust behavior primitives.

Future Work

We have shown that hybrid control can be used to solve a variety of tasks in single [7] and multirobot systems. In ongoing work, we are assessing the use of hybrid control in additional swarm robotics tasks of varying complexity. Particularly, we are studying when hybrid control should be used instead of monolithic evolution, and to which degree a task should be decomposed. The hierarchical control synthesis approach has, so far, only been tested with homogeneous robot systems. An opportunity for this line of research is related to the extension of the approach to heterogeneous multirobot systems. Since hybrid controllers simplify the manual synthesis of high-level control, it becomes possible to have different sub-sets of robots with specialized behaviors, by varying only the behavior arbitrators. In this way, the swarm could benefit from an explicit division of labor suited to the needs of the task.

Acknowledgements. This work was supported by Fundação para a Ciência e a Tecnologia (FCT) under the grants, SFRH/BD/76438/2011, SFRH/BD/89095/2012, PEst-OE/EEI/LA0008/2013, and EXPL/EEI-AUT/0329/2013.

References

1. Şahin, E.: Swarm robotics: from sources of inspiration to domains of application. In: Şahin, E., Spears, W.M. (eds.) Swarm Robotics 2004. LNCS, vol. 3342, pp. 10–20. Springer, Heidelberg (2005)
2. Bayındır, L.: A review of swarm robotics tasks. Neurocomputing **172**(8), 292–321 (2016)
3. Brambilla, M., Ferrante, E., Birattari, M., Dorigo, M.: Swarm robotics: a review from the swarm engineering perspective. Swarm Intell. **7**(1), 1–41 (2013)
4. Nolfi, S., Floreano, D.: Evolutionary Robotics: The biology, Intelligence, and Technology of Self-organizing Machines. MIT Press, Cambridge (2000)
5. Trianni, V.: Evolutionary Swarm Robotics: Evolving Self-organising Behaviours in Groups of Autonomous Robots. SCI, vol. 108. Springer, Heidelberg (2008)
6. Silva, F., Duarte, M., Correia, L., Oliveira, S.M., Christensen, A.L.: Open issues in evolutionary robotics. Evol. Comput. (2016, in press)
7. Duarte, M., Oliveira, S.M., Christensen, A.L.: Evolution of hybrid robotic controllers for complex tasks. J. Intell. Robot. Syst. **78**(3–4), 463–484 (2015)
8. Silva, F., Duarte, M., Oliveira, S.M., Correia, L., Christensen, A.L.: The case for engineering the evolution of robot controllers. In: Proceedings of the International Conference on the Simulation & Synthesis of Living Systems (ALIFE), pp. 703–710. MIT Press, Cambridge (2014)
9. Doncieux, S., Bredeche, N., Mouret, J.B., Eiben, A.E.: Evolutionary robotics: what, why, and where to. Front. Robot. AI **2**(4) (2015)
10. Duarte, M., Costa, V., Gomes, J., Rodrigues, T., Silva, F., Oliveira, S.M., Christensen, A.L.: Evolution of collective behaviors for a real swarm of aquatic surface robots (2015). Preprint, http://arxiv.org/abs/1511.03154
11. Trianni, V., López-Ibáñez, M.: Advantages of task-specific multi-objective optimisation in evolutionary robotics. PLoS ONE **10**(8), e0136406 (2015)

12. Gomez, F., Miikkulainen, R.: Incremental evolution of complex general behavior. Adapt. Behav. **5**(3–4), 317–342 (1997)
13. Nelson, A.L., Barlow, G.J., Doitsidis, L.: Fitness functions in evolutionary robotics: a survey and analysis. Robot. Auton. Syst. **57**(4), 345–370 (2009)
14. Whitley, L.D.: Fundamental principles of deception in genetic search. In: Proceedings of the 1st Workshop on Foundations of Genetic Algorithms, pp. 221–241. Morgan Kaufmann, San Mateo (1991)
15. Doncieux, S., Mouret, J.B.: Beyond black-box optimization: a review of selective pressures for evolutionary robotics. Evol. Intel. **7**(2), 71–93 (2014)
16. Christensen, A.L., Dorigo, M.: Incremental evolution of robot controllers for a highly integrated task. In: Nolfi, S., Baldassarre, G., Calabretta, R., Hallam, J.C.T., Marocco, D., Meyer, J.-A., Miglino, O., Parisi, D. (eds.) SAB 2006. LNCS (LNAI), vol. 4095, pp. 473–484. Springer, Heidelberg (2006)
17. Lehman, J., Stanley, K.: Abandoning objectives: evolution through the search for novelty alone. Evol. Comput. **19**(2), 189–223 (2011)
18. Mouret, J.B., Doncieux, S.: Encouraging behavioral diversity in evolutionary robotics: an empirical study. Evol. Comput. **20**(1), 91–133 (2012)
19. Gomes, J., Urbano, P., Christensen, A.L.: Evolution of swarm robotics systems with novelty search. Swarm Intell. **7**(2–3), 115–144 (2013)
20. Gomes, J., Christensen, A.L.: Generic behaviour similarity measures for evolutionary swarm robotics. In: Proceedings of the 15th Annual Conference on Genetic and Evolutionary Computation Conference (GECCO), pp. 199–206. ACM, New York (2013)
21. Lee, W.P.: Evolving complex robot behaviors. Inf. Sci. **121**(1–2), 1–25 (1999)
22. Larsen, T., Hansen, S.T.: Evolving composite robot behaviour - a modular architecture. In: Proceedings of the International Workshop on Robot Motion and Control (RoMoCo), pp. 271–276. IEEE Press, Piscataway (2005)
23. Becerra, J.A., Bellas, F., Santos, J., Duro, R.J.: Complex behaviours through modulation in autonomous robot control. In: Cabestany, J., Prieto, A.G., Sandoval, F. (eds.) IWANN 2005. LNCS, vol. 3512, pp. 717–724. Springer, Heidelberg (2005)
24. Tunstel, E.: Mobile robot autonomy via hierarchical fuzzy behavior control. In: Proceedings of the International Symposium on Robotics and Manufacturing (WAC), pp. 837–842. ASME Press, New York (1996)
25. Duarte, M., Oliveira, S.M., Christensen, A.L.: Hybrid control for large swarms of aquatic drones. In: Proceedings of the 14th International Conference on the Synthesis & Simulation of Living Systems, pp. 785–792. MIT Press, Cambridge (2014)
26. Rodrigues, T., Duarte, M., Figueiró, M., Costa, V., Oliveira, S.M., Christensen, A.L.: Overcoming limited onboard sensing in swarm robotics through local communication. In: Nguyen, N.T., Kowalczyk, R., Duval, B., van den Herik, J., Loiseau, S., Filipe, J. (eds.) TCCI XX. LNCS, vol. 9420, pp. 201–223. Springer, Heidelberg (2015). doi:10.1007/978-3-319-27543-7_10
27. Duarte, M., Silva, F., Rodrigues, T., Oliveira, S.M., Christensen, A.L.: JBotEvolver: a versatile simulation platform for evolutionary robotics. In: Proceedings of the 14th International Conference on the Synthesis & Simulation of Living Systems, pp. 210–211. MIT Press, Cambridge (2014)
28. Miglino, O., Lund, H.H., Nolfi, S.: Evolving mobile robots in simulated and real environments. Artif. Life **2**(4), 417–434 (1996)

29. Jakobi, N.: Evolutionary robotics and the radical envelope-of-noise hypothesis. Adapt. Behav. **6**(2), 325–368 (1997)

30. Stanley, K., Miikkulainen, R.: Evolving neural networks through augmenting topologies. Evol. Comput. **10**(2), 99–127 (2002)

31. Hamann, H.: Towards swarm calculus: universal properties of swarm performance and collective decisions. In: Dorigo, M., Birattari, M., Blum, C., Christensen, A.L., Engelbrecht, A.P., Groß, R., Stützle, T. (eds.) ANTS 2012. LNCS, vol. 7461, pp. 168–179. Springer, Heidelberg (2012)

EvoSTOC

Direct Memory Schemes for Population-Based Incremental Learning in Cyclically Changing Environments

Michalis Mavrovouniotis$^{(\boxtimes)}$ and Shengxiang Yang

School of Computer Science and Informatics, Centre for Computational
Intelligence (CCI), De Montfort University, The Gateway, Leicester LE1 9BH, UK
{mmavrovouniotis,syang}@dmu.ac.uk

Abstract. The population-based incremental learning (PBIL) algorithm is a combination of evolutionary optimization and competitive
learning. The integration of PBIL with associative memory schemes
has been successfully applied to solve dynamic optimization problems
(DOPs). The best sample together with its probability vector are stored
and reused to generate the samples when an environmental change
occurs. It is straight forward that these methods are suitable for dynamic
environments that are guaranteed to reappear, known as cyclic DOPs.
In this paper, direct memory schemes are integrated to the PBIL where
only the sample is stored and reused directly to the current samples.
Based on a series of cyclic dynamic test problems, experiments are conducted to compare PBILs with the two types of memory schemes. The
experimental results show that one specific direct memory scheme, where
memory-based immigrants are generated, always improves the performance of PBIL. Finally, the memory-based immigrant PBIL is compared
with other peer algorithms and shows promising performance.

1 Introduction

Population based incremental learning (PBIL) is an abstraction of evolutionary
algorithms (EAs), which combines evolutionary optimization with competitive
learning [1]. More precisely, PBIL explicitly maintains the statistics contained in
an EA's population. Similarly to EAs, PBIL algorithms have been successfully
applied to different benchmark problems and real-world applications [2–5]. In
most cases, PBILs are applied to stationary optimization problems. However,
many real-world problems have a dynamic environment in which the objective
function, decision variables, problem instance, constraints, and so on, may vary
over time [6]. Dynamic optimization problems (DOPs) are often more challenging
to address because the moving optimum needs to be tracked.

Similarly with EAs, PBIL faces the same serious challenge as EAs when
addressing DOPs, i.e., premature convergence. In fact, it has been confirmed
that PBIL maintains significantly lower diversity than an EA does [3]. Different
strategies taken from EAs have been integrated into PBILs to address DOPs,

G. Squillero and P. Burelli (Eds.): EvoApplications 2016, Part II, LNCS 9598, pp. 233–247, 2016.
DOI: 10.1007/978-3-319-31153-1_16

including memory schemes [3], hyper-learning scheme [4], multi-population schemes [7] and immigrants schemes [5,8].

In this paper, we focus on DOPs that are subject to cyclic environments, where the environments are guaranteed to re-apper in the future. Many real-world situations have cyclic or approximately cyclic dynamic environments. For example, the traffic jams in the road system are more likely to re-appear during the day. Associative memory schemes have been found suitable for cyclic DOPs, where the best samples associated with environmental information of previously optimized environments are stored and reused when the relevant environments re-appear [3,9]. In this paper, we integrate direct memory schemes with PBIL and investigate their effect. Their difference is that direct memory schemes store only the best samples and reuse them directly to the samples whereas associative memory schemes have an indirect impact on the samples generated.

Using the exclusive-or (XOR) DOP generator that constructs cyclic DOPs proposed in [3], a series of DOPs are systematically constructed as the dynamic test environments and experiments are carried out to investigate the performance of PBILs with direct memory schemes. Based on the experimental results, the effect of direct memory schemes on the performance of PBILs in dynamic environments is analyzed. Among the direct memory schemes analyzed in this paper, the memory-based immigrants scheme has the best performance in almost all test cases. The specific algorithm is then compared with other peer PBILs and EAs and shows competitive performance.

The rest of the paper is organized as follows. Section 2 introduces the standard PBIL algorithm. Section 3 describes the existing PBIL with an associative memory scheme and the proposed PBILs with direct memory schemes. Section 4 describes the construction of cyclic DOPs used for this study. The experimental study is presented in Sect. 5. Finally, Sect. 6 concludes this paper with several observations and discusses relevant future work.

2   Population-Based Incremental Learning

The standard PBIL (SPBIL) algorithm, first proposed by Baluja [1], is a combination of evolutionary optimization and competitive learning. The aim of SPBIL is to generate a real-valued probability vector $\boldsymbol{P}(t) = \{P_1, \ldots, P_l\}$ (l is the binary encoding length), at each generation t, which creates high quality solutions with high probability when sampled. Each element $P_i(i = 1, \ldots, l)$ in the probability vector is the probability of creating an allele "1" in locus i. More precisely, a solution is sampled from the probability vector $\boldsymbol{P}(t)$ as follows: for each locus i, if a randomly generated number $R \in \{0, 1\} < P_i$, it is set to 1; otherwise, it is set to 0.

SPBIL starts from an initial (central) probability vector $\boldsymbol{P}(0)$ with values of each entry set to 0.5. This means when sampling by this initial probability vector random solutions are created because the probability of generating a "1" or "0" on each locus is equal. However, as the search progresses, the values in the probability vector are gradually moved towards values representing high evaluation solutions. The evolution process is described as follows.

For every generation t, a set $S(t)$ of n samples (solutions) are created according to the current probability vector $\boldsymbol{P}(t)$. The set of samples are evaluated according to the problem-specific fitness function. Then, the probability vector is moved towards the best solution $\boldsymbol{x}^{bs}(t)$ of the set $S(t)$ as follows:

$$P_i(t+1) \leftarrow (1-\alpha) \times P_i(t) + \alpha \times \boldsymbol{x}^{bs}(t), \quad i = \{1, \ldots, l\}, \tag{1}$$

where α is the learning rate, which determines the distance the probability vector is moved for each generation.

After the probability vector is updated toward the best sample, in order to maintain the diversity of sampling, it may undergo a bit-wise mutation process [10]. Mutation is applied to the SPBIL studied in this paper since diversity maintenance is important when addressing DOPs [3]. The mutation operation always changes the probability vector toward the central probability vector, where values are set to 0.5, to increase exploration. The mutation operation is carried out as follows. The probability of each locus P_i is mutated, if a random number $R \in \{0, 1\} < p_m$ (p_m is the mutation probability), as follows:

$$P_i'(t) = \begin{cases} P_i(t) \times (1 - \delta_m), & \text{if } P_i(t) > 0.5, \\ P_i(t) \times (1 - \delta_m) + \delta_m, & \text{if } P_i(t) < 0.5, \\ P_i(t), & \text{otherwise}, \end{cases} \tag{2}$$

where δ_m is the mutation shift that controls the amount a mutation operation alters the value in each bit position. After the mutation operation, a new set of samples is generated by the new probability vector and this cycle is repeated.

As the search progresses, the entries in the probability vector move away from their initial settings of 0.5 towards either 0.0 or 1.0. The search progress stops when some termination condition is satisfied, e.g., the maximum allowable number of generations is reached or the probability vector is converged to either 0.0 or 1.0 for each bit position.

PBIL has been applied to many optimization problems with promising results [2]. Most of these applications assume stationary environments, whereas only a few applications considered dynamic environments. To address DOPs with PBILs, the algorithm needs to be enhanced to maintain diversity. Existing strategies, which have been integrated with PBILs and were mainly inspired by EAs, include associative memory schemes [3], hyper-learning schemes [4], multi-population schemes [7] and immigrants schemes [5,8]. In this paper, we integrate direct memory schemes, previously used to genetic algorithms (GAs) [11], to PBILs to address cyclically changing DOPs.

3 Memory Schemes

3.1 PBIL with Associative Memory

Memory schemes have proved to be useful in dynamic environments, especially when the environment changes cyclically. The stored information can be reused when the environment cycles to previously optimized environments. An associative memory scheme was integrated to PBIL in [3] that stores good solutions associated with environmental information.

Within the memory-enhanced PBIL (MPBIL) a memory of size $m = 0.1 \times n$ (n is the population size) is used to store pairs of a sample with its associated probability vector. The associative memory is initially empty and stores pairs until it is full. The update policy of the memory (when it is full) is based on the closest replacement strategy: the memory point with its sample $\boldsymbol{B}_M(t)$ closest to the best population sample of the current generation $\boldsymbol{B}(t)$ in terms of the Hamming distance is replaced if it has the worst fitness; otherwise, the memory remains unchanged. Every time the memory is updated with the sample $\boldsymbol{B}(t)$ its probability vector $\boldsymbol{P}(t)$ is also stored in the memory. The memory is updated at a specific random time as follows. At the end of every memory update at generation t, a random number $R \in [5, 10]$ is generated to determine the next memory update time $t_M = t + R$.

The samples in the memory are re-evaluated every generation. If at least one sample has a change in its fitness, then a change has occurred to the environment. Therefore, the best re-evaluated memory sample with its associated probability vector will replace the current probability vector if the memory sample has better fitness than the current best sample of the current probability vector. If no environmental change is detected, MPBIL progresses as the SPBIL does.

3.2   PBIL with Direct Memory

Direct memory schemes have showed promising performance in DOPs when applied to different GAs [11]. The difference of a direct memory scheme from the associative memory scheme described previously for the MPBIL is that the memory point consists of only the solution. In this paper, we integrate a direct memory scheme into PBIL and introduce three algorithmic variations.

Memory-Enhanced PBIL. The direct memory of the memory-enhanced PBIL (MEPBIL) is initialized with random points (solutions). Similarly with MPBIL, the current best sample $\boldsymbol{B}(t)$ replaces the closest memory point $\boldsymbol{B}_M(t)$ (if no randomly initialized point still exists) if its fitness is better; otherwise, the memory remains unchanged. A memory update occurs based on the stochastic time t_M as described previously in the MPBIL. In addition, the memory is updated whenever a dynamic change occurs. In this way, the most relevant information for the specific environment will be stored in the memory.

When a dynamic change is detected the current probability vector $\boldsymbol{P}(t)$ is not replaced by the best memory probability vector as in MPBIL but the samples stored in the memory are re-evaluated and merged with the current samples of $S(t)$. The best $n - m$ samples will survive in the current $S(t)$. In this way, the probability vector will be learned toward $\boldsymbol{B}(t)$ that is most probably provided by a sample that was stored in the memory. When a dynamic change is not detected, the proposed MEPBIL progresses as SPBIL does.

Memory-Enhanced PBIL+Random Immigrants. Random immigrants were previously integrated to PBIL to enhance their performance in (especially severely changing) DOPs [8]. Random immigrants have the ability to maintain the diversity. Since memory schemes are suitable for slightly or cyclically changing environments and random immigrants are suitable for severely changing environments, a straightforward combination of MEPBIL with random immigrants may combine the merits of both.

The memory-enhanced random immigrant PBIL (MRIPBIL) differs from MEPBIL in only the additional immigration process. The immigration process within MRIPBIL occurs after the probability vector is sampled. More precisely, in every generation, $r_i \times n$ immigrants are generated and replace the worst samples in the current set of samples $S(t)$, where r_i is the replacement ratio and n is the population size. In this way, the generated immigrants will maintain diversity within the samples generated.

Memory-Based Immigrants PBIL. Based on the above consideration, the memory-based immigrants scheme, which has been initially proposed for GAs [11] to address cyclic DOPs, can be integrated with PBILs, denoted as memory-based immigrants PBIL (MIPBIL) in this paper, to better tackle DOPs with reappearing environments. A direct memory scheme is maintained and updated as described previously for the MEPBIL and MRIPBIL algorithms.

Similarly with MRIPBIL above, MIPBIL also includes an immigration process but generates memory-based immigrants instead of random ones. Differently from both MRIPBIL and MEPBIL, MIPBIL does not merge the memory with the current samples but generates memory-based immigrants on every generation as follows. For each generation t, before the mutation operation described in Eq. (2), the best memory point $\boldsymbol{B}_M(t)$ is retrieved and used as the base to generate immigrants. In every generation, $r_i \times n$ memory-based immigrants are generated by mutating bitwise with a probability p_m^i. The generated immigrants replace the worst individuals in the current set of samples $S(t)$. In this way, the samples of the next set $S(t+1)$ will have more direction toward the best point retrieved from the memory.

4   Dynamic Test Environments

The XOR DOP generator [3, 12] can construct dynamic environments from any binary-encoded stationary function $f(\boldsymbol{x})(\boldsymbol{x} \in \{0,1\}^l)$ by a bitwise XOR operator. Since for the specific experimental study memory schemes are involved cyclic DOPs are considered. With the XOR DOP, the cyclicity of the changing environments is controlled as follows. Initially, $2K$ XOR masks $\boldsymbol{M}(0), \ldots, \boldsymbol{M}(2K-1)$ are generated as the base states in the search space randomly. Then, the environment can cycle among these base states in a fixed logical ring. Suppose the environment changes in every τ algorithmic generations, then the individuals at generation t are evaluated as follows:

$$f(\boldsymbol{x}, t) = f(\boldsymbol{x} \oplus \boldsymbol{M}(I_t)) = f(\boldsymbol{x} \oplus \boldsymbol{M}(k\%(2K))), \tag{3}$$

where "$\oplus$" is the XOR operator (i.e., $1 \oplus 1 = 0$, $1 \oplus 0 = 1$, $0 \oplus 0 = 0$), $k = \lceil t/f \rceil$ is the index of the current environment, $I_t = k\%(2K)$ is the index of the base state that the environment is in at generation t and M are the XORing masks.

The $2K$ XORing masks are generated as follows. First, K binary templates $T(0), \ldots, T(K-1)$ are constructed with each template containing exactly $\rho \times n = l/K$ ones. Let $M(0) = 0$ denote the initial state. Then, the other XORing masks are generated iteratively as follows:

$$M(i+1) = M(i) \oplus T(i\%K), \ i = 0, \ldots, 2K - 1. \tag{4}$$

The templates $T(0), \ldots, T(K-1)$ are first used to create K masks till $M(K) = 1$ and then orderly reused to construct another K XORing masks till $M(2K) = M(0) = 0$. The Hamming distance between two neighbour XOR masks is the same and equals $\rho \times n$. Here, $\rho \in [1/l, 1.0]$ is the distance factor, determining the number of base states. A higher value of ρ means severer dynamic changes, whereas a lower value of τ means faster dynamic changes.

In this paper, four 100-bit binary-encoded problems are selected as the stationary problems to generate DOPs. Each problem consists of 25 copies of 4-bit building blocks and has an optimum value of 100. The first one is the *OneMax* function, which aims to maximize the number of ones in a solution. The second one is the *Plateau* function, where each building block contributes four (or two) to the total fitness if its unitation (i.e., the number of ones inside the building block) is four (or three); otherwise, it contributes zero. The third one is the *RoyalRoad* function, where each building block contributes four to the total fitness if its unitation is four; otherwise, it contributes zero. The fourth one is the *Deceptive* function, where the building block is a fully deceptive sub-function. Generally, the difficulty of the four functions for optimization algorithms is increasing in the order from OneMax to Plateau to RoyalRoad to Deceptive.

5 Experimental Study

5.1 Experimental Setup

For each algorithm on a DOP, 30 independent runs were executed on the same environmental changes and 100 environmental changes were allowed for each run. The overall performance is calculated as follows:

$$\bar{F}_{BOG} = \frac{1}{G} \sum_{i=1}^{G} \left(\frac{1}{N} \sum_{j=1}^{N} F_{BOG_{ij}} \right), \tag{5}$$

where G is the number of generations of a run, N is the number of runs and $F_{BOG_{ij}}$ is the best-of-generation fitness of generation i of run j. Moreover, the diversity of the population was recorded every generation. The overall diversity of an algorithm on a DOP is defined as:

$$\bar{T}_{DIV} = \frac{1}{G} \sum_{i=1}^{G} \left(\frac{1}{N} \sum_{j=1}^{N} Div_{ij} \right), \tag{6}$$

where G and N are defined as in Eq. (5) and Div_{ij} is the diversity at generation i of run j, which is defined as:

$$Div_{ij} = \frac{1}{ln(n-1)} \sum_{p=1}^{n} \sum_{q \neq p}^{n} HD(p,q),$$ (7)

where l is the encoding length, n is the population size and $HD(p,q)$ is the Hamming distance between the p-th sample and q-th sample.

Dynamic test environments are generated from the four aforementioned binary-encoded functions, described in Sect. 4, using the XOR DOP generator with τ set to 10 and 50, indicating quickly and slowly changing environments, respectively, and ρ set to 0.1, 0.2, 0.5 and 1.0, indicating slightly, to medium, to severely changing environments, respectively. With the specific setting of ρ, basically the environments cycles among 20, 10, 4, and 2 base states respectively. As a result, eight cyclic dynamic environments (i.e., 2 values of $\tau \times 4$ values of ρ) from each stationary function are generated to systematically analyze the algorithms on the DOPs.

In order to have fair comparisons among PBILs, the population size, memory size and immigrant replacement are set to such that each PBIL has 120 fitness evaluation per generation. For memory-based algorithms the memory size was set to $m = 0.1 \times n$ and for all immigrant-based algorithms the immigrant replacement ratio was set to $r_i = 0.2$ and immigrant mutation probability $p_m^i = 0.01$. For all PBILs in the experiments, the parameters were set to typical values [3] as follows: the learning rate $\alpha = 0.25$, mutation probability $p_m = 0.05$ with the mutation shift $\delta_m = 0.05$, and the elitism of size 1.

5.2   Experimental Study of Memory-Based PBILs

The experimental results of the proposed PBILs with direct memory schemes (e.g., MEPBIL, MRIPBIL and MIPBIL) compared with the existing PBIL with associative memory scheme (e.g., MPBIL) are shown in Fig. 1. The corresponding statistical results are presented in Table 1, where Kruskal–Wallis tests were applied followed by posthoc paired comparisons using Mann–Whitney tests with the Bonferroni correction. The statistical results are shown as "−", "+" or "∼" when the first algorithm is significantly better than the second algorithm, when the second algorithm is significantly better than the first algorithm, or when the two algorithms are insignificantly different, respectively. To better understand the behaviour of the PBILs the population diversity against generations for the first 500 generations is plotted in Fig. 2 for the DOPs with $\rho = 0.2$. From Fig. 1 and Table 1, the following observations can be drawn.

First, among the PBILs with direct memory schemes, MIPBIL has the best performance and outperforms both MEPBIL and MRIPBIL in most DOPs. When random immigrants are hybridized with memory schemes the performance is enhanced only in DOPs with $\rho = 1.0$ but degraded in the remaining DOPs; see the comparisons of MEPBIL $\Leftrightarrow$ MRIPBIL in Table 1. This is normal because random immigrants generate high levels of diversity that help the adaptation

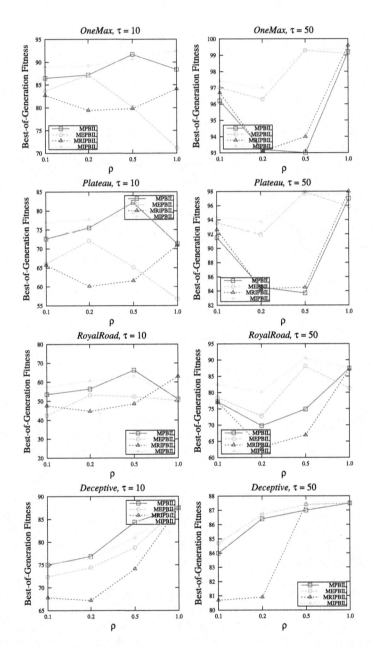

Fig. 1. Experimental results of memory-based PBILs for cyclic DOPs

Table 1. Statistical results of comparing memory-based PBILs for cyclic DOPs

DOPs, $\rho \Rightarrow$	$\tau = 10$				$\tau = 50$			
	0.1	0.2	0.5	1.0	0.1	0.2	0.5	1.0
Algorithms	*OneMax*							
MPBIL ⇔ MEPBIL	−	~	−	−	+	+	+	−
MPBIL ⇔ MRIPBIL	−	−	−	−	+	~	+	+
MPBIL ⇔ MIPBIL	+	+	−	~	+	+	+	+
MIPBIL ⇔ MEPBIL	−	−	−	−	−	−	−	−
MIPBIL ⇔ MRIPBIL	−	−	−	−	−	−	−	+
MEPBIL ⇔ MRIPBIL	−	−	~	+	−	−	−	+
Algorithms	*Plateau*							
MPBIL ⇔ MEPBIL	−	−	−	−	+	+	+	−
MPBIL ⇔ MRIPBIL	−	−	−	~	+	~	+	+
MPBIL ⇔ MIPBIL	+	+	−	+	+	+	+	+
MIPBIL ⇔ MEPBIL	−	−	−	−	−	−	~	−
MIPBIL ⇔ MRIPBIL	−	−	−	−	−	−	−	+
MEPBIL ⇔ MRIPBIL	~	−	−	+	−	−	−	+
Algorithms	*RoyalRoad*							
MPBIL ⇔ MEPBIL	−	−	−	~	~	+	+	−
MPBIL ⇔ MRIPBIL	−	−	−	+	~	−	−	~
MPBIL ⇔ MIPBIL	+	+	−	+	+	+	+	~
MIPBIL ⇔ MEPBIL	−	−	−	−	−	−	−	−
MIPBIL ⇔ MRIPBIL	−	−	−	+	−	−	−	~
MEPBIL ⇔ MRIPBIL	+	−	−	+	−	−	−	+
Algorithms	*Deceptive*							
MPBIL ⇔ MEPBIL	−	−	−	~	+	~	~	~
MPBIL ⇔ MRIPBIL	−	−	−	~	−	−	~	~
MPBIL ⇔ MIPBIL	~	~	−	~	+	~	~	~
MIPBIL ⇔ MEPBIL	−	−	−	~	~	~	~	~
MIPBIL ⇔ MRIPBIL	−	−	−	~	−	−	~	~
MEPBIL ⇔ MRIPBIL	−	−	−	~	−	−	~	~

in severely changing environments but may disturb the optimization process in slightly changing environments. In contrast, memory-based immigrants in MIP-BIL generate guided diversity via transferring knowledge from previous environments. Figure 2 supports our claim since MRIPBIL maintains a higher diversity level than MIPBIL.

Second, MEPBIL and MRIPBIL outperform MPBIL in most DOPs with $\tau = 50$ whereas MPBIL outperforms MEPBIL and MRIPBIL in most DOPs with $\tau = 10$; see the comparisons in Table 1. In contrast, MIPBIL outperforms MPBIL in most DOPs; see the comparisons of MPBIL ⇔ MIPBIL in Table 1.

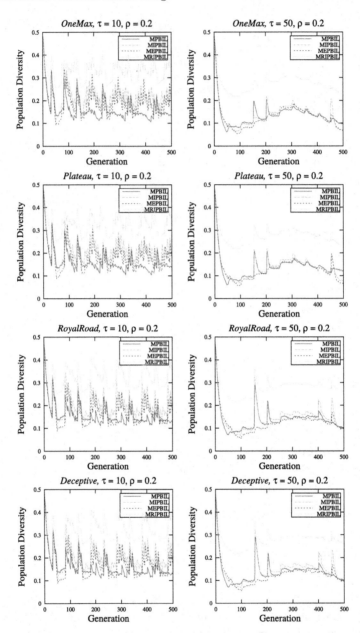

Fig. 2. Dynamic population diversity of PBILs for the first five hundred generations on cyclic DOPs with $\rho = 0.2$

From Fig. 2 it can be observed that MIPBIL maintains higher diversity levels than MPBIL especially in DOPs with $\tau = 10$. Low diversity indicates that MPBIL possibly loses its adaptation capabilities and cannot track the changing environments when they change fast.

Since the memory-based immigrants scheme has the best performance over the other direct memory schemes, we only consider MIPBIL for the remaining experiments.

5.3   Experimental Study of MIPBIL with Other PBILs

The best performing PBIL with the direct memory scheme, i.e., MIPBIL is compared with other existing PBILs: SPBIL [2], random immigrants PBIL (RIPBIL) [3], elitism-based PBIL (EIPBIL) [8] and hybrid immigrants PBIL (HIPBIL) [8]. The experimental results of the aforementioned algorithms are

Table 2. Experimental results of MIPBIL against other PBILs algorithms for cyclic DOPs.

DOPs, $\rho \Rightarrow$	$\tau = 10$				$\tau = 50$			
	0.1	0.2	0.5	1.0	0.1	0.2	0.5	1.0
Algorithms	*OneMax*							
SPBIL	79.6	68.0	59.4	57.1	96.8	92.8	81.8	70.0
RIPBIL	79.3	75.5	74.8	80.7	96.7	92.6	89.0	92.2
EIPBIL	**90.1**	77.7	65.0	61.2	**98.5**	96.4	88.9	80.3
HIPBIL	88.1	77.1	75.1	82.9	98.1	95.7	90.7	94.0
MIPBIL	89.0	**89.2**	**90.6**	**92.5**	97.4	**97.0**	**99.3**	**99.4**
Algorithms	*Plateau*							
SPBIL	60.5	47.0	43.3	50.5	93.0	84.0	61.4	53.6
RIPBIL	60.7	55.4	55.1	66.5	92.7	83.4	76.3	83.9
EIPBIL	**78.0**	58.1	47.3	50.7	**96.4**	91.4	72.7	55.5
HIPBIL	74.2	59.2	56.9	71.4	95.7	89.9	78.4	87.4
MIPBIL	76.9	**77.7**	**79.4**	**81.3**	94.4	**93.7**	**97.5**	**97.6**
Algorithms	*RoyalRoad*							
PBIL	42.6	34.4	37.0	50.0	78.6	59.6	43.8	50.3
RIPBIL	45.1	40.5	43.9	59.2	77.7	60.7	59.5	72.7
EIPBIL	56.3	41.0	39.6	50.0	**85.8**	68.0	47.7	50.4
HIPBIL	52.6	44.3	46.4	**64.8**	84.3	66.2	60.5	76.6
MIPBIL	**57.5**	**60.7**	**63.1**	51.8	82.5	**80.2**	**90.5**	**87.6**
Algorithms	*Deceptive*							
SPBIL	67.1	64.9	69.6	87.5	78.2	75.3	76.9	87.5
RIPBIL	66.7	64.9	70.7	87.5	78.0	75.1	79.2	87.5
EIPBIL	72.6	69.2	72.7	87.5	80.1	78.5	80.7	87.5
HIPBIL	72.6	67.9	76.5	**97.2**	77.5	74.0	82.9	**98.9**
MIPBIL	**75.2**	**76.4**	**81.0**	87.5	**84.7**	**86.7**	**87.4**	87.5

Table 3. Experimental results of MIPBIL against MIGA for cylic DOPs.

DOPs, $\rho \Rightarrow$	$\tau = 10$				$\tau = 50$			
	0.1	0.2	0.5	1.0	0.1	0.2	0.5	1.0
Algorithms	*OneMax*							
MIPBIL (vs)	**89.0**	**89.2**	90.6	92.5	**97.4**	**97.0**	**99.3**	**99.4**
MIGA	82.1	83.5	**92.7**	**96.0**	95.9	93.6	98.4	99.3
Algorithms	*Plateau*							
MIPBIL	**76.9**	**77.7**	79.4	81.3	**94.4**	**93.7**	**97.5**	**97.6**
MIGA	65.4	67.3	**82.3**	**89.7**	91.4	86.5	95.3	**97.6**
Algorithms	*RoyalRoad*							
MIPBIL (vs)	**57.5**	**60.7**	**63.1**	51.8	**82.5**	**80.2**	**90.5**	**87.6**
MIGA	45.2	49.1	61.3	**65.4**	75.7	69.9	83.7	**86.9**
Algorithms	*Deceptive*							
MIPBIL (vs)	**75.2**	**76.4**	81.0	**87.5**	**84.7**	**86.7**	**87.4**	**87.5**
MIGA	66.0	73.4	**82.5**	86.0	81.3	85.2	86.9	87.2

presented in Table 2. Bold values indicate that the algorithm is significantly better than the other algorithms using the same statistical method described previously.

From Table 2, it can be observed that MIPBIL outperforms SPBIL in all DOPs except some cases of the Deceptive function. Also, MIPBIL outperforms the remaining algorithms in most DOPs. This is because MIPBIL can directly move the population to the optimum or close to the optimum of previously optimized environments. Specifically, MIPBIL is underperformed by EIPBIL only in most DOPs with $\rho = 0.1$. This is because when $\rho = 0.1$ more base states (i.e., 20) exist and the memory points stored may not be so accurate to the relevant base states and may misguide the immigrants. As ρ increases the performance of MIPBIL improves whereas EIPBIL degrades which further supports our claim. MIPBIL is underperformed by HIPBIL only in some DOPs with $\rho = 1.0$. This is because when $\rho = 1.0$ the environments switches between two fitness landscapes that are complementary to each other. Therefore, the dualism type of immigrants (e.g., complementary to the best solution) generated in HIPBIL are suitable to cope with these types of DOPs.

5.4   Experimental Results on Pairwise Comparisons of MIPBIL with MIGA

Since the direct memory scheme, i.e., memory-based immigrants, integrated to MIPBIL, was initially introduced and integrated to GAs [11], further pairwise comparisons are performed in this section. Specifically, MIPBIL is compared with memory-based immigrants GA (MIGA) [11]. MIGA is executed on the same DOPs and common parameter settings with MIPBIL above (e.g., $r_i = 0.2$,

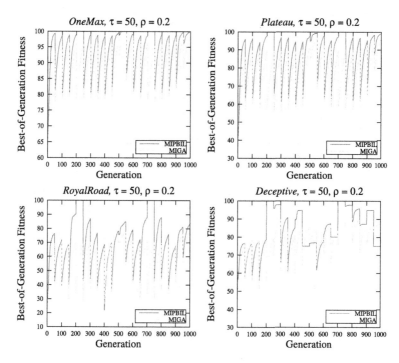

Fig. 3. Dynamic performance of MIPBIL and MIGA for the first twenty environmental changes on DOPs with $\tau = 50$ and $\rho = 0.2$.

$p_m^i = 0.01$ and $m = 0.1 \times n$). The remaining MIGA parameters were set to typical values as follows: generational, uniform crossover with $p_c = 0.6$, flip mutation with $p_m = 0.01$, and fitness proportionate selection with elitism of size 1. The pairwise comparisons regarding the performance are given in Table 3. A bold value indicates that the algorithm is significantly better than the other using Mann–Whitney tests. The dynamic performance of the two algorithms with respect to the best-of-generation fitness against generation on the DOPs with $\tau = 50$ and $\rho = 0.2$ is plotted in Fig. 3.

From Table 3, it can be observed that MIPBIL outperforms MIGA on all DOPs with $\tau = 50$ and on most DOPs with $\tau = 10$ and $\rho = 0.1$ and $\rho = 0.2$. MIGA outperforms MIPBIL in DOPs with $\tau = 10$ and $\rho = 0.5$ and $\rho = 1.0$. This is probably because the mutation operator of MIGA may have a faster effect than the mutation operator of MIPBIL to the adaptation process of the algorithm. The difference lies in that the former is direct (applied to the solutions) whereas the latter is indirect (applied to the probabilistic vector) and may need some time to express its effect.

6   Conclusions

Associative memory schemes have been successfully integrated with PBILs [3]. In this paper, three direct memory schemes are integrated with PBIL and their performance is investigated on cyclically changing DOPs. The direct memory scheme in which memory-based immigrants are generated has the best performance when integrated with PBIL. Therefore, its performance is further compared against other peer algorithms.

From the experimental results, the following concluding remarks can be drawn. First, among the direct schemes integrated with PBIL, the memory-based immigrants scheme has the best performance. Second, although associative memory schemes contain more environmental information from direct memory schemes, the latter may provide faster adaptation in PBILs. Third, PBILs may also benefit when integrated with memory-based immigrants as with GAs [11]. In fact from the experiments it can be observed that MIPBIL outperforms MIGA in many DOPs.

For future work, it would be interesting to apply PBILs to the Knapsack problem that has many applications in the real world.

Acknowledgement. This work was supported by the Engineering and Physical Sciences Research Council (EPSRC) of U.K. under Grant EP/K001310/1.

References

1. Baluja, S.: Population-based incremental learning: A method for integrating genetic search based function optimization and competitive learning. Technical Report CMU-CS-94-163, Carnegie Mellon University, Pittsburgh, PA, USA (1994)
2. Larrañaga, P., Lozano, J. (eds.): Estimation of Distribution Algorithms: A New Tool for Evolutionary Computation. Kluwer, Norwell (2002)
3. Yang, S., Yao, X.: Population-based incremental learning with associative memory for dynamic environments. IEEE Trans. Evol. Comput. **12**(5), 542–561 (2008)
4. Yang, S., Richter, H.: Hyper-learning for population-based incremental learning in dynamic environments. In: IEEE Congress on Evolutionary Computation 2009, CEC 2009, pp. 682–689 (2009)
5. Yang, S., Yao, X.: Experimental study on population-based incremental learning algorithms for dynamic optimization problems. Soft Comput. **9**(11), 815–834 (2005)
6. Jin, Y., Branke, J.: Evolutionary optimization in uncertain environments-a survey. IEEE Trans. Evol. Comput. **9**(3), 303–317 (2005)
7. Yang, S.: Population-based incremental learning with memory scheme for changing environments. In: Proceedings of the 7th Annual Conference on Genetic and Evolutionary Computation. GECCO 2005, 711–718. ACM, New York (2005)
8. Mavrovouniotis, M., Yang, S.: Population-based incremental learning with immigrants schemes for changing environments. In: Proceedings of the 2015 IEEE Symposium on Computational Intelligence in Dynamic and Uncertain Environments (CIDUE), pp. 1444–1451, December 2015

9. Yang, S.: Associative memory scheme for genetic algorithms in dynamic environments. In: Rothlauf, F., Branke, J., Cagnoni, S., Costa, E., Cotta, C., Drechsler, R., Lutton, E., Machado, P., Moore, J.H., Romero, J., Smith, G.D., Squillero, G., Takagi, H. (eds.) EvoWorkshops 2006. LNCS, vol. 3907, pp. 788–799. Springer, Heidelberg (2006)
10. Baluja, S.: An empirical comparison of seven iterative and evolutionary function optimization heuristics. Technical Report CMU-CS-95-193, Carnegie Mellon University, Pittsburgh, PA, USA (1995)
11. Yang, S.: Genetic algorithms with memory- and elitism-based immigrants in dynamic environments. Evol. Comput. 16(3), 385–416 (2008)
12. Yang, S.: Non-stationary problem optimization using the primal-dual genetic algorithm. In: The 2003 Congress on Evolutionary Computation CEC 2003. vol. 3, pp. 2246–2253, December 2003

Simheuristics for the Multiobjective Nondeterministic Firefighter Problem in a Time-Constrained Setting

Krzysztof Michalak[1]([✉]) and Joshua D. Knowles[2]

[1] Department of Information Technologies, Institute of Business Informatics,
Wroclaw University of Economics, Wroclaw, Poland
krzysztof.michalak@ue.wroc.pl
[2] School of Computer Science, University of Birmingham,
Edgbaston, Birmingham, UK
j.knowles.1@cs.bham.ac.uk

Abstract. The firefighter problem (FFP) is a combinatorial problem requiring the allocation of 'firefighters' to nodes in a graph in order to protect the nodes from fire (or other threat) spreading along the edges. In the original formulation the problem is deterministic: fire spreads from burning nodes to adjacent, unprotected nodes with certainty.

In this paper a nondeterministic version of the FFP is introduced where fire spreads to unprotected nodes with a probability P_{sp} (lower than 1) per time step. To account for the stochastic nature of the problem the simheuristic approach is used in which a metaheuristic algorithm uses simulation to evaluate candidate solutions. Also, it is assumed that the optimization has to be performed in a limited amount of time available for computations in each time step.

In this paper online and offline optimization using a multipopulation evolutionary algorithm is performed and the results are compared to various heuristics that determine how to place firefighters. Given the time-constrained nature of the problem we also investigate for how long to simulate the spread of fire when evaluating solutions produced by an evolutionary algorithm. Results generally indicate that the evolutionary algorithm proposed is effective for $P_{sp} \geq 0.7$, whereas for lower probabilities the heuristics are competitive suggesting that more work on hybrids is warranted.

Keywords: Graph-based optimization · Nondeterministic firefighter problem · Simheuristics

1 Introduction

The firefighter problem (FFP) is a discrete-time optimization problem in which spreading of fire is modelled on a graph and the goal is to select nodes that should be protected in order to prevent fire from spreading. The same formalism can be used for analyzing threats in computer networks and a spread of diseases in

© Springer International Publishing Switzerland 2016
G. Squillero and P. Burelli (Eds.): EvoApplications 2016, Part II, LNCS 9598, pp. 248–265, 2016.
DOI: 10.1007/978-3-319-31153-1_17

humans as well as in livestock. There are other similar problems complementary to the FFP For example, in the area of research on broadcasting and computer networks a question arises how many edges in the graph (rather than nodes) should be prevented from transmitting to stop the spread of an infection [2]. The paper [9] studies evolving (i.e. growing) graphs and analyzes their properties. While the aforementioned paper does not discuss any means of preventing the growth, some similarities can be seen between the evolving graphs and the subgraph composed of the nodes on fire in the FFP.

Originally, the firefighter problem was proposed by Hartnell in 1995 [6] as a single-objective, deterministic problem. Since then the research on the FFP has followed three main directions. The first area is the analysis of various theoretical aspects of the problem itself. For example, some researchers analyzed boundary conditions for which it is possible to save the graph [4]. The second line of work is the application of classical optimization methods, such as the linear integer programming [3]. Only recently a third area emerged in which metaheuristic algorithms are applied to the FFP. The paper [1], presented at the EvoCOP 2014 conference was, according to its authors, the first attempt to use a metaheuristic approach (ACO) to solving the FFP. In a paper [11] published later the same year the multiobjective version of the FFP was proposed and solved using and evolutionary algorithm (EA). Two more papers on the FFP were published at EvoCOP in 2015: [7] where variable neighborhood search (VNS) approach was applied to the single-objective FFP and a solution representation suitable for this algorithm was proposed, and [13] where the multiobjective version of the FFP was tackled using a multipopulation algorithm Sim-EA with migration based on similarity between search directions assigned to subpopulations.

All papers mentioned above concerned the deterministic version of the FFP. In a paper [5] the FFP is studied on randomly constructed graphs with randomly chosen starting points, but the spreading of fire is deterministic. In this paper we introduce and study the nondeterministic version of the FFP which involves uncertainty in the spreading of fire in the graph.

The rest of this paper is structured as follows. Section 2 provides a definition of both the deterministic and the nondeterministic version of the FFP. In Sect. 3 the optimization method, an EA combined with simulation-based evaluation, is described. Section 4 presents experiments including comparisons with some simple but effective heuristics, and discusses the results. Section 5 concludes the paper.

2 Problem Definition

The **deterministic** version of the firefighter problem can be formalized in the following way. Let $G = \langle V, E \rangle$ be an undirected graph with N_v vertices. Each of the nodes of G can be in one of the states from a set $L = \{'B', 'D', 'U'\}$ with the interpretation $'B' = $ burning, $'D' = $ defended and $'U' = $ untouched. The state of the graph at time t is denoted S_t and for any vertex v the state of this vertex at time t is $S_t[v]$. Apart from the graph (stored, for example as an adjacency matrix $A_{N_v \times N_v}$) the initial state S_0 of the graph and the number N_f of firefighters assigned per a time step are provided in each FFP instance. Most often the initial state S_0 is constructed by setting the state of vertices from a given non-empty set $\emptyset \neq S \subset V$ to 'B' and the remaining ones to 'U'. Therefore, we can consider an FFP instance as an ordered triple $\langle G, S_0, N_f \rangle$.

The spread of fire is simulated in discrete time steps $t = 0, 1, \ldots$, with the graph set to the initial state S_0 at $t = 0$. In each time step $t > 0$ two events occur. First, a predefined number N_f of still untouched nodes become defended by firefighters (i.e. change their state from 'U' to 'D'). Then the fire spreads along the edges of the graph G from the nodes labelled 'B' to the neighbouring nodes labelled 'U'. Nodes defended by firefighters before fire gets to them (i.e. marked 'D') remain protected until the end of the simulation. Conversely, nodes that catch on fire are considered lost and firefighters are not assigned to them. The simulation ends when fire cannot spread anymore because all nodes adjacent to the burning ones are defended (we say that fire is contained) or when all undefended nodes are burning.

A very often used representation of **solutions** of the FFP is a permutation-based representation, even though other representations are also considered [7]. In the permutation-based representation an individual solution is encoded as a permutation π of numbers $1, \ldots, N_v$. When firefighters are assigned, the first N_f numbers for which the corresponding vertices are untouched ('U') are taken from π and the state of these selected vertices is changed to 'D'. The goal in the **single-objective FFP** is to maximize the number of non-burning nodes ('D' or 'U') at the simulation end. Note, that we are only interested in the state of the nodes. Edges in the graph just define the topology and are not themselves subject to burning nor protecting by firefighters.

In the paper [11] the **multiobjective version of the FFP** was proposed in which there are m values $v_j(v)$, $j = 1, \ldots, m$ assigned to each node v in the graph. The objectives f_j, $j = 1, \ldots, m$ attained by a given solution π are calculated by simulating the spreading of fire from the initial state S_0 until the fire stops spreading. When the final state S_t is reached the objectives are calculated as: $f_j = \sum_{v \in V : S_t[v] \neq 'B'} v_j(v)$, where $v_j(v)$ is the value of the node v according to the j-th criterion.

In this paper we tackle the multiobjective version of the FFP described above, but contrary to the paper [11] we introduce the nondeterminism to the problem.

2.1   Nondeterministic FFP as a Dynamic Optimization Problem

The **nondeterministic** version of the FFP used in this paper can be described similarly as the deterministic one with one key difference that there is an uncertainty in the way the fire spreads (we assume, however, that the initial set S of burning nodes is fixed). For describing the nondeterministic spreading of fire a simple model is used in which, in each time step t, for every edge $e = \langle v_1, v_2 \rangle \in E$ such that $S_t[v_1] = 'B'$ and $S_t[v_2] = 'U'$ the state of the vertex v_2 is set to 'B' with a constant probability $P_{sp} \leq 1$. Formally, assume that the state of the graph is represented as a vector of length N_v of states from the set L and the edges are represented as an adjacency matrix $A = [a_{ij}]_{N_v \times N_v}$ containing $\{0, 1\}$ elements with $a_{ij} = 1$ denoting that there exists an edge between vertices v_i and v_j. Then a procedure that transforms the current state S_t at a time step t to the state S_{t+1} at the next time step can be implemented as shown in Algorithm 1. In this procedure rand() is a function that returns a value randomly drawn from the uniform probability distribution on the $[0, 1)$ range.

Algorithm 1. Nondeterministic spreading of fire in the graph (one time step).

IN: S_t - the state of the graph at a time step t
OUT: S_{t+1} - the state at a time step $t + 1$ (to be calculated by this procedure)

$S_{t+1} := S_t$ // Copy the state to the next time step

for $i := 1, \ldots, N_v$ **do**   // Allow fire to spread
    if $S_t[i] == {}'B'$ **then**
        for $j := 1, \ldots, N_v$ **do**
            if $(A[i, j] == 1)$ and $(S_t[j] == {}'U')$ and $(\text{rand}() < P_{sp})$ **then**
               $S_{t+1}[j] := {}'B'$
           end if
        end for
    end if
end for
  return S_{t+1}

It is worth noticing that in the deterministic version of the FFP the spread of fire in the graph is dynamic, but the optimization problem is, in fact, static. This follows from the fact that once a solution (i.e. a permutation) is selected the spreading of fire can be exactly simulated from the initial state to the very end. Therefore, the optimizer can always work with the initial state and can obtain an exact evaluation regardless if it is a single-objective problem or a multiobjective one. To the contrary, the nondeterminism makes the problem truly dynamic. Because it is not possible to predict the spread of fire with certainty, the solution chosen as the best one may change as time progresses depending on which nodes actually caught on fire and which did not.

As with any dynamic optimization problem two typical approaches are the **offline** and **online** optimization. In the offline approach we try to find the best permutation π of firefighters and then, as fire spreads, we assign firefighters using the same π at each time step. In the online approach the optimizer can take into consideration how the fire did actually spread and which nodes are burning at each time step $t > 0$. Therefore, a different permutation π_t can be produced at a time t step and used for assigning firefighters in this particular time step. In this paper we assume, that the already assigned firefighters cannot be reallocated, so the nodes that were defended remain defended and the new permutation π_t only affects the assignment of firefighters done after this permutation was generated.

3   Simheuristics

The prohibitive computational cost of running exact solvers on larger instances of some combinatorial problems can suggest the use, instead, of heuristics, or metaheuristics such as EAs. We adopt this approach here and use simulation of the spreading fire to enable the candidate solutions proposed by the EA to be evaluated. In previous work on the deterministic version of the FFP the evaluation of a permutation π was performed by simulating the spread of fire in the graph while simultaneously assigning firefighters according to the ordering determined by the permutation π. This approach can also be used for the non-deterministic version of the FFP. In fact, the approach called "simheuristics" – combining simulation with metaheuristic optimization was proposed in a recent survey [8] as a proper approach to nondeterministic optimization. Of course, in the case of nondeterministic problems each run of the simulation can yield different values of the optimization criteria. Therefore, in this paper we adopt an approach in which we perform a number N_{sim} of simulations for a given solution (permutation) π and then average the results. In order to speed up computations, in the experiments described in this paper the simulation routine was implemented for a GPU massively-parallel architecture using the CUDA technology. From the N_{sim} different values of the objectives obtained in these runs we calculate a mean value for each of the objectives $\overline{f_j}$, $j = 1, \ldots, m$. Therefore, the implementation of the optimization algorithm is split into two parts: an EA that runs on the CPU and a simulation routine that runs on the GPU in N_{sim} parallel threads (see Fig. 1).

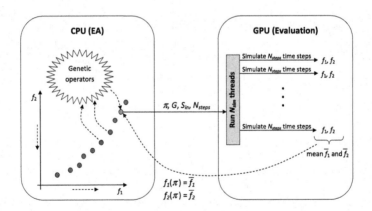

Fig. 1. Implementation of the simheuristic on a CPU/GPU machine.

Each thread simulates the spreading of fire starting from a given initial state S_{in} until a predefined number of time steps N_{steps} is completed or until fire can no longer spread. The working of a single simulation thread is presented in Algorithm 2.

Algorithm 2. The working of a single simulation thread.

IN: S_{in} - an initial state of the graph for the simulation
          π - a solution to test

$t := 0$ // Simulation of the spreading of fire
$S_0 := S_{in}$
while not SpreadingFinished(S_t) and $t < N_{steps}$ **do**
     $S_{t+1} :=$ AssignFirefighters(S_t, π)
     $S_{t+1} :=$ SpreadFire(S_{t+1});
     $t := t + 1$
end while

Set $f_j := 0$ for $j = 1, \dots, m$ // Evaluation of the final state
for $v \in V$ **do**
     if $S_t[v] \neq {}'B'$ **then**
         for $j := 1, \dots, m$ **do**
             $f_j := f_j + v_j(v)$
         end for
     end if
end for

The **SpreadingFinished**(S_t) function used in Algorithm 2 checks if the fire spreading has finished for a given graph state S_t. The spreading of fire is considered finished if there are no untouched nodes adjacent to the burning ones, that is $\neg \exists i, j : (S_t[v_i] = {}'B' \wedge S_t[v_j] = {}'U' \wedge A[i,j] = 1)$, where A is the adjacency matrix of the graph G. The **AssignFirefighters** procedure assigns firefighters to those N_f untouched nodes that are placed first in the permutation π. The **SpreadFire** procedure performs one time step of the nondeterministic fire spreading according to Algorithm 1.

In this paper the Sim-EA algorithm [12] is used that was applied to the multiobjective FFP in a previous paper [13] and was found to outperform a well-known multiobjective optimization algorithm MOEA/D [10]. The Sim-EA is a decomposition-based approach in which several populations $P_1, \dots, P_{N_{sub}}$ perform optimization with respect to scalar objectives obtained by aggregating the original ones using different weight vectors $\lambda^{(1)}, \dots, \lambda^{(N_{sub})}$. In Sim-EA specimens can migrate between populations according to various migration strategies.

In the paper [13] several migration strategies were tested and the "rank" strategy that worked best is used this paper. For each destination subpopulation $P_d, d = 1, \dots, N_{sub}$ all subpopulations P_s, where $s \neq d$ are ranked according to the dot product of weight vectors $\lambda^{(d)} \cdot \lambda^{(s)}$. The source population is selected using the roulette wheel selection with probabilities proportional to the ranks of the subpopulations. Then, N_{imig} best specimens from the source population are taken and merged into the population P_d. In the merge phase each migrated specimen is matched against the currently weakest specimen in the destination population P_d. The new specimen replaces the currently weakest specimen in the destination population P_d if it has a higher value of the objectives aggregated using the weight vector $\lambda^{(d)}$. The main loop of the Sim-EA algorithm is presented in Algorithm 3. It is very similar to the one used in papers [12] and [13], but a different stopping condition is used in this paper. In the previous works the optimization was run for a preset number of generations N_{gen}. In this paper we use a time limit T_{max} in which the main loop of the algorithm has to finish, because we assume, that decisions have to be made in a given amount of time. Also, we are interested in how to use the available time effectively, by either making longer simulation runs (large N_{steps}) or making shorter simulation runs (small N_{steps}) allowing more generations for the EA. Because the EA runs on a regular CPU and simulations are performed in parallel on a GPU it is hard to find a common measure of the amount of computations done on both devices other than the running time limit.

Algorithm 3. The main loop of the Sim-EA algorithm (see also [12] and [13])

IN: $P_1, P_2, \ldots, P_{N_{sub}}$ - populations, one for each search direction $\lambda^{(d)}$
        S_{in} - state of the graph for which to optimize
OUT: $P_1, P_2, \ldots, P_{N_{sub}}$ - populations after evolution

for $d := 1, \ldots, N_{sub}$ **do**   // Initial evaluation
    Evaluate(S_{in}, P_d, $\lambda^{(d)}$)
end for
while not StoppingConditionMet() **do**
    for $d := 1, \ldots, N_{sub}$ **do**   // Genetic operators
        P' := GeneticOperators(P_d)
        Evaluate(S_{in}, P', $\lambda^{(d)}$)
        $P_d := P_d \cup P'$
    end for

    for $d := 1, \ldots, N_{sub}$ **do**   // Source populations
        s := SelectSourcePopulation(d)
        P'_d := the N_{imig} best specimens from P_s
    end for
    for $d := 1, \ldots, N_{sub}$ **do**   // Migration
        for $x \in P'_d$ **do**
            Evaluate(S_{in}, $\{x\}$, $\lambda^{(d)}$)
            w := the weakest specimen in P_d
            $P_d := P_d - \{w\}$
            b := BinaryTournament($w, x, \lambda^{(d)}$)
            $P_d := P_d \cup \{b\}$
        end for
    end for

    for $d := 1, \ldots, N_{sub}$ **do**   // Elitist selection
        e := the best specimen in P_d
        P_d := Select($P_d \backslash \{e\}$, $N_{pop} - 1$)
        $P_d := P_d \cup \{e\}$
    end for
end while

The main loop of the Sim-EA uses the following procedures:

StoppingConditionMet() – A function that determines if the stopping condition is satisfied, such as the number of generations completed, maximum running time, etc.

GeneticOperators(P) – Applies the crossover and mutation to specimens in P and returns the new specimens. In papers [11,13] a mechanism for adjusting probabilities of the application of several different crossover and mutation operators was introduced. The same mechanism is used in this paper. The same 10 crossover and 5 mutation operators were used as in the previous papers. The crossover operators were: Cycle Crossover (CX), Linear Order Crossover (LOX), Merging Crossover (MOX), Non-Wrapping Order Crossover (NWOX), Order Based Crossover (OBX), Order Crossover (OX), Position Based Crossover (PBX), Partially Mapped Crossover (PMX), Precedence Preservative Crossover (PPX) and Uniform Partially Mapped Crossover (UPMX). The mutation operators were: displacement mutation, insertion mutation, inversion mutation, scramble mutation and transpose mutation.

Evaluate(S_{in}, P, λ) – Evaluates specimens from a given set using N_{sim} parallel simulations performed on a GPU as shown in Fig. 1 starting at the state S_{in}. In the simulation estimates of the average values $\overline{f_1}, \ldots, \overline{f_m}$ of m objectives are obtained. The fitness of the specimens is then calculated by aggregating these objectives using a given weight vector λ.

SelectSourcePopulation(d) – Selects a population P_s to migrate specimens from into P_d. Various strategies can be used for selecting the source population P_s. In the paper [13] several strategies were tested and the one that worked best with the FFP turned out to be to select the source population using the roulette wheel selection procedure based on ranking calculated using the dot product of weight vectors $\lambda^{(d)} \cdot \lambda^{(s)}$.

BinaryTournament(s_1, s_2, λ) – compares two specimens s_1 and s_2 according to the fitness calculated by aggregating the m objectives of the specimens using a given weight vector λ. Returns the winning specimen.

Select(P, n) – selects n specimens from a given population P using binary tournament selection method.

Because this paper concerns a dynamic optimization problem, the optimization can be performed either in an offline or in an online mode. In the offline mode one long optimization run is performed based on the initial state of the graph. The best permutation π is selected and then, as fire spreads, firefighters are assigned according to this permutation in increments of N_f per a time step. Because in this paper a multiobjective problem is concerned, we select a different permutation $\pi^{(d)}$, $d = 1, \ldots, N_{sub}$ for each optimization direction $\lambda^{(d)}$. In the offline mode we simulate the spreading of fire for each optimization direction $\lambda^{(d)}$ separately starting from the initial state of the graph and assigning firefighters using $\pi^{(d)}$. The final evaluation E_d along each direction $\lambda^{(d)}$ is equal to the sum of the values of the non-burning nodes weighted using $\lambda^{(d)}$ when the simulation ends. The offline optimization is presented in Algorithm 4.

Algorithm 4. The optimization in the offline mode.

---

for $d := 1, \ldots, N_{sub}$ **do**
    $P_d := \text{InitPopulation}(N_{pop})$
end for
$\text{Evolve}(\{P_1, P_2, \ldots, P_{N_{sub}}\}, S_0, T_{max})$

for $d := 1, \ldots, N_{sub}$ **do**
    $\pi^{(d)} := \text{SelectBestSolution}(P_d)$

    $t := 0$ // Simulation of the spreading of fire
    $S_t^{(d)} := S_0$
    while not $\text{SpreadingFinished}(S_t^{(d)})$ **do**
        $S_{t+1}^{(d)} := \text{AssignFirefighters}(S_t^{(d)}, \pi^{(d)})$
        $S_{t+1}^{(d)} := \text{SpreadFire}(S_{t+1}^{(d)})$;
        $t := t + 1$
    end while

    $E_d := 0$ // Evaluation of the final state for the d-th subproblem
    for $v \in V$ **do**
        if $S_t^{(d)}[v] \neq \text{'B'}$ **then**
            for $j := 1, \ldots, m$ **do**
                $E_d := E_d + \lambda_j^{(d)} v_j(v)$
            end for
        end if
    end for
end for

In the online mode the optimization is performed at each time step for a short period of time. The best currently known permutation is selected and firefighters are assigned using this permutation when fire spreads in the current time step. For the multiobjective problem a different permutation $\pi_t^{(d)}$ is selected for each optimization direction $\lambda^{(d)}$ at each time step t. Because of that we have to keep the current state $S_t^{(d)}$ of the simulation at the time step t for each search direction $\lambda^{(d)}$. The final evaluation E_d along each direction $\lambda^{(d)}$ is equal to the sum of the values of the non-burning nodes in the state $S_t^{(d)}$ weighted using $\lambda^{(d)}$ when the simulation ends. The optimization in the online mode is presented in Algorithm 5.

The Algorithms 4 and 5 use the following procedures:

InitPopulation(n) - Creates a given number n of new specimens initialized as random permutations.

Evolve$(\{P_1, P_2, \ldots, P_{N_{sub}}\}, S_t, T_{max})$ - Runs the main loop of the Sim-EA algorithm described in Algorithm 3 with the maximum running time T_{max} as the stopping criterion. This run of the algorithm optimizes solutions based on a given graph state S_t.

SelectBestSolution(P_d) - Selects the best solution from a given population P_d with respect to fitness calculated using the weight vector $\lambda^{(d)}$.

Algorithm 5. The optimization in the online mode.

$t := 0$
for $d := 1, \ldots, N_{sub}$ **do**
    $P_d := \text{InitPopulation}(N_{pop})$
    $S_t^{(d)} := S_0$
end for

while $\exists d$ not SpreadingFinished$(S_t^{(d)})$ **do**
    Evolve$(\{P_1, P_2, \ldots, P_{N_{sub}}\}, S_t^{(d)}, T_{max})$
    for $d := 1, \ldots, N_{sub}$ **do**
        if not SpreadingFinished$(S_t^{(d)})$ **then**
            $\pi_t^{(d)} := \text{SelectBestSolution}(P_d)$
            $S_{t+1}^{(d)} := \text{AssignFirefighters}(S_t^{(d)}, \pi_t^{(d)})$
            $S_{t+1}^{(d)} := \text{SpreadFire}(S_{t+1}^{(d)})$
        end if
    end for
    $t := t + 1$
end while

for $d := 1, \ldots, N_{sub}$ **do**
    $E_d := 0$ // Evaluation of the final state for the d-th subproblem
    for $v \in V$ **do**
        if $S_t^{(d)}[v] \neq \text{'B'}$ **then**
            for $j := 1, \ldots, m$ **do**
                $E_d := E_d + \lambda_j^{(d)} v_j(v)$
            end for
        end if
    end for
end for

SpreadingFinished(S_t) - Checks if the fire spreading has finished for a given graph state S_t. The spreading of fire is considered finished if there are no untouched nodes adjacent to the burning ones, that is $\neg\exists i, j : S_t[v_i] = \text{'B'} \land S_t[v_j] = \text{'U'} \land A[i, j] = 1$, where A is the adjacency matrix of the graph G.

AssignFirefighters(S_t, π) - Modifies the state S_t by assigning firefighters to those N_f untouched nodes that are placed first in the permutation π.

SpreadFire(S_t) - Performs one time step of the nondeterministic fire spreading according to Algorithm 1.

4 Experiments and Results

The experiments were aimed at investigating three issues: comparing the online and offline optimization, comparing the two with some simple heuristics that determine how to place firefighters and determining the influence of the N_{steps} parameter on the quality of the obtained results. The larger the N_{steps} parameter the longer the simulation used for evaluating specimens in the EA. Therefore, we assumed that the entire main loop of the EA can run for at most T_{max} seconds and when N_{steps} is smaller more generations of the EA can fit within the same time limit.

The time limit for the online optimization was set to $T_{max} = 60$ s per a time step and to $T_{max} = 300$ s for the offline optimization. The equipment used for experiments was the Intel Q6600 CPU running at 2.4 GHz with 4 GB of RAM with a 470 GTX GPU with 1.25 GB of RAM. However, the memory sizes did not play an important role in the experiments, because the actual memory usage was far lower than the available maximum. The number of simulations run in parallel was set to $N_{sim} = 200$. It is worth mentioning, that even on moderately advanced 470 GTX GPUs such number of threads can easily run in parallel, so the running time of the simulations cannot be significantly decreased by just lowering the N_{sim} number.

Data sets used in the experiments were created in the same way as used in a previous paper on the FFP [13]. In these data sets the graph G was generated by randomly determining, for each pair of vertices v_i, v_j, if there exists an edge $\langle v_i, v_j \rangle$. The probability of generating an edge was set to $P_{edge} = 2.5/N_v$, where N_v was the number of vertices. This value was used in order to ensure that the mean number of edges adjacent to a vertex was similar for all the instances.

Costs $v_j(v)$, $j = 1, \ldots, m$ assigned to vertices of the graph G were generated by drawing pairs of random values with the uniform probability on a triangle formed in $\mathbb{R}^2$ by points $[0, 0]$, $[0, 100]$, $[100, 0]$. With this method of cost assignment individual costs fall in the range $[0, 100]$ but also the sum of costs associated with a single vertex cannot exceed 100. Therefore it is not possible to maximize both objectives at the same time, so the ability of the algorithm to find good trade-offs can be tested.

The number of initially burning nodes in each graph was set to $N_s = 1$ and the number of firefighters to assign in each time step was set to $N_f = 2$. Because of the higher computational cost of the experiments involving several values of the P_{sp} and N_{steps} parameters the instances used in this paper were smaller than in [13] with $N_v = 30, 40, 50, 75, 100$ and 125. For each graph size N_v, 30 test instances $I_{N_v}^{(1)}, \ldots I_{N_v}^{(30)}$ were prepared following the procedure described above to allow comparing the algorithms on different, but fixed, test cases.

The nondeterminism is involved in how the fire spreads in the graph, therefore the same graph structure can be used for different values of P_{sp}. In this paper five values $P_{sp} = 0.3, 0.5, 0.7, 0.9$ and 1.0 were used. For the length of the simulation $N_{steps} = 2, 4, 6, 8, 10$ and ∞ were used. The last value of $N_{steps} = \infty$ causes the simulation to run until the fire can no longer spread regardless of how many time steps it takes.

The number of subpopulations was $N_{sub} = 20$ with each subpopulation size $N_{pop} = 100$. The number of specimens migrating between the populations was set to $N_{imig} = 0.1 \cdot N_{pop} = 10$. Similarly as in the paper [13] a set of 10 crossover operators and 5 mutation operators was used in the experiments with an autoadaptation mechanism used for adjusting probabilities of the usage of the operators. The number of new specimens generated by the crossover operator N_{cross} was equal to the population size N_{pop} and the probability of mutation was set to $P_{mut} = 0.05$. The minimum probability of selecting a particular operator in the auto-adaptation mechanism was set to $P_{min} = 0.02$ for crossover auto-adaptation and to $P_{min} = 0.05$ for mutation auto-adaptation.

For each mode of optimization (online and offline) and for each P_{sp}, N_{steps} pair 30 repetitions of the optimization were performed. In each repetition the E_d values, calculated as shown in Algorithms 4 and 5 for $d = 1, \ldots, N_{sub}$ were stored.

The performance of the optimization algorithms was compared to heuristics that suggest where to place firefighters. Three different heuristics were tested:

- **Max degree** - Assign firefighters to those nodes first that have higher degrees among untouched ('U') nodes.
- **Max degree (adjacent)** - Assign firefighters to those nodes first that have higher degrees among those untouched ('U') nodes that are adjacent to burning ('B') nodes.
- **BFS** - Perform a breadth-first search (BFS) in the graph starting at the burning nodes that finds possible paths from the burning ('B') nodes to the untouched ('U') ones. For each untouched node v determine the smallest number of steps s it takes to get to this node from a burning one. Assign a value of $P_{BFS} = (P_{sp})^s$ to node v. Assign firefighters to those nodes first that have higher values of P_{BFS}.

These heuristics were used in each time step to select the best N_f locations for firefighters. If in one time step more than N_f nodes had the maximum score according to the selected heuristic, N_f nodes were selected randomly from them. An evaluation of the performance of the heuristics was done in a similar way to the evaluation of the offline optimizer (see Algorithm 4). For each optimization direction $\lambda^{(d)}$, $d = 1, \ldots, N_{sub}$, a simulation of fire was performed with a selected heuristic used for allocating firefighters. When simulation finished, the score E_d was calculated by adding cost values assigned to non-burning nodes weighted using $\lambda^{(d)}$. Similarly as with the optimization algorithms, the tests with the heuristics were repeated 30 times. In the k-th of the 30 repetitions of the tests the same test instance $I_{N_v}^{(k)}$ was used with both modes of optimization, all heuristics and all P_{sp} and N_{steps} values. Therefore, a range of methods was tested on a diversified set of 30 test instances, but each method worked in the same starting conditions.

Median values of the E_d score (see Algorithms 4 and 5) are presented in Tables 1, 2, 3, 4, 5 and 6. The best result obtained by a given type of methods (online, offline, heuristic) for a given P_{sp} is marked by an arrow '→'. Table 7 shows the results of a statistical comparison of the optimizers performed using the Wilcoxon test at $\alpha = 0.05$. This comparison was performed between the 30 measurements obtained for the best value of N_{steps} for each of the optimizers. For smaller instances ($N \leq 50$) both optimization approaches are often comparable, even though the online method was significantly better four times (the offline one just once). For larger instances ($N \geq 75$) the online optimizer outperformed the offline one 12 times (with 3 cases undecided, none in favour of the offline optimizer).

Table 1. Median values of the E_d score obtained for 30 nodes.

Method		$P_{sp} = 0.3$	$P_{sp} = 0.5$	$P_{sp} = 0.7$	$P_{sp} = 0.9$	$P_{sp} = 1.0$
Online	$N_{steps} = 2$	1364.81	1228.16	959.29	703.56	676.03
optimization	$N_{steps} = 4$	1386.63	1253.57	→ 1096.32	→ 914.42	→ 855.57
	$N_{steps} = 6$	→ 1388.47	1254.53	1064.32	889.36	826.10
	$N_{steps} = 8$	1382.90	→ 1276.79	1081.58	887.28	801.36
	$N_{steps} = 10$	1363.90	1241.03	1061.86	871.30	823.18
	$N_{steps} = \infty$	1382.90	1269.30	1088.25	882.06	828.57
Offline	$N_{steps} = 2$	1189.17	817.99	653.77	566.83	534.29
optimization	$N_{steps} = 4$	1329.74	1156.53	892.96	831.48	791.11
	$N_{steps} = 6$	→ 1341.22	1176.64	984.11	→ 888.73	866.42
	$N_{steps} = 8$	1323.47	1160.76	968.56	880.54	→ 869.50
	$N_{steps} = 10$	1322.26	→ 1187.83	→ 1010.22	886.86	859.31
	$N_{steps} = \infty$	1330.42	1155.04	965.01	885.53	819.07
Heuristics	Max. degree	850.32	676.07	549.14	416.60	385.08
	Max. degree (adjacent)	→ 1352.98	→ 1188.69	→ 907.59	→ 722.75	→ 703.93
	BFS	1294.78	768.77	752.02	562.32	488.12

Table 2. Median values of the E_d score obtained for 40 nodes.

Method		$P_{sp} = 0.3$	$P_{sp} = 0.5$	$P_{sp} = 0.7$	$P_{sp} = 0.9$	$P_{sp} = 1.0$
Online	$N_{steps} = 2$	1795.27	1545.76	903.35	720.89	661.38
optimization	$N_{steps} = 4$	1803.58	→ 1677.30	→ 1294.07	→ 1002.20	→ 923.94
	$N_{steps} = 6$	→ 1806.78	1675.89	1114.45	913.54	872.59
	$N_{steps} = 8$	1800.09	1541.41	1064.09	895.61	872.05
	$N_{steps} = 10$	1795.85	1378.15	1049.11	901.01	877.04
	$N_{steps} = \infty$	1759.37	1420.91	1047.97	896.04	886.23
Offline	$N_{steps} = 2$	1470.07	925.41	723.90	617.66	574.65
optimization	$N_{steps} = 4$	1762.00	1310.05	975.70	846.58	843.00
	$N_{steps} = 6$	→ 1765.74	1512.08	→ 1101.16	→ 968.91	921.43
	$N_{steps} = 8$	1764.88	→ 1561.17	1051.66	938.20	920.22
	$N_{steps} = 10$	1729.15	1373.84	1055.34	948.79	→ 939.76
	$N_{steps} = \infty$	1720.21	1418.45	1032.26	951.91	904.45
Heuristics	Max. degree	1120.14	724.68	583.10	481.13	442.67
	Max. degree (adjacent)	1766.37	→ 1771.33	→ 1019.21	→ 813.96	→ 809.71
	BFS	→ 1848.82	894.25	825.46	552.55	539.19

Another aspect of the optimization is the number of the steps N_{steps} for which the simulation is carried out. The shorter the simulation the more generations can the EA perform in the same amount of time. Figure 2 shows how many times the best result was obtained for $N_{steps} = 2, 4, 6, 8, 10, \infty$. Clearly, letting

Table 3. Median values of the E_d score obtained for 50 nodes.

Method		$P_{sp} = 0.3$	$P_{sp} = 0.5$	$P_{sp} = 0.7$	$P_{sp} = 0.9$	$P_{sp} = 1.0$
Online	$N_{steps} = 2$	1300.12	626.98	503.00	431.68	408.95
optimization	$N_{steps} = 4$	→ 1386.02	679.02	615.17	→ 579.97	562.21
	$N_{steps} = 6$	1385.33	→ 875.73	→ 670.68	567.47	→ 563.33
	$N_{steps} = 8$	1269.63	795.62	634.16	575.82	556.82
	$N_{steps} = 10$	1313.69	784.41	634.71	577.18	553.11
	$N_{steps} = \infty$	1071.02	771.25	640.98	570.84	560.65
Offline	$N_{steps} = 2$	850.63	601.22	471.63	412.41	390.39
optimization	$N_{steps} = 4$	1081.48	718.70	610.37	572.84	552.81
	$N_{steps} = 6$	→ 1128.72	→ 799.05	→ 654.90	601.09	604.99
	$N_{steps} = 8$	1093.90	750.79	625.35	598.93	613.88
	$N_{steps} = 10$	1029.41	714.61	626.82	→ 604.06	611.61
	$N_{steps} = \infty$	921.29	698.84	629.26	601.70	→ 614.32
Heuristics	Max. degree	716.63	477.24	399.16	336.53	296.22
	Max. degree (adjacent)	→ 1419.40	→ 803.09	→ 577.10	→ 437.66	→ 440.70
	BFS	1275.95	581.44	473.22	337.26	342.84

Table 4. Median values of the E_d score obtained for 75 nodes.

Method		$P_{sp} = 0.3$	$P_{sp} = 0.5$	$P_{sp} = 0.7$	$P_{sp} = 0.9$	$P_{sp} = 1.0$
Online	$N_{steps} = 2$	956.88	614.59	487.15	415.99	389.99
optimization	$N_{steps} = 4$	981.13	620.49	531.32	571.25	550.45
	$N_{steps} = 6$	992.92	774.51	→ 685.25	→ 629.67	→ 596.58
	$N_{steps} = 8$	959.63	806.06	674.45	612.49	595.02
	$N_{steps} = 10$	1090.14	795.37	677.82	608.71	596.54
	$N_{steps} = \infty$	→ 1092.12	→ 807.45	671.97	610.85	593.34
Offline	$N_{steps} = 2$	922.30	640.93	522.01	454.48	424.78
optimization	$N_{steps} = 4$	→ 1042.61	700.90	608.54	567.38	587.20
	$N_{steps} = 6$	961.70	→ 715.41	→ 633.43	→ 585.37	600.81
	$N_{steps} = 8$	931.17	691.37	602.31	573.06	604.34
	$N_{steps} = 10$	895.80	676.03	598.90	579.47	→ 606.50
	$N_{steps} = \infty$	868.76	680.01	598.87	569.24	604.17
Heuristics	Max. degree	849.68	570.76	440.48	343.04	341.51
	Max. degree (adjacent)	→ 2272.92	→ 1114.03	→ 617.77	→ 526.89	→ 485.55
	BFS	2269.66	664.75	491.11	400.65	381.74

the simulation run without limit can deteriorate the results of the optimization. For larger instances ($N_v = 75, 100$ and 125) the unlimited simulation length worked best for $P_{sp} = 0.3$ and 0.5, but for larger values of P_{sp} limiting the duration of simulations worked better. Although not conclusive, this observation may

Table 5. Median values of the E_d score obtained for 100 nodes.

Method		$P_{sp} = 0.3$	$P_{sp} = 0.5$	$P_{sp} = 0.7$	$P_{sp} = 0.9$	$P_{sp} = 1.0$
Online	$N_{steps} = 2$	908.03	608.99	487.98	422.16	398.28
optimization	$N_{steps} = 4$	914.67	599.72	490.98	493.24	518.26
	$N_{steps} = 6$	867.02	709.03	694.02	611.86	→ 583.91
	$N_{steps} = 8$	947.36	840.08	694.69	609.91	581.56
	$N_{steps} = 10$	1090.21	839.29	→ 699.85	→ 613.16	576.68
	$N_{steps} = \infty$	→ 1154.59	→ 843.16	697.52	609.95	577.14
Offline	$N_{steps} = 2$	926.14	676.89	545.51	477.67	446.36
optimization	$N_{steps} = 4$	→ 946.40	665.64	551.58	512.98	500.06
	$N_{steps} = 6$	887.11	661.15	→ 597.94	→ 558.89	563.46
	$N_{steps} = 8$	881.87	→ 691.41	589.34	535.96	→ 564.41
	$N_{steps} = 10$	882.68	663.10	583.07	536.55	561.50
	$N_{steps} = \infty$	904.78	669.16	584.03	539.25	562.49
Heuristics	Max. degree	871.62	582.74	467.80	379.39	375.07
	Max. degree (adjacent)	→ 3024.79	→ 862.67	→ 600.93	→ 515.56	→ 443.76
	BFS	974.80	648.55	505.87	419.71	385.88

Table 6. Median values of the E_d score obtained for 125 nodes.

Method		$P_{sp} = 0.3$	$P_{sp} = 0.5$	$P_{sp} = 0.7$	$P_{sp} = 0.9$	$P_{sp} = 1.0$
Online	$N_{steps} = 2$	963.07	637.80	519.84	440.04	397.65
optimization	$N_{steps} = 4$	868.58	600.06	501.59	461.53	482.77
	$N_{steps} = 6$	906.73	692.04	723.36	→ 660.97	→ 630.15
	$N_{steps} = 8$	917.48	893.64	756.26	658.30	625.60
	$N_{steps} = 10$	1108.00	907.41	→ 766.36	653.44	623.34
	$N_{steps} = \infty$	→ 1230.37	→ 912.10	751.03	656.34	621.58
Offline	$N_{steps} = 2$	→ 961.28	688.38	571.53	491.56	462.26
optimization	$N_{steps} = 4$	953.17	662.25	532.82	483.99	491.69
	$N_{steps} = 6$	891.00	653.68	→ 601.89	→ 566.59	559.24
	$N_{steps} = 8$	905.52	701.63	597.22	538.33	559.82
	$N_{steps} = 10$	943.15	→ 711.44	594.74	536.91	→ 563.81
	$N_{steps} = \infty$	936.90	701.85	597.07	539.40	559.68
Heuristics	Max. degree	917.63	587.32	500.42	414.06	377.07
	Max. degree (adjacent)	3957.17	→ 844.74	→ 672.27	→ 537.26	→ 503.02
	BFS	→ 3975.47	703.07	551.18	444.86	415.94

indicate that for lower values of P_{sp} it is harder to predict the short-time movement of fire and thus averaged behaviour observed in the longer run becomes more important. The optimization algorithms were also compared with heuristic methods described in Sect. 4. Clearly, the best of the tested heuristics is the one that considers placing the firefighters adjacently to the burning nodes (see Fig. 3). Statistical comparison of optimization algorithms with heuristics shown that the "Max degree (adjacent)" heuristic is a very effective strategy when the

Table 7. The results of a statistical comparison of the online and offline optimization modes: 'N' - online better, 'F' - offline better, '=' - no statistical difference.

N_v	$P_{sp} = 0.3$	$P_{sp} = 0.5$	$P_{sp} = 0.7$	$P_{sp} = 0.9$	$P_{sp} = 1.0$
30	=	N	N	=	=
40	N	=	=	=	=
50	=	N	=	=	F
75	=	N	N	N	=
100	N	N	N	N	=
125	N	N	N	N	N

probability of the spreading of fire is low ($P_{sp} \leq 0.5$). It was significantly better than the online optimizer in 7 out of 12 cases, significantly worse in 3 cases, with 2 cases undecided (p-value > 0.05). Compared to the offline optimizer it was significantly better in 11 cases, with only one case undecided. On the other hand, in the tests for $P_{sp} \geq 0.7$ the heuristic was significantly worse than the online optimizer 15 times out of 18 (3 cases undecided). Compared to the offline optimizer it was significantly worse 13 times and significantly better once (4 cases undecided).

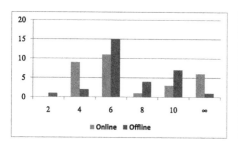

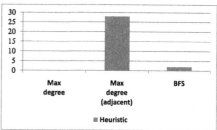

Fig. 2. The number of times each value of N_{steps} produced the best result.

Fig. 3. The number of times each heuristic produced the best result.

These results might also indicate the fact that for lower values of P_{sp} it is harder to predict the short-time movement of fire. In such situation heuristics may be more effective than trying to assess what location of firefighters will be best based on simulations with a high level of uncertainty.

5   Conclusion

In this paper the multiobjective nondeterministic firefighter problem is studied. For solving this problem the Sim-EA multipopulation EA is applied that was shown in [13] to outperform the MOEA/D algorithm on the FFP. Solving the nondeterministic FFP requires dynamic optimization and therefore the effectiveness of the algorithm in the online and offline optimization modes is compared. For small problem instances both approaches are comparable, but for larger ones the online approach seems better. Also, it was tested how long to run the simulation that evaluates candidate solutions. Limiting the length of simulation runs often improves the results, but for low P_{sp} longer simulations may be better. Evolutionary optimization was compared with several heuristic approaches that use simple rules for allocating firefighters. The heuristic that selects nodes with a high degree placed adjacently to the already burning ones worked by far the best. It significantly outperformed the online optimizer in several cases for $P_{sp} \leq 0.5$ and the offline optimizer in all cases except one. For $P_{sp} \geq 0.7$, however, the heuristic approach performed poorly. The observations concerning the length of simulation runs and comparison of optimization and heuristics seem to indicate that for low P_{sp} it is hard to obtain good evaluations of solutions using simulations. Because different types of methods work best for different probabilities of the spreading of fire, further work on the nondeterministic FFP may concern hybrid methods combining heuristics with online optimization.

References

1. Blum, C., Blesa, M.J., García-Martínez, C., Rodríguez, F.J., Lozano, M.: The firefighter problem: application of hybrid ant colony optimization algorithms. In: Blum, C., Ochoa, G. (eds.) EvoCOP 2014. LNCS, vol. 8600, pp. 218–229. Springer, Heidelberg (2014)
2. Comellas, F., Mitjana, M.: Broadcasting in small-world communication networks. In: Kaklamanis, C., Kirousis, L. (eds.) 9th International Colloquium on Structural Information and Communication Complexity, pp. 73–85. Carleton Scientific, Waterloo (2002)
3. Develin, M., Hartke, S.G.: Fire containment in grids of dimension three and higher. Discrete Appl. Math. **155**(17), 2257–2268 (2007)
4. Feldheim, O.N., Hod, R.: 3/2 firefighters are not enough. Discrete Appl. Math. **161**(1–2), 301–306 (2013)
5. García-Martínez, C., et al.: The firefighter problem: empirical results on random graphs. Comput. Oper. Res. **60**, 55–66 (2015)
6. Hartnell, B.: Firefighter! an application of domination. In: 20th Conference on Numerical Mathematics and Computing (1995)
7. Hu, B., Windbichler, A., Raidl, G.R.: A new solution representation for the firefighter problem. In: Ochoa, G., Chicano, F. (eds.) EvoCOP 2015. LNCS, vol. 9026, pp. 25–35. Springer, Heidelberg (2015)
8. Juan, A.A.: A review of simheuristics: extending metaheuristics to deal with stochastic combinatorial optimization problems. Oper. Res. Perspect. **2**, 62–72 (2015)
9. Kumar, R., et al.: Stochastic models for the web graph. In: Proceedings of the 41st Annual Symposium on Foundations of Computer Science, FOCS 2000, pp. 57–65. IEEE Computer Society (2000)
10. Li, H., Zhang, Q.: Multiobjective optimization problems with complicated pareto sets, MOEA/D and NSGA-II. IEEE Trans. Evolut. Comput. **13**(2), 284–302 (2009)

11. Michalak, K.: Auto-adaptation of genetic operators for multi-objective optimization in the firefighter problem. In: Corchado, E., Lozano, J.A., Quintián, H., Yin, H. (eds.) IDEAL 2014. LNCS, vol. 8669, pp. 484–491. Springer, Heidelberg (2014)
12. Michalak, K.: Sim-EA: an evolutionary algorithm based on problem similarity. In: Corchado, E., Lozano, J.A., Quintián, H., Yin, H. (eds.) IDEAL 2014. LNCS, vol. 8669, pp. 191–198. Springer, Heidelberg (2014)
13. Michalak, K.: The Sim-EA algorithm with operator autoadaptation for the multiobjective firefighter problem. In: Ochoa, G., Chicano, F. (eds.) EvoCOP 2015. LNCS, vol. 9026, pp. 184–196. Springer, Heidelberg (2015)

Benchmarking Dynamic Three-Dimensional Bin Packing Problems Using Discrete-Event Simulation

Ran Wang[1]([✉]), Trung Thanh Nguyen[1], Shayan Kavakeb[1], Zaili Yang[1], and Changhe Li[2]

[1] Liverpool Logistics Offshore and Marine Research Institute (LOOM),
School of Engineering, Technology and Maritime Operations,
Liverpool John Moores University, Liverpool L3 3AF, UK
R.Wang@2015.ljmu.ac.uk, {T.T.Nguyen,S.Kavakeb,Z.Yang}@ljmu.ac.uk
[2] School of Computer Science, China University of Geosciences, Wuhan, China
Changhe.lw@gmail.com

Abstract. In this paper a framework is developed to generate benchmark problems for dynamic three-dimensional (3D) bin packing problems (BPPs). This framework is able to generate benchmark problems for different variants of BPPs by taking into account potential uncertainty in real-world BPPs, which are uncertainties in dimensions, costs, weights of upcoming items. This paper has three main contributions. First, a benchmark generator framework is developed for the first time using an open source discrete-event simulation platform. This framework generates benchmark problems for BPPs by reproducing uncertainty in real-world BPPs. Second, this framework can be integrated with any dynamic BPP algorithm so that the optimisation algorithm can be run alongside the simulation to solve dynamic BPPs. Third, various performance measures from the literature are included in the framework to evaluate the optimisation algorithms from different perspectives. Thanks to the 3D visualisation feature of this framework, the optimisation results can also be observed visually. Finally, empirical experiments on a real-world BPP are conducted to verify these contributions.

Keywords: Benchmarking · Bin packing problem · Dynamic optimisation · Simulation

1 Introduction

Three-dimensional (3D) bin packing problems (BPPs) are NP-hard optimisation problems in which a number of boxes are packed into one or multiple 3D bins. Depending on the characteristics of the problem, different objectives can be defined for BPPs such as input minimisation or output maximisation. Input minimisation aims to find the minimum total cost or minimum number of bins. The size of bins can be either identical [1–3] or varied [4–8]. The aim of output

© Springer International Publishing Switzerland 2016
G. Squillero and P. Burelli (Eds.): EvoApplications 2016, Part II, LNCS 9598, pp. 266–279, 2016.
DOI: 10.1007/978-3-319-31153-1_18

maximisation is maximising the volume or number of packed boxes given the limited number of bins [9–12]. Some BPPs have multiple objectives: minimising the cost while packing items with preferences [8]; minimising the cost and also packing items with the same destinations together [13]. In terms of size, items may be identical, weakly heterogeneous (i.e. many items but a few item types), or strongly heterogeneous (i.e. a few items but many item types). Regarding constraints on BPPs, Bortfeldt [14] introduced various constraints on potential containers, items, loading and allocation.

In the literature the BPPs are normally considered in an ideal situation with no uncertainty. In reality, however, uncertainty is a frequent feature of real-world BPPs. Hence, we need to take uncertainty into account in academic research. To address uncertainty to a variety of BPPs, it is necessary to have data set for each problem based on different constraints and features of the problem. In this paper, we propose a new approach of using discrete-event simulation to generate benchmark problems and to evaluate dynamic optimisation algorithms including evolutionary algorithms. To the best of our knowledge this is the first time such an approach is proposed. It is expected that the new approach will help researchers generate dynamic problems and evaluate their algorithms easier than before.

This paper is structured as follows. Section 2 reviews benchmark problems for BPPs, possible uncertainties in BPPs and the performance measures used in the literature. Section 3 describes the structure and functions of the proposed framework. It then proposes a group of test cases for BPPs. Section 4 carries out an experiment for a real-world BPP using the proposed framework including a dynamic BPP algorithm. The experimental results are then provided. Section 5 concludes this paper.

2   Literature Review

2.1   Benchmarks

Real-world optimisation problems are subject to uncertainty and unknown changes. To approach these problems, dynamic algorithms have been developed to the problems as time goes by. As changes occur in an optimisation problem, the dynamic optimisation algorithm need to react to the changes to produce a new optimal solution in regards to the changes. Due to the complexity of uncertainty in real-world problems, it is difficult to evaluate dynamic optimisation algorithms effectively. In academic research on dynamic benchmark problems, attempts are made to capture changes that reflect different characteristics of real problems. These benchmarks can then be used as a basis to evaluate algorithms. However, generating such benchmark problems is not trivial due to difficulty in capturing and reproducing uncertainty in real-world problems. This difficulty arises due to the lack of enough knowledge about uncertainty and also due to the difficulty in formulating/modelling uncertainty.

In general, as in Nguyen et al. [15], a decent benchmark problem for uncertain optimisation problems should encompass the following characteristics: (1) the flexibility of setting objectives, problem uncertainties and dimensions;

(2) simplicity and efficiency of problem implementation and evaluation; (3) allowing conjectures to real-world problems. In the literature, there have been some attempts to provide benchmark problems for uncertain optimisation problems. Nguyen et al. [15] reviewed these benchmark problems based on the following properties: (1) time-linkage; (2) predictability of changes; (3) objective types and numbers; (4) constraint types and frequency; (5) factors changing other parameters. This survey paper reveals that most existing academic benchmark problems focused only on some of the above-mentioned criteria due to complexity of these problems. Nevertheless, there have been few publications that attempted to consider some of these criteria by taking into account changeable constraints or the situation where future events depend on previous solutions [16–24]. These attempts, however, still have some limitations to capture properly the uncertainty in real-world problems. Thus, these limitations in benchmark problems make it very difficult to evaluate robust/dynamic algorithms properly and hence it is not very clear that the developed robust/dynamic optimisation algorithms in the literature can effectively tackle real-world optimisation problems. This indicates a gap of knowledge between real-world optimisation problems and academic research.

2.2   Uncertainty

This section reviews uncertainty in BPPs by introducing a real case from an industrial partner in the first place. In this problem, there is a set of items with different sizes to be packed into containers whose sizes (20, 40, 45 feet) and structures (closed, open or flat rack) depend on the sizes of the items. Items include boxes, anchors, tubes, and some other odd shape items fixed on a palette. The cost of hiring containers varies depending on their sizes and structures. Assuming the number of containers is unlimited, our objective is to minimise the total hiring cost of containers in order to pack all needed items. This problem, similar to other real-world optimisation problems, is subject to uncertainty. For example, some items can be tied together into a bundle. How many items, and how they can be tied together into one bundle, is uncertain. The size of each bundle is also uncertain, depending on how items are tied together. There is also an uncertainty of the number of items that arrive at any single time, as detailed below.

One of the most common uncertainties in literature is the uncertain characteristic of upcoming items. It was mentioned in Peng and Zhang [25] that the volumes of items and the capacities of bins are only approximately known due to economic restrictions or technical limitations. Also in Perboli et al. [26] the item profits were considered as random variables because of the handling operations. Another uncertainty proposed in Crainic [27] is the unknown future demand of items. It may generate extra cost or loss profit that the planned containers are insufficient. Thus, due to the variety of uncertainty, it is difficult to generate test case depending on different problem requirements.

2.3   Performance Measures

According to the literature, many performance measures used in existing experiments are specific for each algorithm. It means that while tackling the same problem, it is hard to compare the performance of different algorithms. With the purpose of providing a framework that different algorithms of BPP can be compared, common performance measures are essential. The volume utilisation[1] is mentioned in several experiments [28–30] which is either total volume utilisation or the average. In addition, the running time of the algorithm is another measurement that is commonly referenced [13,31].

3   BPP Framework

In this section we propose a new framework which can be used to generate and test dynamic instances of the BPPs. This framework is developed based on an open-source simulation software named JaamSim. To the best of our knowledge this is the first time a discrete-event simulation framework is used as a platform for benchmarking dynamic optimisation algorithms. We believe discrete-event simulation could be a good solution for benchmarking purposes. First, the event-based nature of discrete-event simulation makes it possible to generate different problem instances very easily by just creating different events and adjusting how the system changes its state upon an event. Second, generating dynamics is the nature of simulation - all simulation software are intrinsically equipped with some random/uncertainty generator. This makes simulations naturally suitable for generating dynamic benchmark problems. Third, the visualisation feature of simulation can help researchers visually observe the behaviour of dynamic problems as well as algorithms much more easily. Fourth, many simulation software have drag-and-drop features, which would make the process of creating dynamic problems more user-friendly. Fifth, discrete-event simulation are naturally suitable for complex systems. This makes it suitable for creating more challenging benchmark problems. Finally, the flexibility of open source makes it possible to easily integrate different algorithms and extend the framework to other type of problems.

3.1   Features of the Framework

Objectives and Uncertainty. The framework provides three objectives that optimisation algorithms can choose to optimise: (1) minimise the number of bins; (2) minimise the cost of bins; (3) maximise the profit of packed items.

Given the objectives above, the framework can generate uncertainty in: size of items, weight of items, profit of items, and cost of bins. The uncertain values can be generated under any distribution (e.g. uniform distribution, normal distribution). Both two-dimensional (2D) and 3D (cuboid items) BPP can be visualised. The framework can generate problems with single bin, multiple identical bins, or multiple bins of different type.

---

[1] The volume utilisation of one bin is the total volume of packed items in this bin divided by the volume of the bin.

Offline BPP Versus Online BPP. The framework supports solving the BPP in both static (offline) and dynamic (online) ways. In the static case, all items are available beforehand, and the optimisation algorithm can freely choose any item to load into a bin. In the dynamic case, the bins need to be packed when time goes by, and items arrive at different time. Whenever one or some item(s) arrive, the algorithm needs to find the best way to pack the items to a bin, then waits for the next set of items to come, and so on.

Algorithms Integration. The framework is developed in Java, therefore any algorithm that can be executable/called from Java can be simulated and evaluated in this framework. The experimental study carried out in this paper (Sect. 4) shows an example of integration of the proposed framework with one online and one static algorithms which were developed in Java and C programming languages, respectively. To illustrate how the framework can be used to test algorithms that solve the static BPP, we use a static bin packing algorithm [3] which solves the BPP in an offline way, assuming that all items are available beforehand. This algorithm uses heuristic approaches on initially sorted items by nonincreasing volume. The algorithm also assumes that unlimited identical bins are given, and bins have fixed orientation. If one bin is full then it is closed, a new bin is set as open to receive items. The algorithm was written in C, then was compiled into an executable file, which is then called by the simulation framework. We choose this algorithm because, although it is not the latest method, it is one of the few available 3D bin packing algorithms whose source code is accessible and detailed algorithm description is available. Because the purpose of this paper is to provide a proof of concept, we feel the decision of choosing this algorithm is justified.

To illustrate how the framework can be used to test dynamic algorithms that solve the dynamic BPP in an online way, we implement a new online algorithm (Algorithm 1). It should be noted that the dynamic BPP is very new to the academic community, and while there has been a few research that proposes solving algorithms [32–34], these research have not provided any experimental details to prove that these algorithms work. Because of that, here we just provide a simple algorithm as a proof of concept. The algorithm is written in Java, and it works by packing upcoming items layer by layer under the same assumption as the static algorithm.The algorithm assumes that there is no information of upcoming items and hence the problem needs to be solved online. At the time an item comes, the information of the item, the current bin and a packing location are passed to the algorithm as parameters and the algorithm goes on packing the new item into the current available bins. Items are packed in a bin in "layers" (see Figs. 1 and 2).

Algorithm 1. OnlineBinPacking(a_i, b, p)

```
while (b_j) { //if the j^th bin exists
    if a_{i,x} > b_{j,x} || a_{i,y} > b_{j,y} || a_{i,z} > b_{j,z} {
        //if any of the dimensions of the item is larger than the bin
        return }//this item can not be packed
    if p_x + a_{i,x} > b_{j,x} {
        //if the space through x coordinate is not enough
        //move the packing location to the origin of next bin, close the current bin
        update p and j to next bin
    } else if p_y + a_{i,y} > b_{j,y} {
        //if the space through y coordinate is not enough
        //move the packing location to the next layer
        update p to next layer through x coordinate
    } else if p_z + a_{i,z} > b_{j,z} {
        //if the space through z coordinate is not enough
        //move the packing location to the next column
        update p to next layer through y coordinate
    } else {
        //space is enough, pack the item
        pack the item and update p,i and j
        return

    }
}
return
```

where p_x,p_y,p_z are the packing location, $a_{i,x}, a_{i,y}, a_{i,z}$ are the width, height and length of the ith item in the list of items, $b_{j,x}, b_{j,y}, b_{j,z}$ are the width, height and length of the jth bin in the list of bins.

The size of each layer is the maximum size of the packed items. The algorithm will check whether the current vertical layer in a bin has enough space for this item. If there is enough space on the current layer, the item is packed to the location and

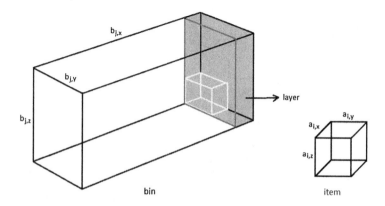

Fig. 1. Example of item, bin, and layer in 3D

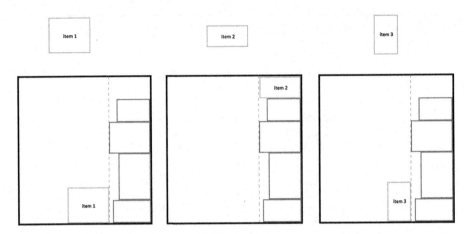

Fig. 2. Example of packing in a layer in 2D using Algorithm 1. Four items are packed in the first column. For the newly arrived item 1, the space left in the first column is not enough, so it is placed to the second column. Because the volume of item 2 is smaller than the free space, it is placed on the top of those items in the first column. Item 3 has the same volume as item 2 but with different orientation. It is placed to the second column since our algorithm is under the assumption of fixed orientation.

return the updated packing location. If there is no enough space on this layer, it will check the next layer till this bin is full and then open a new one.

The simulation process of online bin packing is outlined as follows. This is also shown in Fig. 3.

1. Setting up inputs to generate test cases. The format of input for minimising the number of identical bins is shown in Table 1.
2. Once a test instance is generated, the online algorithm is called while passing the information of this item as parameters (width, height, length). The test cases are saved as a text file as well in the format below.

> $n\ b_x, b_y, b_z$
> $a_{1,x}\ a_{1,y}\ a_{1,z}$
> $a_{2,x}\ a_{2,y}\ a_{2,z}$
> ...
> ...
> ...
> $a_{n,x}\ a_{n,y}\ a_{n,z}$
> where n is the total number of items, x, y, z are the width, height and length.

3. Then the packing location is returned by the algorithm so that the framework could visualise.
4. In the meanwhile, the packing result is exported to a file with a format as follows: bin type, bin no., item width, item height, item length, packing position coordinate x, y, z.

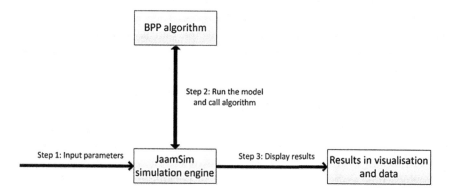

Fig. 3. The process flow of simulation

Table 1. The format of input for bin number minimisation

Parameters	Explanation
String file_name	The output file name of test cases. It should be the same name read in algorithm
int bin_x, int bin_y, int bin_z	The size of bins
int item_num	The total number of items to be packed
int x_dis, int y_dis, int z_dis	The distribution of each dimension (width, height, length). 1 is normal distribution, the default is uniform distribution
int x_min, int y_min, int z_min	It is the mean for normal distribution or the lower bound of the range for uniform distribution
int x_max, int y_max, int z_max	It is the standard deviation for normal distribution or the upper bound of the range for uniform distribution

Performance Measures. As mentioned in Sect. 2.3, performance measures are necessary to be introduced due to the lack of common measures. To evaluate the effectiveness of the algorithms comparing to other algorithms, we use the following measures: average utilisation and number of bins. The average utilisation shows how much the bin capacity is used on average. It provides a reference to see how close the items are packed in each bin. The average utilisation of the bins is the total volume of packed items dividing by the total volume of used bins. Moreover, the number of bins is used as another performance measure regarding the effectiveness of algorithms. It represents how an algorithm performs with an objective of input minimisation such as number of bins minimisation or cost of bins minimisation.

Performance measures regarding efficiency of algorithms generally means the process time of an algorithm, for example, how fast or slow the algorithm identified the optimum solution. The criteria at the moment the framework supports is the running time, which is the total process time of an algorithm in our framework. This criteria can be easily collected and it is valuable in comparing performances of different algorithms on computers of same standard. Other criteria like CPU usage could be implemented in the future.

3.2   Generating Test Problems

This subsection proposes test cases that we suggest to use as default. However, our framework is not limited to them. Users can customise test cases based on their own needs. Here we considered three levels of test cases: easy, medium, and hard which represent the level of difficulty of packing items. The time that an algorithm takes normally depends on the difficulty of the problem. Due to the lack of a proper 3D benchmark, the following instances are extended from Lodi et al. [35] which provides instances in 2D. Another dimension is added for bins and items. They are displayed in Table 2 based on following types of cuboids that are defined in terms of the width W, height H and length L of the bins.

Type 1: w_j uniformly random in $[\frac{2}{3}W, W]$; h_j uniformly random in $[1, \frac{1}{2}H]$; l_j uniformly random in $[1, \frac{1}{3}L]$;

Type 2: w_j uniformly random in $[1, \frac{1}{2}W]$; h_j uniformly random in $[\frac{2}{3}H, H]$; l_j uniformly random in $[\frac{1}{2}L, L]$;

Type 3: w_j uniformly random in $[\frac{1}{2}W, W]$; h_j uniformly random in $[\frac{1}{2}H, H]$; l_j uniformly random in $[1, \frac{1}{2}L]$;

Type 4: w_j uniformly random in $[1, \frac{1}{2}W]$; h_j uniformly random in $[1, \frac{1}{2}H]$; l_j uniformly random in $[\frac{2}{3}L, L]$.

Table 2. Example of the problem instances generated by the framework

Data set No	Category	Bin size (W*H*L)	Total number of items	Item $(w_j * h_j * l_j)$
I_20	Easy	30*30*30	20	Uniformly random in [1, 10]
I_40	Easy	30*30*30	40	Uniformly random in [1, 10]
I_60	Easy	30*30*30	60	Uniformly random in [1, 10]
I_80	Hard	30*30*30	80	Uniformly random in [1, 10]
I_1000	Hard	30*30*30	1000	Uniformly random in [1, 10]
II_20	Easy	100*100*100	20	Uniformly random in [1, 35]
II_40	Medium	100*100*100	40	Uniformly random in [1, 35]
II_60	Medium	100*100*100	60	Uniformly random in [1, 35]
II_80	Hard	100*100*100	80	Uniformly random in [1, 35]
II_1000	Hard	100*100*100	1000	Uniformly random in [1, 35]
III_20	Medium	100*100*100	20	Uniformly random in [1, 100]
III_40	Medium	100*100*100	40	Uniformly random in [1, 100]
III_60	Hard	100*100*100	60	Uniformly random in [1, 100]
III_80	Hard	100*100*100	80	Uniformly random in [1, 100]
III_1000	Hard	100*100*100	1000	Uniformly random in [1, 100]
IV_40	Hard	100*100*100	40	Type 1 with probability 70 %, Type 2, 3, 4 with probability 10 % each
IV_1000	Hard	100*100*100	1000	Type 1 with probability 70 %, Type 2, 3, 4 with probability 10 % each

The input used for generating the data set is listed below. The fifth parameter, n is the number of instances which must be an integer. In Java, the specific function with the same name will be called depending on the number of input parameters. We have two combinations of parameters below. The first one generates instances by choosing a specific distribution (data set I, II, III), and the instances generated by the second one are based on the size of bins (data set IV). The n below represents the number of items to be generated.

parameters: String file_name, int bin_x, int bin_y, int bin_z,
            int item_num, int x_dis, int x_min, int x_max, int y_dis, int
            y_min, int y_max, int z_dis, int z_min, int z_max
            OR
            String file_name, int bin_x, int bin_y, int bin_z, int item_num

I: "test_cases1.txt", 30, 30, 30, n, 0, 1, 10, 0, 1, 10, 0, 1, 10
II: "test_cases2.txt", 100, 100, 100, n, 0, 1, 35, 0, 1, 35, 0, 1, 35
III: "test_cases3.txt", 100, 100, 100, n, 0, 1, 100, 0, 1, 100, 0, 1, 100
IV: "test_cases4.txt", 100, 100, 100, n

4   Case Study

4.1   Peer Algorithms

The optimal solution that a dynamic BPP algorithm could find in the online dynamic case would always be worse than or at best equal to the optimal solution found in the static case. Due to that, to evaluate the efficiency of a dynamic BPP algorithm, we can compare its solution with that of an established static BPP algorithm.

To demonstrate this type of comparison, in this experiment we are going to compare our online algorithm (Algorithm 1 in Sect. 3.1) with the static algorithm (Sect. 3.1) to evaluate the effectiveness/efficiency of Algorithm 1.

4.2   Experimental Design

The simulation runs on an Intel Core 2, 3.06 GHz computer with 4.0 GB RAM. The test set in Table 2 is applied to both algorithms. Different number of items are set for each group of instances. In each group, we run the algorithm for ten replications. In order to evaluate the performance of algorithms, we use the number of bins used, the average utilisation and the running time (see Sect. 3.1).

4.3   Performance Analysis

Table 3 shows the average results of ten replications. The average numbers of bins determined by the online and static algorithms do not have a significant difference when the number of items are low or the sizes of items are small in proportion to bin sizes, i.e. I_20 and II_20. However, for I_40, I_60, II_40, II_60, III_20 and III_40 the static solution provides a smaller number of bins with

Table 3. Bin packing results. N/A represents that an algorithm has not finished the job by the time of submission.

Data set No.	Total No. of items	Online			Static		
		No. of bins	Average utilisation	Running time (s)	No. of bins	Average utilisation	Running time (s)
I_20	20	1	11.994 %	0.001	1	11.994 %	0.129
I_40	40	1.1	24.054 %	0.001	1	25.752 %	0.136
I_60	60	2	19.037 %	0.001	1	38.077 %	0.116
I_80	80	5	37.43 %	0.001	N/A	N/A	N/A
I_1000	1000	22.3	29.16 %	0.001	N/A	N/A	N/A
II_20	20	1	11.619 %	0.001	1	11.619 %	0.226
II_40	40	1.4	17.442 %	0.001	1	22.273 %	0.146
II_60	60	2	16.92 %	0.001	1	33.83 %	0.19
II_80	80	3.2	13.71 %	0.001	N/A	N/A	N/A
II_1000	1000	26	22.46 %	0.002	N/A	N/A	N/A
III_20	20	8.8	29.77 %	0.001	4.5	58.43 %	2.001
III_40	40	15.7	29.633 %	0.001	6.6	70.107 %	14.214
III_60	60	24.3	32.13 %	0.001	N/A	N/A	N/A
III_80	80	37	33.11 %	0.001	N/A	N/A	N/A
III_1000	1000	382	32.77 %	0.001	N/A	N/A	N/A
IV_40	40	2	25.54 %	0.001	N/A	N/A	N/A
IV_1000	1000	81	42.44 %	0.001	N/A	N/A	N/A

a larger average utilisation rate. This reflects the fact that in most cases the solution found online would be worse than the solution found offline in a static way.

In terms of the running time, the online algorithm has significantly shorter process times in comparison with the static algorithm. According to the results, the static algorithm takes much longer to achieve a packing plan for hard level of test cases. By the time we submit this paper it has not even been able to find a solution for all the large scale cases. The online algorithm, on the contrary, is much faster and can solve all the problems, including the hard, large-scale instances. For example, in the results of I_1000, II_1000, III_1000 and IV_1000 shown in Table 3, in the online algorithm it takes only 0.002 s at most. This solving time includes both computational time of the proposed online algorithm and the simulation time to visually display the loading process. It means the simulation is also fast and our framework is capable of handling problems with a large number of items which is common in the real-world BPPs. Figure 4 shows an example of how the process of packing bins online is displayed in 3D in our proposed framework.

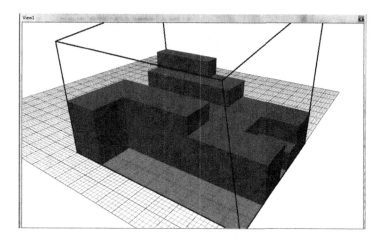

Fig. 4. A 3D view of the online bin packing process, as displayed by our framework.

5   Conclusion

This paper for the first time proposes a framework to generate benchmark problems for dynamic BPPs using discrete-event simulation. The developed benchmarks provides a test bed for evaluation of the applicability of dynamic optimisation algorithms to real-world BPPs. We also developed an online bin packing algorithm which was integrated in the framework. A set of test cases were generated using the framework to evaluate the online algorithm. The results of the online algorithm on these test cases were then compared with those of an static algorithm from the literature.

The future directions for further improvement of the proposed framework are as follows. First, in this paper only limited algorithms were evaluated, but this framework is capable of being integrated with other algorithms such as evolutionary algorithms. Second, state-of-the-art online bin packing algorithms from the literature can be reproduced in the framework to achieve competitive results. Third, the impact of uncertainty on each dimension of items can be investigated. In addition, currently, we only used this framework for BPPs, however this framework can be extended to many other optimisation problems e.g. the travelling salesman problem and job shop scheduling. This can be investigated in our future study.

Acknowledgement. This work was supported by a Dean's scholarship from the Faculty of Engineering and Technology, Liverpool John Moores University, a British Council UK-ASEAN Knowledge Partnership grant and a British Council Newton Institutional Links grant.

References

1. Crainic, T.G., Perboli, G., Tadei, R.: Ts 2 pack: A two-level tabu search for the three-dimensional bin packing problem. Eur. J. Oper. Res. **195**(3), 744–760 (2009)
2. Feng, X.: Hybrid genetic algorithms for the three-dimensional multiple container packing problem. Flex. Serv. Manufact. J. **27**(2), 451–477 (2013)
3. Martello, S.: Algorithm 864: general and robot-packable variants of the three-dimensional bin packing problem. ACM Trans. Math. Softw. **33**(1), 1–12 (2007)
4. de Almeida, A.: A particular approach for the three-dimensional packing problem with additional constraints. Comput. Oper. Res. **37**, 1968–1976 (2010)
5. Alvarez-Valdes, R.: A grasp/path relinking algorithm for two- and three-dimensional multiple bin-size bin packing problems. Comput. Oper. Res. **40**, 3081–3090 (2013)
6. Che, C.H.: The multiple container loading cost minimization problem. Eur. J. Oper. Res. **214**, 501–511 (2011)
7. Eley, M.: A bottleneck assignment approach to the multiple container loading problem. OR Spectr. **25**, 54–60 (2003)
8. Tian, T., Zhu, W., Lim, A., Wei, L.: The multiple container loading problem with preference. Eur. J. Oper. Res. **248**, 84–94 (2015)
9. Lim, A.: The single container loading problem with axle weight constraints. Int. J. Prod. Econ. **144**(1), 358–369 (2013)
10. Liu, J.: A novel hybrid tabu search approach to container loading. Comput. Oper. Res. **38**(4), 797–807 (2011)
11. Junqueira, L.: Three-dimensional container loading models with cargo stability and load bearing constraints. Comput. Oper. Res. **39**, 74–85 (2012)
12. Costa, M.G., Captivo, M.E.: Weight distribution in container loading: a case study. Int. Trans. Oper. Res. (2014)
13. Ceschia, S., Schaerf, A.: Local search for a multi-drop multi-container loading problem. J. Heuristics **19**(2), 275–294 (2013)
14. Bortfeldt, W.: Constraints in container loading - a state-of-the-art review. Eur. J. Oper. Res. **229**, 1–20 (2012)
15. Nguyen, T.T., Yang, S., Branke, J.: Evolutionary dynamic optimization: A survey of the state of the art. Swarm Evol. Comput. **6**, 1–24 (2012)
16. Nguyen, T.T., Yao, X.: Continuous dynamic constrained optimization - the challenges. IEEE Trans. Evol. Comput. **16**(6), 769–786 (2012)
17. Jin, Y., Sendhoff, B.: Constructing dynamic optimization test problems using the multi-objective optimization concept. In: Raidl, G.R., et al. (eds.) EvoWorkshops 2004. LNCS, vol. 3005, pp. 525–536. Springer, Heidelberg (2004)
18. Richter, H.: Memory Design for Constrained Dynamic Optimization Problems. In: Chio, C., Cagnoni, S., Cotta, C., Ebner, M., Ekárt, A., Esparcia-Alcazar, A.I., Goh, C.-K., Merelo, J.J., Neri, F., Preuß, M., Togelius, J., Yannakakis, G.N. (eds.) EvoApplicatons 2010, Part I. LNCS, vol. 6024, pp. 552–561. Springer, Heidelberg (2010)
19. Weicker, K., Weicker, N.: Dynamic rotation and partial visibility. In: Proceedings of the 2000 Congress on Evolutionary Computation, pp. 1125–1131 (2000)
20. Nguyen, T.T., Yao, X.: Dynamic time-linkage problems revisited. In: Giacobini, M., et al. (eds.) EvoWorkshops 2009. LNCS, vol. 5484, pp. 735–744. Springer, Heidelberg (2009)
21. Nguyen, T.T., Yao, X.: Benchmarking and solving dynamic constrained problems. In: IEEE Congress on Evolutionary Computation, CEC 2009, pp. 690–697. IEEE (2009)

22. Nguyen, T.T.: Continuous dynamic optimisation using evolutionary algorithms. PhD thesis, University of Birmingham (2011)
23. Nguyen, T.T., Yao, X.: Dynamic time-linkage evolutionary optimization: definitions and potential solutions. In: Alba, E., Nakib, A., Siarry, P. (eds.) Metaheuristics for Dynamic Optimization. SCI, vol. 433, pp. 379–405. Springer, Heidelberg (2013)
24. Kavakeb, S., Nguyen, T.T., Yang, Z., Jenkinson, I.: Evolutionary fleet sizing in static and uncertain environments with shuttle transportation tasks - the case studies of container terminals. IEEE Computational Intelligence Magazine (in press, 2016)
25. Peng, J., Zhang, B.: Bin packing problem with uncertain volumes and capacities (2012)
26. Perboli, G., Tadei, R., Baldi, M.M.: The stochastic generalized bin packing problem. Discrete Appl. Math. **160**(7), 1291–1297 (2012)
27. Crainic, T.G.: Bin packing problems with uncertainty on item characteristics: an application to capacity planning in logistics. Soc. Behav. Sci. **111**, 654–662 (2014)
28. Lim, A., Zhang, X.: The container loading problem. In: Proceedings of the 2005 ACM Symposium on Applied Computing, pp. 913–917. ACM (2005)
29. Zhu, W., Lim, A.: A new iterative-doubling greedy-lookahead algorithm for the single container loading problem. Eur. J. Oper. Res. **222**(3), 408–417 (2012)
30. Jiang, J., Cao, L.: A hybrid simulated annealing algorithm for three-dimensional multi-bin packing problems. In: 2012 International Conference on Systems and Informatics (ICSAI), pp. 1078–1082. IEEE (2012)
31. Egeblad, J., Pisinger, D.: Heuristic approaches for the two-and three-dimensional knapsack packing problem. Comput. Oper. Res. **36**(4), 1026–1049 (2009)
32. Burcea, M., Wong, P.W.H., Yung, F.C.C.: Online multi-dimensional dynamic bin packing of unit-fraction items. In: Spirakis, P.G., Serna, M. (eds.) CIAC 2013. LNCS, vol. 7878, pp. 85–96. Springer, Heidelberg (2013)
33. Epstein, L., Levy, M.: Dynamic multi-dimensional bin packing. J. Discrete Algor. **8**(4), 356–372 (2010)
34. Wong, P.W.H., Yung, F.C.C.: Competitive multi-dimensional dynamic bin packing via l-shape bin packing. In: Bampis, E., Jansen, K. (eds.) WAOA 2009. LNCS, vol. 5893, pp. 242–254. Springer, Heidelberg (2010)
35. Lodi, A., Martello, S., Vigo, D.: Heuristic and metaheuristic approaches for a class of two-dimensional bin packing problems. INFORMS J. Comput. **11**(4), 345–357 (1999)

Genetic Programming Algorithms
for Dynamic Environments

João Macedo[1]([✉]), Ernesto Costa[2], and Lino Marques[1]

[1] Institute of Systems and Robotics, Coimbra, Portugal
{jmacedo,lino}@isr.uc.pt
[2] Center of Informatics and Systems of the University of Coimbra,
University of Coimbra, Coimbra, Portugal
ernesto@dei.uc.pt

Abstract. Evolutionary algorithms are a family of stochastic search heuristics that include Genetic Algorithms (GA) and Genetic Programming (GP). Both GAs and GPs have been successful in many applications, mainly with static scenarios. However, many real world applications involve dynamic environments (DE). Many work has been made to adapt GAs to DEs, but only a few efforts in adapting GPs for this kind of environments. In this paper we present novel GP algorithms for dynamic environments and study their performance using three dynamic benchmark problems, from the areas of Symbolic Regression, Classification and Path Planning. Furthermore, we apply the best algorithm we found in the navigation of an Erratic Robot through a dynamic Santa Fe Ant Trail and compare its performance to the standard GP algorithm. The results, statistically validated, are very promising.

Keywords: Evolutionary algorithms · Genetic programing · Dynamic environments

1 Introduction

Evolutionary Algorithms (EA) are a family of stochastic search heuristics inspired by the principle of natural selection and genetics that has been successfully applied in problems of learning, optimisation and design. Genetic Algorithms (GA) and Genetic Programming (GP) are two subfamilies of EAs. While GAs evolve solutions for a given problem, the purpose of GPs is to evolve programs that, when executed, produce solutions for specific problems.

Many of the successful applications of EAs assume that the environments are static, i.e., environments whose conditions remain constant throughout time. However in many real world applications, the environment is dynamic, altering its constraints or the problem itself over time. An example of such situation is the navigation problem of a mobile robot, where the constraints to the navigation change both in time and in space.

An EA works by iteratively evolving a population of individuals with operators that resemble sexual and asexual reproduction. Usually, during evolution,

© Springer International Publishing Switzerland 2016
G. Squillero and P. Burelli (Eds.): EvoApplications 2016, Part II, LNCS 9598, pp. 280–295, 2016.
DOI: 10.1007/978-3-319-31153-1_19

the best individuals reproduce more than the worse. This causes a gradual convergence of the entire population to individuals that are similar to its best. When the environment changes, it is likely that the former best individual does not perform well anymore. Furthermore, as the population has lost its diversity, the algorithm will have an added difficulty to evolve solutions that are strongly different from the existing ones.

Some work was already done in adapting GAs to Dynamic Environments. However, the same can not be said for GPs. Our goal is to fill that gap, making GPs capable of working with good performance in such environments.

We start by transposing some of the available techniques for GAs to GPs, and testing them in three dynamic benchmark problems. We then proceed to create novel mechanisms, based on the combination of some pre-existing concepts, and compared their performance to the previous ones. Finally, we apply a standard GP and the best technique to a real robot and compared their performance in navigating a dynamic Santa Fe Ant Trail.

The rest of this paper is organised as follows: Sect. 2 presents a brief state of the art in adapting GAs and GPs for Dynamic Environments, Sect. 3 describes the proposed GP algorithms for Dynamic Environments, as well as the experiments made to asses their usefulness, Sect. 4 describes the experiments made on a robotics, Sect. 5 presents the results of those experiments, and Sect. 6 makes the conclusion remarks and suggests future work.

2   State of the Art

In this section we briefly describe the work that have been done to deal with dynamic environments for GA's and GP's separately. We also present a method for classifying the different types of dynamic environments, inspired by [17]. For further information on these topics, in the context of GA's, the interested reader may consult Nguyen et al. [12], Yang et al. [16] or Branke [1].

2.1   Genetic Algorithms

Generally, Dynamic Environments are dealt with by either maintaining the population's diversity or by increasing it as a response to an environmental change. In Genetic Algorithms, this has been achieved through a number of methods. For the sake of clarity, we classify such methods into three classes, depending on their manipulation of the algorithm's parameters, use of memory or introduction of immigrants.

Parametric Methods. The Triggered Hypermutation [2,3] is a parametric method that consists in considerably increasing the base mutation rate of the algorithm when an environmental change is detected. Another method, know as Variable Local Search [23] works similarly to the Triggered Hypermutation. However, when the environment changes, the mutation rate is only slightly increased, thus favouring exploitation rather than exploration. In [9] is proposed a method that adapts the crossover and mutation rates depending on the fitness of both the single individuals and the entire population. Although this method was initially proposed for avoiding premature convergence, it could also perform well in dynamic environments, as it leverages exploration and exploitation around good quality individuals.

Memory Methods. The Memory based algorithms are specially useful in scenarios where future environments are similar or equal to those of the past. They can make use of implicit or explicit memory. The implicit memory methods are characterised by embedding memory mechanisms within each individual. In [7,11] this is achieved by means of multiploidy. On the other hand, explicit memory methods are characterised by having a special location where information is stored. In [17], four explicit memory based algorithms are compared. In these algorithms, the memory is used not only for storing good individuals from past environments, but also for detecting environmental changes. Apart from that, each algorithm makes use of it differently. The Memory-Enhanced Genetic Algorithm (MEGA) reacts to environmental changes by creating a new population, containing the best performing individuals from memory and the previous population, in the new context. The Memory-Immigrants Genetic Algorithm (MIGA) uses the best individual from the memory on each generation to create, by means of mutation, immigrants that are injected into the population. The Associative-Memory Genetic Algorithm (AMGA) stores in memory information about the environments along with their best individuals. This algorithm reacts to changes by using the environmental information associated to the best individual from memory to create new individuals that are injected into the population. Finally, the Variable-Size Memory Evolutionary Algorithm (VMEA) works similarly to MEGA, with the difference that it adapts the size of the memory and population, while maintaining the total number of individuals constant.

Immigrant Methods. The methods based on immigrants usually aim at maintaining a reasonable population diversity throughout the generations, rather than increasing it when the environment changes. A simple method, Random Immigrants, consists in randomly creating a set of individuals and injecting them into the population. Yang and Tinós [26] proposed the hybrid immigrants method, that combines individuals created by mutation of the best from the population and randomly generated individuals.

Other Methods. Simões and Costa [18], proposed a method that uses Transformation, a novel variation operator that replaces crossover. This method, inspired by bacterial processes, consists in the incorporation into the individuals' genome, of gene segments present in the environment. In [19], the authors compared two variants of the transformation method to a triggered hypermutation and a random immigrants approach. They concluded that with low severity environmental changes, the transformation performs better than the others, while in high severity scenarios, the triggered hypermutation is better.

2.2   Genetic Programming

The majority of work made on Genetic Programming has focused on static environments. However, there is a few literature in adapting this family of algorithms to dynamic environments, that we categorise depending on their manipulation of the algorithms' parameters, use of memory or other methods.

Parametric Methods. In [15], Riekert et al. proposed an algorithm that adapts the size of the elite, the crossover and mutation rates and uses culling. Their study showed that this algorithm is able to outperform a standard GP and a neural network in three dynamic classification problems. In [27], the adaptation of the crossover and mutation rates was explored with the goal of preventing premature convergence by increasing diversity if a new best solution is not found after a number of generations. Vanneschi and Cuccu [22] explored the adaptation of the population size in dynamic environments, reducing it when the fitness of the population is good, in order to save resources. On the other hand, when the fitness drops, the population grows to increase the exploration capabilities.

Memory Methods. In [24] Wagner et al. proposed an algorithm for time series forecasting in Dynamics Environments, using a sliding, variable length, time window and an explicit memory where good individuals from the past are stored for future use.

Other Methods. O'Neill et al., [13], used Dynamic Environments for speeding up the convergence of a standard GP algorithm in a symbolic regression problem. By varying the target polynomial functions, the authors concluded that the use of Dynamic Environments not only leads to a faster convergence of the individuals to a good quality area of the search space, but also contributes to the discovery of higher quality solutions.

2.3   Dynamic Environments

Dynamic environments can be defined according to the time (when) and nature (how) of the environmental changes. The presentation below follows [17].

When. The success of a GP algorithm depends greatly on the moments when the environment changes. Not only it is important to know if the changes happen frequently or rarely, but also if they can be predicted or are completely unexpected. This time component of the Dynamic Environments can be characterised according to: (a) the **frequency of change**, i.e., the number of times the environment changes in a run; (b) the **type of change period**, i.e., the way the period between changes varies over time (periodic, patterned, nonlinear, random); and (c) the **predictability of the change**, i.e., the ability to predict when the next change will take place.

How. The environmental changes depend on (a) the **severity of the change** i.e., how different the new environment is from the previous one; (b) the **types of environmental changes** i.e., the way an environment changes (cyclic, cyclic with noise, probabilistic, random); and (c) the **predictability of the new environment** i.e., the ability to predict how the new environment will be, based on the current and past ones.

Difficulty Scenarios. Based on the characteristics presented above it is possible to define scenarios with different difficulty levels. An easy scenario would have changes with low frequency and taking place periodically or in patterned moments. These modifications would have low severity, altering the environment cyclically, possibly with noise. A medium scenario would change more frequently, at periodic, patterned or nonlinear moments. The changes could be cyclic with noise or probabilistic, having average severity. Finally, in a hard scenario the changes would take place very frequently, at random or nonlinear moments. They would provoke random modifications in the scenario, with high severity.

3   Proposed Algorithms

We propose a set of tree based GP algorithms for dealing with Dynamic Environments which are divided into two categories. They are compared to each other and to a standard GP (SGP) as described in [8], which is used as baseline.

3.1   Simple Techniques

The first category contains techniques that are a direct adaptation of algorithms already proposed for genetic algorithms. We implemented this algorithms as it is our belief that they would also perform better than the SGP in Dynamic Environments.

Triggered Hypermutation (TH). This algorithm is an adaptation of the one proposed in [2]. In our implementation, it uses subtree mutation with a base mutation rate of 10 %. When an environmental change is detected, the mutation rate is increased to 50 %, staying at that value for a period of 25 generations.

Immigrants (I). The immigrants algorithm we propose is an adaptation of ERIGA [26]. It combines random and elitist immigrants in adaptive quantities, with the restriction that the total amount of immigrants is constant.

Fixed Memory (FM). We also propose an algorithm that uses an explicit memory of fixed length, where the best individual of the population at the time of the update is stored. When an environmental change is detected, the best individual from memory replaces the worst from the population, provided that it is better than it. This mechanism is an adaptation of MEGA, proposed by Yang [26].

3.2   Hybrid Techniques

In this section we present algorithms that result largely from the combination of the simple techniques, except for the Transformation, that is an adaptation of the algorithm proposed in [18], but is presented here for chronological reasons.

Transformation. As we said, this algorithm is a direct adaptation of one already proposed for preventing premature convergence in GAs. Our belief, is that as it promotes population diversity, it will be able to handle Dynamic Environments better than the SGP.

Combination of Two Techniques. The Transformation Memory combines Transformation with Memory. It is similar to the Fixed Memory algorithm, apart from the variation operator. It is our belief that it will combine the better results of the transformation with the ability to resume evolution provided by the memory. This reasoning served as inspiration to the following combinations:

- The Hypermutation Memory (HM), which combines the Triggered Hypermutation with Fixed Memory;
- Immigrants Memory (IM), which combines the ERIGA inspired approach with Fixed Memory;
- Random Immigrants Memory (RIM), which combines the Fixed Memory with immigrants that are only randomly generated.

Combination of More Techniques. In this section we present a set of algorithms that combine three techniques with multi populations. The first of these algorithms is the Hypermutation Memory Transformation SGP (HMTS), which uses three subpopulations, being two of them evolved in isolation. The first subpopulation is evolved with Triggered Hypermutation Memory, the second subpopulation is evolved with Transformation and the third subpopulation is evolved with the standard GP algorithm. For producing each offspring, three candidate parents are selected by means of tournament, from each of the populations. The crossover operator is then applied to the best two individuals. On each generation, the best individual outputted is the best individual from the entire population. Is also that individual that is introduced into the memory, thus being an exception to the isolated evolution of the first subpopulation. The other two variants of this algorithm are the Transformation Memory Hypermutation SGP (TMHS) and the Fixed Memory Transformation Hypermutation (FMTH), the difference between them being the algorithms that evolve each subpopulation.

3.3   Experimental Results

These algorithms were applied in three dynamic benchmark problems, a Symbolic Regression, a Classification and the Santa Fe ant Trail, each of them having three scenarios of increasing difficulty. These benchmark problems were chosen because they belong to classes relevant to the area [25]. In this Section we briefly describe the experiments made, directing the interested reader to [10] for further details.

Symbolic Regression. This benchmark problem consists in using Symbolic Regression to approximate a set of polynomial functions of the form: $x + x^2 + x^3 + x^4 + x^5 + x^6 + x^7 + x^8$. In order to make this problem dynamic we modify a number of addictions to subtractions. The Root Mean Square Error was used to assess the quality of each individual. For this benchmark problem, we defined the following scenarios:

- In the **Easy** scenario, the environmental changes take place periodically, every 200 generations. They are cyclical in nature, and with relatively low severity, consisting in alternating between two target functions that differ in only one operator.
- In the **Medium** scenario, the modifications are more severe and frequent. The environmental changes take place at pre-determined moments, following a pattern of 100 - 120 - 80 - 100 generations and consist in modifying three operators. Furthermore, the number of possible environments is limited to 5. On each environmental change, the next environment shall be chosen probabilistically, depending on the current one.
- In the **Hard** scenario, the changes will not only be more frequent, as they will take place at any random moment, with a minimum change period of 50 generations. After that number of generations there will be a 50 % probability of the environment being modified in the current moment. Furthermore, there will be no predefined number of environments. Instead, each operator will have a 50-50 chance of being inverted, i.e. positive members becoming negative and vice versa, thus causing random modifications to the environment. Initially, the environment shall be defined by the original function: $x + x^2 + x^3 + x^4 + x^5 + x^6 + x^7 + x^8$, as in the other scenarios.

Classification. In this section we describe the binary classification problem we used as benchmark. In it, each example belongs to one of two classes, that are separated by a 10-dimensional hyperplane. The task of the classifier is to attribute a class to each example, and its performance is measured with the F1-score.

The decision hyperplane is of the form:

$$H(x) = \sum_{i=1}^{9} (a_i * x_i) + c$$

This problem was made dynamic by modifying the decision hyperplane, more specifically, the coefficients a_i. We defined three scenarios, with increasing difficulty:

- In the **Easy** scenario the modifications take place periodically, every 200 generations, and are cyclical in nature, consisting in alternating between two environments that only differ in one coefficient.
- In the **Medium** scenario, the environmental changes take place periodically, every 100 generations. They are cyclical with noise, consisting in a succession of 5 environments, that differ from each other in two coefficients, which have an added noise.
- In the **Hard** scenario, the environment will be modified at random moments, provided that there are a minimum of 50 generations between changes. At each change, 5 coefficients are randomly selected and altered.

Santa Fe Ant Trail. The third benchmark problem is from the class of path planning. It consists in evolving controllers for navigating an artificial ant through a two dimensional grid world, with the goal of collecting all existing food. The original map, depicted in Fig. 1, contains 89 food pellets, disposed in the filled cells. The quality of a controller can be measured by the percentage of food pellets collected in at most 400 movements. We chose to use a percentage to allow a fair comparison between environments, as they change by omitting food pellets.

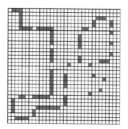

Fig. 1. Santa Fe Ant Trail. The filled cells contain food pellets.

This problem can be made dynamic by simply removing some food pellets from the environment. We proposed three scenarios for making it dynamic, with increasing difficulty:

- The **Easy** scenario is modified periodically, every 200 generations. Each change consists in cyclically alternating between two environments, the original and one created by omitting 7 food pellets from it.
- In the **Medium** scenario, the changes take place in moments that follow a pattern, taking place between in periods of 100 - 120 - 80 -100 generations. They are of a probabilistic nature, being possible to go from one environment to any of the other four, each transition having an associated probability. The environment set consists of the original environment and four others, generated by omitting 12 food pellets from the original one.

– In the **Hard** scenario, there are five possible environments, the original and four others, created by removing 17 food pellets from the original. The changes are of a random nature, being possible to choose any next scenario regardless of the current one. Furthermore, the changes take place at random moments, provided that there are a minimum of 50 generations between them.

3.4   Experimental Setup

For each benchmark problem and for each scenario we did 30 runs. The fitness of the best individual from the population at each generation was collected. Those values were used to compute two metrics, the offline performance and the best of generation. The offline performance measures the ability of the algorithm to produce good solutions before the environment changes. More formally, the offline performance of an algorithm's run is given by:

$$OfflinePerformance = \frac{\sum_{t=1}^{nGenerations}(u_e^*(a_t))}{nGenerations}$$

where $nGenerations$ is the number of generations in a run, and $u_e^*(a_t)$ is the fitness of the best individual in the current environment.

The best of generation (BOG) metric measures the average fitness of the best individual over all generations in a run. More formally, it is defined by:

$$BOG = \frac{\sum_{g=1}^{nGenerations}(f(b_g))}{nGenerations}$$

where $f(b_g)$ is the fitness of the best individual from generation g.

In order to assess whether the proposed algorithms were any better than the existing ones, the simple techniques were compared with the SGP, the hybrid techniques were compared with the simple techniques that constitute them, and the transformation was compared with all simple techniques and the SGP. Furthermore, all algorithms were compared with each other in an attempt to find and overall best. These comparisons were made with statistical tests, specifically with Friedman's Anova and the Wilcoxon Signed Ranks test, at a 95 % confidence interval, and using the Bonferroni correction for continuity. Due to space limitations we do not present all the results here, and we direct the interested reader to [10]. In Table 1 we present the total number of times an algorithm was significantly superior, or inferior, to the others, over all scenarios of all benchmark problems, at this confidence level.

From this table we conclude that the algorithm HMTS performs very well, being the one that is more times superior to others, and never inferior. For the same reasons, we may say that the worst algorithms are the RIM, that is the most times inferior to others, and the SGP, being the least times superior to others. In the second part of this paper we apply the HMTS to the evolution of the controller of a real robot, along with the SGP and a Random Walk for baseline.

Table 1. Number of times an algorithm is significantly superior or inferior to any other. The highest values are highlighted

Algorithm	Superior	Inferior
SGP	6	40
TH	10	19
Immigrants	13	26
FM	15	11
Transformation	9	39
HM	28	3
IM	9	25
RIM	17	**50**
TM	20	9
HMTS	**35**	0
TMHS	33	0
FMTH	28	1

4 Experiment in Evolutionary Robotics

In this section we describe the application of two GP algorithms to the navigation of a real robot.

4.1 Problem

The application problem at hand consists of the adaptation of the Santa Fe Ant Trail benchmark problem to the real world. Instead of a toroidal grid, the maze will be delimited by walls. Furthermore, similarly to the Lausanne Trail [21], there will not be a grid, but the proportions and displacement of the food pellets will be maintained. Thus, the robot is able to move freely in the world. A consequence of moving to the real world is the presence of noise, both in the perceptions, i.e., robot localisation, orientation, distance to walls, and in actions, i.e., turning and moving. Furthermore, this problem is made dynamic using the scenarios described in Sect. 3.3. The evolved controllers will have to deal with both noise and dynamic environments.

The Santa Fe Ant Trail problem is a well known problem within the GP community. In [4], Doucette and Heywood proposed a novelty-based fitness function to evaluate GP individuals navigating the Santa Fe trail. They tested two methods that are based only on novelty, a combination of novelty and quality, i.e., amount of food eaten, and quality alone. They found that while the quality-based and combination functions did not provide significant differences, they performed significantly better than the ones based on novelty alone. On the other hand, the introduction of novelty allowed for better generalisation capabilities.

To the best of our knowledge there is only one adaptation of this problem to robotics. Teuscher et al. [21] used small robots to navigate the Lausanne Trail, an adaptation of the Santa Fe trail to the real world. The trails were printed to A0 paper, with black squares indicating the food pellets. The controllers for those robots were evolved with an EA, with local tournament selection. The main drawback they found in their system was the robot's inability to detect loops, as they did not possess any localisation mechanism and were not able to delete the discovered food pellets.

Sugawara et al. [20] proposed a framework for simulating the behaviour of foraging robots based on virtual pheromone. They experimented different combinations of number of robots and pheromone evaporation rates in tasks of foraging uniformly distributed food, which has some similarities to the Santa Fe Ant Trail, as well as foraging food that is concentrated in a specific location.

Fioriti et al. [5] addressed the problem of foraging in Dynamic Environments using Lévy search. They experiment on environments with a single moving target, positioned accordingly to a radial probability density function, comparing the Lévy search with the random walk.

4.2   Algorithms

We will compare two tree based GP algorithms, the SGP, as defined by Koza [8] and the HTMS, presented in Sect. 3.2, which we consider the best approach for dynamic environments. We also implemented a Random Walk algorithm (RW) for providing a baseline. This latter algorithm chooses random positions in the map and navigates to them using the same actions as the controllers evolved by the GP algorithms.

4.3   Parametrisation

The main parameters are defined in Table 2. Moreover, the variation operators used are sub-tree crossover, sub-tree mutation and node mutation. All algorithms will be given 400 steps to solve the trails, and use the same function and terminal sets:

- Function Set = {IfObstacleAhead, IfTargetAhead, Progn2, Progn3}
- Terminal Set = {Move, Right, Left}

Where IfObstacleAhead and IfTargetAhead are boolean functions that check if there is an obstacle or target directly ahead of the robot, Progn2 and Progn3 are two progression functions that execute two or three actions in order. Move is the action of moving the equivalent of one cell forward and Left and Right are the actions of turning 90 degrees in each direction. The fitness of each individual will be measured as the percentage of food pellets collected by it. A change in the environment will be considered to have occurred, when the fitness of the best individual of the previous generation is now different.

Table 2. GP parameters

Parameter	Value
Population Size	400
Number of generations	5000
Crossover probability	80 %
Mutation probability	10 %
Maximum tree depth	7
Elite size	25
Tournament size	50
Population initialisation	Ramped half and half
Base mutation probability	10 %
Hypermutation probability	50 %
Hypermutation period	25 generations
Memory size	40
Transformation probability	80 %
Gene pool size	160
Subpopulation size	133

4.4 Resources

To test the algorithms we use an Erratic Robot, having one Hokuyo range finder with $180°$ aperture angle and $4 m$ range and a camera on its front facing downwards at a $30°$ angle. The robot's position and orientation are given by its odometry. Using these sensors we can provide perceptions such as determining whether there is a target ahead of the robot and if it is safe to move forward or turn without hitting a wall. The algorithms were tested in Stage [6], a simulator of the ROS framework [14] known for being realistic enough to allow an almost direct transition from simulation to real robots.

5 Experimental Results

We measure the quality of the algorithms with two metrics, the offline performance and the best of generation, presented in Sect. 3.4. Figures 2, 3 and 4 show the fitness of the best individual, averaged over 30 runs, for each algorithm. An analysis of these graphs leads us to the conclusion that the HMTS is always better than the SGP, being able to gain quality faster, not loosing much fitness when the environment changes, and achieving overall better results. Furthermore, both GP algorithms clearly outperform the RW. In the next section we proceed to statistically validate the results of only the GP algorithms, as we feel that there is not a need to statistically compare them with the Random Walk.

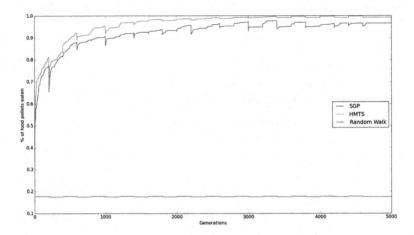

Fig. 2. Performance of the SGP, HMTS and Random Walk algorithms in the easy scenario of the Santa Fe Ant Trail.

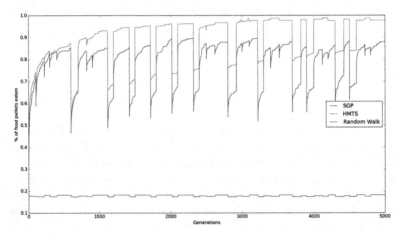

Fig. 3. Performance of the SGP, HMTS and Random Walk algorithms in the medium scenario of the Santa Fe Ant Trail.

5.1 Statistical Tests

We apply a pairwise test to the distribution of the results of the 30 runs, determining, at a 95 % confidence interval, which is the best for this application.

For comparing the performances of the algorithms we employed the Wilcoxon Signed Ranks test. This test was chosen because we only have two algorithms to compare, the data are dependent and do not follow the same distribution.

The results of the Wilcoxon Signed Ranks test are shown in Table 3. The test was made in a way that positive ranks favour the HMTS and negative ranks favour the SGP. From it, we can conclude that the HMTS is an improvement

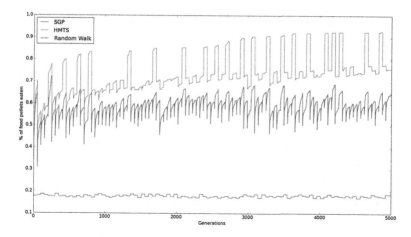

Fig. 4. Performance of the SGP, HMTS and Random Walk algorithms in the hard scenario of the Santa Fe Ant Trail.

over the standard algorithm, never being worse than it, and being significantly better than it in the medium and hard scenarios, regardless of the metric used.

Table 3. Wilcoxon Signed Ranks test

		Asymp. Sig	Z	Sum of positive ranks	Sum of negative ranks
BOG	Easy	0.318	−0.998	281	184
	Medium	0.001	−3.425	399	66
	Hard	0.000	−4.782	465	0
Offline P.	Easy	0.229	−1.203	291	174
	Medium	0.000	−4.103	432	33
	Hard	0.000	−4.782	465	0

6 Conclusions and Future Work

In this paper we presented a study where a set of GP algorithms were proposed for coping with Dynamic Environments. Those algorithms were tested in a set of dynamic benchmark problems, each having three scenarios of increasing difficulty. Having shown good results in those experiments, we proceeded to employ the best technique found and compare it with a standard GP and a Random Walk algorithm in the navigation of a real robot in a dynamic Santa Fe Ant Trail problem, in different difficulty scenarios. The experimental results showed that this algorithm is robust and better than the SGP in the two harder scenarios and not being significantly different from it in the easiest one.

In the future, work has to be done to analyse the characteristics of each algorithm, such as population diversity throughout the generations and exploration and exploitation capacities. This study would allow for better understanding on how to combine techniques in order to construct algorithms that excel at both exploration and exploitation, while reusing past knowledge. Research should also be made concerning the creation of other memory mechanisms that take advantage of past knowledge while preventing the convergence to local optima previously found.

Preliminary testing of the HMTS with a real Erratic Robot already showed promising results.

Acknowledgements. This work was partially supported by the TIRAMISU project (www.fp7-tiramisu.eu) under grant agreement FP7/SEC/284747 and the MassGP project (www.novaims.unl.pt/massgp) under grant agreement PTDC/EEI-CTP/2975/2012.

References

1. Branke, J.: Evolutionary Optimization in Dynamic Environments. Kluwer Academic Publishers, Norwell (2001)
2. Cobb, H.G.: An investigation into the use of hypermutation as an adaptive operator in genetic algorithms having continuous, time-dependent nonstationary environments. Technical report, DTIC Document (1990)
3. Cobb, H.G., Grefenstette, J.J.: Genetic algorithms for tracking changing environments. Technical report, DTIC Document (1993)
4. Doucette, J., Heywood, M.I.: Novelty-based fitness: an evaluation under the Santa Fe Trail. In: Esparcia-Alcázar, A.I., Ekárt, A., Silva, S., Dignum, S., Uyar, A.Ş. (eds.) EuroGP 2010. LNCS, vol. 6021, pp. 50–61. Springer, Heidelberg (2010)
5. Fioriti, V., Fratichini, F., Chiesa, S., Moriconi, C.: Lévy foraging in a dynamic environment-extending the Lévy search. Int. J. Adv. Robot Syst. **12**(98) (2015)
6. Gerkey, B., Vaughan, R.T., Howard, A.: The player/stage project: tools for multi-robot and distributed sensor systems. In: Proceedings of the 11th International Conference On Advanced Robotics (ICAR), pp. 317–323 (2003)
7. Goldberg, D.E., Smith, R.E.: Nonstationary function optimization using genetic algorithms with dominance and diploidy. In: ICGA, pp. 59–68 (1987)
8. Koza, J.: Genetic Programming: On the Programming of Computers by Means of Natural Selection, vol. 1. MIT press, Cambridge (1992)
9. Liu, C., Liu, H., Yang, J.: A path planning method based on adaptive genetic algorithm for mobile robot. J. Inf. Comput. Sci. **8**(5), 808–814 (2011)
10. Macedo, J.: Genetic programming algorithms for dynamic environments. Master's thesis, University of Coimbra, Portugal, September 2015
11. Ng, K.P., Wong, K.C.: A new diploid scheme and dominance change mechanism for non-stationary function optimization. In: Proceedings of the 6th International Conference on Genetic Algorithms, pp. 159–166. Morgan Kaufmann Publishers Inc (1995)
12. Nguyen, T.T., Yang, S., Branke, J.: Evolutionary dynamic optimization: a survey of the state of the art. Swarm Evol. Comput. **6**, 1–24 (2012)

13. O'Neill, M., Nicolau, M., Brabazon, A.: Dynamic environments can speed up evolution with genetic programming. In: Proceedings of the 13th Annual Conference Companion on Genetic and Evolutionary Computation, pp. 191–192. ACM (2011)

14. Quigley, M., Conley, K., Gerkey, B., Faust, J., Foote, T., Leibs, J., Wheeler, R., Ng, A.Y.: Ros: an open-source robot operating system. In: ICRA Workshop on Open Source Software (2009)

15. Riekert, M., Malan, K.M., Engelbrect, A.P.: Adaptive genetic programming for dynamic classification problems. In: IEEE Congress on Evolutionary Computation, CEC 2009, pp. 674–681. IEEE (2009)

16. Shengxiang, Y., Yao, X.: Evolutionary Computation for Dynamic Optimization Problems, vol. 1. Springer, Berlin (2013)

17. Simões, A.: Improving memory-based evolutionary algorithms for dynamic environments. Ph.D. thesis, University of Coimbra (2010)

18. Simões, A., Costa, E.: On biologically inspired genetic operators: transformation in the standard genetic algorithm. In: Proceedings of the Genetic and Evolutionary Computation Conference, pp. 584–591 (2001)

19. Simões, A., Costa, E.: Using genetic algorithms to deal with dynamic environments: a comparative study of several approaches based on promoting diversity. In: Proceedings of the Genetic and Evolutionary Computation Conference GECCO, vol. 2 (2002)

20. Sugawara, K., Kazama, T., Watanabe, T.: Foraging behavior of interacting robots with virtual pheromone. In: Proceedings of 2004 IEEE/RSJ International Conference on Intelligent Robots and Systems, (IROS 2004), vol. 3, pp. 3074–3079. IEEE (2004)

21. Teuscher, C., Sanchez, E., Sipper, M.: Romero's Odyssey to Santa Fe: from simulation to real life. Robot. Manufact. Syst. Recent Results Res. Dev. Appl. **10**, 262–267 (2000)

22. Vanneschi, L., Cuccu, G.: A study of genetic programming variable population size for dynamic optimization problems. In: IJCCI, pp. 119–126 (2009)

23. Vavak, F., Fogarty, T.C., Jukes, K.: A genetic algorithm with variable range of local search for tracking changing environments. In: Ebeling, W., Rechenberg, I., Voigt, H.-M., Schwefel, Hans-Paul (eds.) PPSN 1996. LNCS, vol. 1141, pp. 376–385. Springer, Heidelberg (1996)

24. Wagner, N., Michalewicz, Z., Khouja, M., McGregor, R.R.: Time series forecasting for dynamic environments: the DyFor genetic program model. IEEE Trans. Evol. Comput. **11**(4), 433–452 (2007)

25. David, R., McDermott, J., Castelli, M., Manzoni, L., Goldman, B.W., Kronberger, G., Jaśkowski, W., O'Reilly, U.-M., Luke, S.: Better GP benchmarks: community survey results and proposals. Genet. Program. Evolvable Mach. **14**(1), 3–29 (2013)

26. Yang, S., Tinós, R.: A hybrid immigrants scheme for genetic algorithms in dynamic environments. Int. J. Autom. Comput. **4**(3), 243–254 (2007)

27. Yin, Z., Brabazon, A., O'Sullivan, C., O'Neil, M.: Genetic programming for dynamic environments. In: Proceedings of the International Multiconference on Computer Science and Information Technology, pp. 437–446 (2007)

A Memory-Based NSGA-II Algorithm for Dynamic Multi-objective Optimization Problems

Shaaban Sahmoud and Haluk Rahmi Topcuoglu[(✉)]

Computer Engineering Department, Marmara University, 34722 Istanbul, Turkey
ssahmoud@marun.edu.tr, haluk@marmara.edu.tr

Abstract. Dynamic multi-objective optimization problems (DMOPs) have been rapidly attracting the interest of the research community. Although static multi-objective evolutionary algorithms have been adapted for solving the DMOPs in the literature, some of those extensions may have high running time and may be inefficient for the given set of test cases. In this paper, we present a new hybrid strategy by integrating the memory concept with the NSGA-II algorithm, called the MNSGA-II algorithm. The proposed algorithm utilizes an explicit memory to store a number of non-dominated solutions using a new memory updating technique. The stored solutions are reused in later stages to reinitialize part of the population when an environment change occurs. The performance of the MNSGA-II algorithm is validated using three test functions from a framework proposed in a recent study. The results show that performance of the MNSGA-II algorithm is competitive with the other state-of-the-art algorithms in terms of tracking the true Pareto front and maintaining the diversity.

Keywords: NSGA-II · Dynamic multi-objective optimization problems · Memory/search algorithms · Hybrid genetic algorithms

1 Introduction

In a multi-objective optimization problem, there is more than one objective where at least two objectives are in conflict with one another. Multi-objective evolutionary algorithms (MOEAs) have been applied to solve various multi-objective optimization problems in different domains during the last twenty years.

MOEAs evolve a population of candidate solutions to find a set of optimal solutions instead of a single optimal solution. This set of optimal solutions is called Pareto optimal front (POF) in the fitness space, and it is called Pareto optimal set (POS) in the decision space. The non-dominated sorting strategy was used to determine the best solutions in the population. The non-dominated solutions can be defined as follows: each solution is compared with all other solutions in the population, and solutions which are not dominated by any other solutions are called non dominated solutions. There are a large number of Multi-objective Evolutionary Algorithms (MOEAs) that have been proposed to solve MOPs including NSGA [1], NSGA-II [2], SPEA [3], MOPSO [4–8] where they mainly target for the stationary environments.

© Springer International Publishing Switzerland 2016
G. Squillero and P. Burelli (Eds.): EvoApplications 2016, Part II, LNCS 9598, pp. 296–310, 2016.
DOI: 10.1007/978-3-319-31153-1_20

Although MOEAs work very well when the environment is stationary, most of the real-world MOPs have parameters that change over time [9, 10]. In a dynamic multi-objective optimization problem (DMOP), one or more elements including the objective functions or the problem constraints change in time. According to the classification of DMOPs presented by Farina et al. [11], the dynamism may become as one or more of the types listed below:

- The Pareto optimal set (POS) changes over time while the Pareto optimal front (POF) remains stationary.
- Both the POF and POS change over time.
- The POF changes over time while the POS remains stationary.
- Both the POF and POS remain stationary.

This classification does not take into account the sources of dynamism in the problem such that if the dynamism happens because of changing in constraints or changing in objectives. Therefore another classification depending on other criteria may be found in literature to classify the DMOPs. As in single DOPs in order to solve DMOPs, we need to track the moving set of optimal solutions once change is detected. It will be time consuming and inefficient to restart the optimization process with new population when a new change is detected.

In this paper, we propose a memory-based algorithm to solve DMOPs by enhancing the well known NSGA-II algorithm. The explicit memory can be very useful for improving the performance of the NSGA-II to deal with dynamic environments. The experimental results show that if we design a fair comparison by equalizing the cost (i.e. the running time) of the algorithms, the memory based NSGA-II algorithm presented in this study outperforms the state-of-the-art dynamic multi-objective evolutionary algorithms.

The rest of this paper is organized as follows. Section 2 presents a brief overview on dynamic multi-objective optimization problem and it provides a classification of dynamic multi-objective evolutionary algorithms. Section 3 presents our proposed approach, the MNSGA-II algorithm. A comparative study and related discussions based on a set of DMOP test instances are given in Sect. 4. Finally Sect. 5 summarizes the paper.

2 Dynamic Multi-objective Evolutionary Algorithms

In a dynamic multi-objective optimization problem (DMOP), there exist two or more conflicting objectives, where the objectives, constraints or parameters of the problem change over time. Whenever a change happens and is detected in the environment of a DMOP, usually at least the Pareto optimal set (POS) or the Pareto optimal front (POF) may changed, as shown in previous section. The dynamic POF is the set of nondominated solutions with respect to the objective space, while the dynamic POS is the set of nondominated solutions with respect to the decision space, at the given time. Since a change in the DMOP affects the existing solutions of the DMOP, the goal is to track the dynamic POF. Therefore the new POF and the new POS must be found as fast as possible before the next change happening.

There are a number of dynamic multi-objective evolutionary algorithms that consider the diversity in the population in order to track the dynamic POF, which can be broadly classified into three categories [12]: diversity introduction, diversity maintenance and multiple populations. Prediction-based and memory-based techniques are the alternatives of the techniques that are based on diversity of the population. In the following section, we briefly summarize dynamic MOEAs in four categories, which is adapted from the classification given in [12].

Approaches That Control the Diversity. This category of algorithms either introduces diversity during restart or they preserve diversity throughout the execution of the algorithm. All population or part of it can be reinitialized randomly whenever a change is detected in the environment [10]. A more efficient and widely used method is randomly reinitializing a fixed or changed percent of population in each generation regardless of the detected change in the environment, as in the random immigrants method [13]. Additionally, the diversity level of a population can be updated when a change occurs by using the mutation operator. The mutation rate can be increased or decreased according to the problem changes and the diversity of the population [14, 15].

The non-dominated sorting genetic algorithm (NSGA-II) proposed in [2] is one of the most efficient algorithms in solving multi-objective optimization problems. NSGA-II solved the main issues of NSGA [1] and introduced an improved version with the $O(MN^2)$ computational complexity. There are a number of diversity maintenance based extensions on NSGA-II in order to apply for the DMOPs. The DNSGA-II-A algorithm [10] replaces fixed percent of the population with randomly created new solutions whenever there is a change in the problem. This helps the algorithm to introduce new (random) solutions whenever there is a change in the problem and so the diversity will be maintained while running. The DNSGA-II-B algorithm [10] replaces fixed percent of the population with mutated solutions of existing solutions chosen randomly, where the mutation rate is increased whenever environment change is detected. In this way, the new solutions that will be introduced in the population are near the existing population.

In [16], two NSGA-II based algorithms to solve DMOPs are proposed. The first algorithm DNSGA-II-Rm replaces 20 % of the population in each generation to maintain the diversity in the population. The second algorithm, the DNSGA-II-HM algorithm, works by increasing the fixed small mutation rate for one generation after a change, otherwise the algorithm works with the normal mutation value. Based on the experimental study, hyper-mutation based algorithm performs better than all other traditional approaches [16].

Multi-population. A multi-population approach maintains multiple subpopulations concurrently, where each of them may have a separate task (i.e., searching the global optimum or tracking any change) and may handle separate area of the search space. There are multi-population algorithms that use Particle Swarm Optimization (PSO) in order to solve DMOPs. In [17], the author presents a heuristic algorithm to solve multi-objective problems using sharing in swarm optimization algorithms and then the work was extended to solve DMOPs. Helbig and Engelbrecht present DVEPSO algorithm [18] to solve DMOPs. DVEPSO was extended from multi-swarm PSO-based algorithm, called the vector evaluated particle swarm optimization (VEPSO) algorithm. There are alternative strategies proposed in the literature to improve the performance of the DVEPSO algorithm such as managing the archive of the algorithm after a change in the environment occurred [19], and using suitable boundary constraint management approach [20].

Another multi-population approach is the Dynamic Competitive-Cooperative Coevolutionary Algorithm (dCOEA) that combines the competitive and cooperative techniques of Coevolutionary algorithms together to solve the MOPs and DMOPs [21]. It uses stochastic competitors and a temporal memory to track the Pareto front in dynamic environments. In the proposed algorithm, individuals from different sub-populations evolve separately and collaborate with other subpopulations to solve the problem. The competitive and cooperative processes may be executed iteratively or with a certain frequency entered to the system. The dCOEA algorithm outperforms other state-of-the-art algorithms for a set of solving test instances of DMOPs [22].

Memory-Based Approaches. Memory based approach in dynamic optimization uses implicit or explicit memory schemes to store the detected best solutions and to reuse this information at a later stage when environment change happens. This approach is very effective when periodic changes occurred, which is widely used in single objective problems [23–25]. In another study, an artificial immune system uses a repository in order to store the generated nondominated solutions [26]. There are only a few memory-based multi-objective optimization algorithms presented in the literature.

Prediction-Based Approaches. There are prediction based algorithms by exploiting the past information in order to solve DMOPs [27, 28]. These algorithms try to estimate the POF or the POS of next change from the history of past sequence locations of the global optimum using a forecasting method. After applying the prediction methods, there are many methods that can be used to seed the population when a change in the objective landscape arrives in order to make the convergence to the new global optimum faster. For example new solutions are created by varying the solution in the last time window with a "predicted" Gaussian noise, or new solutions are sampled around the predicted locations using modified mutation, crossover or fitness functions [9]. The main problem of the algorithms in this category is the highly dependence on the predictors' training process. Selecting inconvenient training data or an inconvenient training method could cause frequent errors in prediction phase.

3   Memory-Based NSGA-II Algorithm (MNSGA-II)

To enhance the ability of NSGA-II algorithm to track the non-dominated solutions in dynamic environment, we use explicit memory in the algorithm to store the best solutions in each generation, as in the memory/search algorithm [25]. The memory/search algorithm is a memory-based algorithm which divides a population of individuals into two sub-populations: a *memory population* and a *search population*.

A separate *explicit memory* is also considered, where the memory population exploits the explicit memory by maintaining minimum jumps, and the search population submits new peaks to the memory.

Our proposed algorithm uses two populations with the NSGA-II which are search and main populations as shown in Fig. 1. The search population is randomly reinitialized whenever an environment change occurs. It is only used to provide the memory with the best non-dominated solutions, so it should not be updated from memory. On the other hand, the main population is the population considered in the original NSGA-II algorithm. Whenever an environment change is detected the main population updates itself from memory.

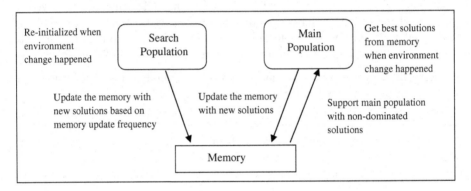

Fig. 1. Scheme of the Memory/Search Algorithm [25]

The outline of the MNSGA-II algorithm is given in Fig. 2. Instead of using a single main population as in the memory/search algorithm, we consider parent and child populations in the MNSGA-II algorithm. The MNSGA-II algorithm starts by initializing the search, parent and child populations randomly, and by using an empty memory. In each generation the algorithm checks if there is a change in the environment. If a change occurs, the search population is re-initialized randomly. After that parents and child populations are merged with memory populations to construct "*population R*" with $2N + K$ elements, where N is the size of the parent or child population and K is the memory size. If there is no change in the environment, parent and child populations are merged to construct "population R" with 2N elements.

Updating the Memory. In a memory-based evolutionary algorithm, the memory updating process is very important and it can directly affect the performance of the algorithm. For updating the memory, it is required to set how the best solution is selected and how to replace solutions when the memory is full. For DMOPs, we have a set of best solutions rather than one best solution as in the single objective problems.

```
begin
    InitializeRandomly(S(N), P(N), Q(N))     // Initialize Search, parent and child population
    Initialize(Memory(k))                    //Initially memory is empty.
    while termination condition not satisfied do
        If change happened Then
            ReinitializeRandomly(S(N))           // Re-initialize randomly all the Search Population
            R = Merge(P(N), Q(N),Memory(k))  // Merge memory, parent and child populations (2N+k)
        else
            R = Merge(P(N), Q(N))                // Merge parent and child populations (2N)

        F = FastNonDominatedSort(R)          // Sort the population to F = F1,F2 ...
        P(N) = Φ and j = 1
        // while the parent population not filled add  population set of rank j to the parent population
        while | P | + | Fⱼ | < N  do
            P = P + Fⱼ
            j = j +1
        end
        DensitySort(Fⱼ)              // Sort in descending order according to the density
        P  = P + Fⱼ[1: ( N − | P|) ]   // Fill the remaining positions in P according to density
        Q  = NormalGenetic(P)      // generate new population using normal Genetic algorithm

        if update frequency condition satisfied Then
            //Get S Best non-dominated solutions
            Best[S] = MergeAndSelectBest(Pmemory(N), PSearch (N))
            // Add Best[S] non-dominated solutions to memory
            UpdateMemory(Best[S], Memory(k))
        end  // if end
    end   // while end
end  // Program end
```

Fig. 2. Memory-based NSGA-II Algorithm for DOPs.

Therefore it may not be appropriate to select only one solution and add it to memory. We have three options to update the memory:

(a) Add all non-dominated solutions to the memory;
(b) Add the non-dominated solutions that represent the sides of the POF to the memory, where the number of these solutions will be the same as the number of objectives in the problem;
(c) Add selected solutions from the set of non-dominated solutions based on other heuristic methods.

In our proposed algorithm the third option with minimum distance search is used to select solutions that will be added to memory. Figure 3 shows the steps of updating the memory.

The size of population R must be reduced to include only N solutions. The non-domination sorting strategy of NSGA-II is used for this reduction. Firstly, the

R population is sorted using fast non-dominated sort. Secondly, the next iteration parent population is constructed from the faces one by one and respectively. The faces are continuing taken until there is no space for next full face let it F_i. Then F_i is sorted in descending order according to the density measure proposed in NSGA-II which uses a crowding distance definition. The remaining positions in the next iteration parent population are filled respectively from sorted F_i.

```
S = size(Memory)              // get memory size
Index = 0                     // memory index
P(N) = currentPop()           // get current population
SSet = getFaceOne(P(N))       // get the Non-Dominated Solutions


For all SSet elements
    If Index < S Then
            // if memory is not full add this solution
            Memory = Memory + SSet[i]
            Index = Index +1
    Else
        // get nearest solution from memory to SSet[i]
        So = getNearestSol(Memory, SSet[i] )
        // if new solution is better than nearest one to it, replace them
        If  isBetter(SSet[i], So) Then
         Memory = replace( SSet[i],So)
        End
    End
End
```

Fig. 3. The steps of the memory updating in MNSGA-II algorithm.

The memory in our proposed algorithm starts empty and at every x generation it will be updated, where x is a predefined value that represents the memory update frequency [25]. When MNSGA-II is started, the memory will be empty and the non-dominated individuals will be selected from the current main and search populations. If the memory is not full, the non-dominated solutions are directly stored in the memory. But when the memory becomes full, the nearest solution from memory to each non-dominated individual is searched. If the new non-dominated solution is better that its nearest solution from memory, the two solutions are replaced; otherwise, the memory stays without change. To find the nearest solution, the minimum Euclidean distance in decision space is used. Using this method the memory of the MNSGA-II algorithm will be able to store the non-dominated solutions of past generations in different regions in decision space. The memory contents will help the algorithm to find new solutions very fast, after an environment change occurs. If the environment changes are cyclic, best performance can be expected from the algorithm. On the other hand, if the new POF are not tracked before, it will find it quickly because the algorithm already keeps the best solutions in distributed areas in the decision space.

4   Experimental Study

In this section we present a comprehensive comparison of our MNSGA-II algorithm, with five multi-objective dynamic evolutionary algorithms by using a set of test functions from a recently proposed framework. The algorithms in the comparison study are the random immigrant NSGA-II (NSGA-II-RI) [16], the hyper-mutation NSGA-II (NSGA-II-HM) [16], DNSGA-II-A [10], DNSGA-II-B [10], and the Competitive-Cooperative Coevolutionary algorithm (dCOEA) [21].

4.1   Test Problems and Comparison Metrics

Jiang and Yang [22] proposed a framework to construct scalable dynamic test problems, where dynamism can be easily added and controlled. This framework can be used to implement dynamic test problems from all the four dynamic problem types as described in the introduction. In this paper, three dynamic test problems from this framework will be used, which are the SJY1, the SJY2 and the SJY3 problems. The time t in each of the three problems is defined as,

$$t = \frac{1}{n_t} * \frac{\tau}{fr} \tag{1}$$

Where n_t represents the severity of change, τ is the iteration count and fr is the frequency of change. In the experiments we assume that the change time is known, therefore there is no change detection phase.

- **SJY1 test problem:** The POF is linear and it remains stationary over time where POS changes [22]. This problem can be used to test the tracking ability of dynamic multi-objective evolutionary algorithms. This problem is defined as:

$$f_{i=1:M}(x,t) = (1 + S(x,t)) * \left(\frac{x_i}{x_1 + x_2 + \ldots + x_M} \right)$$
$$S(x,t) = \sum_{x_i \in X_{II}} (x_i - G(t))^2$$
$$G(t) = \sin(0.5\pi t)$$

where $x_{i=1:M} \in X_I$, $x_{i=M+1:n} \in X_{II}$, the POF of this problem is defined as: $f_1 + f_2 + \ldots + f_M = 1$.
And the POS of this problem is defined as: $x_i = G(t) \, \forall x_i \in X_{II}$

- **SJY2 test problem:** The POF is a convex hyper-surface and it remains stationary over time where POS changes [22]. This test problem can be used to test the diversity and the percent of covering the POF. This problem is defined as:

$$f_{i=1:M}(x,t) = (1+S(x,t)) * \left(\frac{x_i}{\sqrt[M-1]{\prod_{j \neq i, x_j \in X_I} x_j}} \right)$$

$$swap_t(x_i \in X_I, x_{M-i} \in X_I), i = i : n_p$$

$$S(x,t) = \sum_{x_i \in X_{II}} (x_i - \frac{\sum_{j=1}^{n_p} x_j}{n_p})^2$$

Where $n_p = rand(1, M/2)$ and $x_{i=1:M} \in X_I$, $x_{i=M+1:n} \in X_{II}$, the POF of this problem is defined as: $f_1 * f_2 * \ldots * f_M = 1$
And the POS of this problem is defined as:

$$x_i = \frac{\sum_{j=1}^{n_p} x_j}{n_p}, \forall x_j \in X_I, \forall x_i \in X_{II}$$

- **SJY3 test problem:** In this problem, both POF and POS change over time. This problem tests the tracking ability of dynamic multi-objective evolutionary algorithms because it adds some difficulty in relocating the new POS after each change [22]. It is defined as:

$$f_{i=1:M}(x,t) = (1+S(x,t)) * \left(\frac{x_i}{\sqrt[M-1]{\prod_{j \neq i, x_j \in X_I} x_j}} \right)^{H(t)}$$

$$swap_t(x_i \in X_I, x_{M-i} \in X_{II}), i = i : n_p$$

$$S(x,t) = \sum_{x_i \in X_{II}} (x_i - 5)^2$$

where $n_p = rand(1, \frac{M}{2})$, $H(t) = 0.5 + 2|\sin(0.5\pi t)|$.
And $x_{i=1:M} \in X_I$, $x_{i=M+1:n} \in X_{II}$, the POF of this problem is defined as:

$$f_1 * f_2 * \ldots * f_M = 1.$$

And the POS of this problem is defined as: $x_i = 5, \forall x_i \in X_{II}$

Performance Metrics. Two metrics are used in order to measure both the diversity and convergence of the algorithms:

- The mean inverted generational distance (mIGD): This metric is derived from IGD as discussed in [29], which can be calculated as:

$$mIGD = \frac{1}{g_n} * \sum_{t=1}^{g_n} IGD_t, \tag{2}$$

where g_n is the number of generations

- The mean inverted generational distance just before the change (mIGDB): This metric is developed to compute the performance of the DMOE algorithms just before the next change. It is used because mIGD computes the total average generational distance for all generations together and could not compute the last generational distance value just before the change. The term mIGDB can be calculated as:

$$mIGDB = \frac{1}{C_n} * \sum_{t=1}^{C_n} IGD_t, \tag{3}$$

Where C_n is the number of changes and IGD_t is calculated just before the next change

The use of IGD metrics needs the generation of randomly uniformly distributed solutions on the POF. In this work 500 points on the POF were generated for SYJ1 and SJY2 test problems. For SJY3, because the POF changes as the t changes, 500 points were generated for each different t.

4.2 Comparative Study

To implement a fair framework for comparison, the NSGA-II parameters are fixed for all dynamic algorithms with population size of 200, the crossover probability of 0.7 and a mutation probability of $1/n$ (where n is the string length for binary-coded). To confirm fairness in the experiments, all algorithms variables are represented using binary representation. The single-point crossover and bitwise mutation for binary-coded were used.

For the proposed MNSGA-II algorithm, a memory of size 100 and update frequency of 10 are considered, based on the results of preliminary experiments. Five different algorithms are used in our comparison study. The first and the second algorithms are DNSGA-II-A and DNSGA-II-B, where a replacement value of 20 % is used for both algorithms as suggested in [10]. The first algorithm replaces 20 % of the population randomly whenever a change happens, where the second algorithm mutates 20 % of randomly selected individuals whenever a change happens.

The third and fourth algorithms are the DNSGA-II-RI and the DNSGA-II-HM algorithms [16]. The third algorithm is a random immigrant algorithm which replaces 20 % of the population randomly in each generation. The fourth algorithm is a hyper-mutation algorithm which increases the mutation for one generation after environment change; and otherwise the mutation remains $1/n$, where n is the string length. For the dCOEA algorithm, the population size set to 200 as the other algorithms. The subpopulation size was 20 and bit representation with 30 bit was used.

The number of generations is fixed to be $10*fr$, where fr is the frequency of change. This ensures that 10 changes will happen in each experiment. All algorithms are tested using four different frequencies of changes, which are 10, 20, 30 and 50. The number of variables is fixed to be 10 and 20 and the number of objectives was set to three for all experiments conducted.

Table 1. Performance of the six algorithms for the SJY1 problem.

Number of variables	Frequency of change	DNSGA-II-A		DNSGA-II-B		DNSGA-II-HM		DNSGA-II-Rim		dCOEA		MNSGA-II	
		mIGDB	mIGD	mIGDB	mIGD	mIGDB	mIGD	mIGDB	mIGD	mIGDB	mIGD	mIGDB	mIGD
10	10	0.1584	0.2417	0.1575	0.2434	0.1222	0.1349	0.1521	0.2175	**0.0657**	**0.0763**	0.1145	0.1264
	20	0.0870	0.1608	0.0970	0.1668	0.0721	0.0811	0.0882	0.1541	**0.0472**	**0.0621**	0.0633	0.0729
	30	0.0679	0.1316	0.0719	0.1326	0.0534	0.0611	0.0642	0.1230	**0.0463**	**0.0559**	0.0535	0.0611
	50	0.0506	0.1044	0.0523	0.1134	0.0467	0.0513	0.0496	0.0983	**0.0416**	**0.0494**	0.0472	0.0520
20	10	0.8700	1.1307	0.8700	1.1307	0.5923	0.6452	0.8976	1.1279	**0.1376**	**0.1328**	0.6027	0.6568
	20	0.4389	0.7329	0.4389	0.7329	0.2932	0.3377	0.5098	0.7720	**0.1043**	**0.1039**	0.2950	0.3366
	30	0.2813	0.5517	0.2813	0.5517	0.1643	0.2027	0.3172	0.5659	**0.0878**	**0.0950**	0.1499	0.1914
	50	0.1550	0.3659	0.1550	0.3659	0.0838	0.1128	0.1813	0.3864	**0.0663**	**0.0806**	0.0892	0.1184

Table 2. Performance of the six algorithms for the SJY2 problem.

Number of variables	Frequency of change	DNSGA-II-A		DNSGA-II-B		DNSGA-II-HM		DNSGA-II-Rim		dCOEA		MNSGA-II	
		mIGDB	mIGD	mIGDB	mIGD	mIGDB	mIGD	mIGDB	mIGD	mIGDB	mIGD	mIGDB	mIGD
10	10	1.3748	1.9443	1.2118	1.8443	1.2504	1.9110	2.1213	2.7190	0.9633	0.9753	**0.7230**	**0.8393**
	20	0.8935	1.3443	0.8945	1.5642	0.7769	1.3447	1.5473	2.0780	0.8641	0.8764	**0.5892**	**0.6236**
	30	0.7395	1.1393	0.7456	1.1563	0.6167	0.9602	1.3544	1.8475	0.8489	0.8479	**0.5762**	**0.5888**
	50	0.7943	1.0007	0.7913	1.1127	0.5520	0.7581	1.3002	1.5882	0.6799	0.7547	**0.5415**	**0.5462**
20	10	10.0852	15.2759	10.0852	14.5642	4.7281	7.1789	16.0380	20.3555	**1.5988**	**1.5748**	4.1459	5.6707
	20	3.7515	8.3075	3.1235	8.5494	1.9764	2.9203	5.3453	9.6182	**1.2229**	**1.2466**	1.2520	2.0412
	30	1.7770	4.9044	1.8150	4.5584	0.8928	1.3431	3.0311	6.5787	0.9811	**1.0373**	**0.8176**	1.1098
	50	1.3827	3.8694	1.3547	3.1294	0.6968	0.8469	2.0806	4.9672	**0.6798**	**0.7451**	0.6915	0.7512

Table 3. Performance of the six algorithms for the SJY3 problem.

Number of variables	Frequency of change	DNSGA-II-A		DNSGA-II-B		DNSGA-II-HM		DNSGA-II-Rim		dCOEA		MNSGA-II	
		mIGDB	mIGD	mIGDB	mIGD	mIGDB	mIGD	mIGDB	mIGD	mIGDB	mIGD	mIGDB	mIGD
10	10	11.783	12.488	11.927	12.543	11.843	12.627	12.398	13.119	**11.684**	**11.851**	11.729	12.448
	20	11.684	12.212	11.686	12.095	11.503	11.972	12.174	12.779	11.580	**11.736**	**11.442**	12.007
	30	11.567	11.945	11.563	11.904	11.494	11.857	12.112	12.611	11.486	**11.628**	**11.416**	11.811
	50	11.401	11.604	11.470	11.707	**11.334**	**11.570**	12.118	12.381	11.460	11.587	11.401	11.672
20	10	15.992	22.169	14.717	20.288	15.342	20.871	16.996	22.736	**11.756**	**11.954**	15.310	20.880
	20	13.241	17.683	12.298	16.010	12.256	16.573	14.046	18.790	**11.601**	**11.775**	12.184	16.861
	30	11.661	15.210	11.830	14.764	11.443	14.785	13.044	16.852	11.568	**11.716**	**11.223**	14.560
	50	11.019	13.717	11.628	13.237	**10.562**	12.679	12.064	14.898	11.538	**11.668**	10.725	13.047

Tables 1, 2 and 3 show the results of the SYJ1, SJY2 and SJY3 problems respectively, where bold number represent the best values of each case. The dCOEA and the MNSGA-II algorithms are the best two algorithms in all experiments except three cases that NSGA-II-HM got the best result. Based on the results given in Table 1, the dCOEA algorithm gives the minimum mIGD and mIGDB values in all experiments. On the other hand, the MNSGA-II algorithm is the second best algorithm in SJY1 problem when comparing it to the other algorithms. The DNSGA-II-Rim also works well in some cases but it is still not good as the MNSGA-II algorithm.

Based on the results of SJY2 problem given in Table 2, our proposed algorithm, the MNSGA-II algorithm provides the best values in both metrics when the number of variables is 10. When the number of variables is 20, the dCOEA algorithm gains the best results in small frequency of change numbers. For higher frequency of change cases, the MNSGA-II algorithm starts to outperform the dCOEA algorithm. The DNSGA-II-HM algorithm works well and provides results close the results of the MNSGA-II algorithm in both metrics and especially in high change frequencies.

The results of the SJY3 problem are similar to the results of the SJY2 problem. For the mIGDB metric, the MNSGA-II and dCOEA algorithms are the best algorithms except in two cases where the DNSGA-II-HM algorithm generates better results. The dCOEA algorithm obtained the best values in most cases for the mIGD metric. Based on the results of the three problems, the dCOEA is an efficient algorithm and works better than all the other algorithms in many cases where MNSGA-II algorithm becomes the second best algorithm. The MNSGA-II algorithm outperforms all other remaining algorithms in most cases and outperforms the dCOEA algorithm in some cases.

Table 4. Time and number of function evaluations of the six algorithms.

Number of variable	Frequency of change	DNSGA-II-A		DNSGA-II-B		DNSGA-II-HM		DNSGA-II-Rim		dCOEA		MNSGA-II	
		Time (s)	Func. Eval.	Time (s)	Func. Eval.	Time (s)	Func. Eval.	Time (s)	Func. Eval.	Time (s)	Func. Eval.	Time (s)	Func. Eval.
10	10	0.38	2220	0.38	2220	0.38	2020	0.4	4020	1.18	3307	0.61	4200
	20	0.73	4220	0.73	4220	0.75	4020	0.73	8020	3.67	7351	1.17	8220
	30	1.01	6220	1.01	6220	1.05	6020	1.07	12020	6.51	11409	1.71	12220
	50	1.80	10220	1.80	10220	1.84	10020	1.86	20020	10.60	19523	2.92	20220
20	10	0.51	2220	0.51	2220	0.52	2020	0.55	4020	1.27	5071	0.88	4200
	20	1.06	4220	1.06	4220	1.06	4020	1.09	8020	3.92	11138	1.75	8220
	30	1.51	6220	1.51	6220	1.51	6020	1.51	12020	6.57	17175	2.60	12220
	50	2.61	10220	2.61	10220	2.61	10020	2.69	20020	12.20	29245	4.15	20220

4.2.1 Performance Evaluation of Algorithms by Equalizing Running Times

For performance evaluation of dynamic optimization algorithms, the number of generations is the common measure for setting the frequency of change values. During the execution of previous experiments, although the dCOEA algorithm outperforms the other algorithms for some of the test cases, it takes significantly longer time than the other algorithms for all cases. Table 4 provides the running times (in seconds) and the number of function evaluations of all algorithms in the comparison study. The average execution times between two consecutive changes in the environment are computed for all algorithms for each frequency of change value.

The results clearly show that all algorithms except the dCOEA algorithm take approximately similar amount of time for execution; and the dCOEA algorithm requires significantly longer time. For example, for the case 20 variables with change frequency of 50 generations, the dCOEA algorithm takes average of 12.20 s between two environment changes, where MNSGA-II algorithm can be executed in only 4.15 s. The MNSGA-II works three times faster than the dCOEA algorithm for the given case. The longer execution time of the dCOEA algorithm is due to the high computations in the competitive and cooperative mechanisms in the algorithm.

In this subsection, we repeat the experimental study by setting equal execution time between two consecutive changes in the environment for the best two algorithms from

Table 5. Performance of the MNSGA-II and the dCOEA algorithms for SJY1, SJY2 and SJY3 problems.

Number of variables	Frequency of change (seconds)	SJY1 Problem				SJY2 Problem				SJY3 Problem			
		MNSGA-II		dCOEA		MNSGA-II		dCOEA		MNSGA-II		dCOEA	
		mIGDB	mIGD	mIGDB	mIGD	mIGDB	mIGD	mIGDB	mIGD	mIGDB	mIGD	mIGDB	mIGD
10	1.18	**0.0596**	**0.0697**	0.0657	0.0763	**0.5714**	**0.6959**	0.9633	0.9753	**11.442**	12.007	11.684	**11.851**
	3.67	**0.0468**	**0.0520**	0.0472	0.0621	**0.5581**	**0.6704**	0.8641	0.8764	**11.401**	**11.672**	11.580	11.736
	6.51	**0.0458**	**0.0475**	0.0463	0.0559	**0.5473**	**0.6108**	0.8489	0.8479	**11.325**	**11.488**	11.486	11.628
	10.6	**0.0388**	**0.0414**	0.0416	0.0494	**0.5351**	**0.5919**	0.6799	0.7547	**11.403**	**11.441**	11.460	11.587
20	1.27	0.251	0.2157	**0.1376**	**0.1328**	**1.2520**	2.0412	1.5988	**1.5748**	12.184	16.861	**11.756**	**11.954**
	3.92	**0.0892**	0.1184	0.1043	**0.1039**	**0.6915**	**0.7512**	1.2229	1.2466	**10.725**	13.047	11.601	**11.775**
	6.57	**0.0526**	**0.0635**	0.0878	0.0950	**0.5462**	**0.9521**	0.9811	1.0373	**10.009**	**11.198**	11.568	11.716
	12.2	**0.0519**	**0.0606**	0.0663	0.0806	**0.5125**	**0.7326**	0.6798	0.7451	**10.001**	**10.633**	11.538	11.668

the previous section, which are dCOEA and MNSGA-II algorithms. In other words, the frequency of change will be measured using the time between two changes in seconds rather than using number of generations.

In order to perform this experiment, the average time needed to execute 10, 20, 30 and 50 generations for dCOEA algorithm is computed. Then, we run the MNSGA-II for the same amount of time before the next change occurs. When the frequency of change is equal to 10 generations, the dCOEA takes 1.18 s to execute. So the frequency of change becomes 1.18 s for the MNSGA-II algorithm rather than 10 generations and because MNSGA-II is faster than dCOEA, it will run more than 10 generations in this time period.

Table 5 shows the result of this experiment for the given three test cases. It is clear that if we do a fair comparison and assign equal running times for the MNSGA-II and the dCOEA algorithms between two environmental changes, the MNSGA-II algorithm outperforms dCOEA for most considered cases. For the mIGDB metric, the MNSGA-II algorithm outperforms the dCOEA algorithm in 22 out of 24 considered cases, where it outperforms the dCOEA algorithm for 18 out of 24 cases based on the mIGD metric. The results show that MNSGA-II much better than dCOEA when the number of variables is 10. When the number of variables becomes 20, dCOEA works better than MNSGA-II in fast environment changes, where MNSGA-II outperforms dCOEA when environment changes happen in long periods of time.

This result highlights a very important remark in designing DMOEAs that researchers must take attention to it; in dynamic environments the execution time of the algorithm is important exactly as the performance of the algorithm. Finally, the results show that the proposed MNSGA-II algorithm can be a good alternative to state-of-the-art algorithms in solving the dynamic multi-objective optimization problems.

5   Conclusion

In this paper, we propose a new memory-based algorithm that uses an explicit memory to store a number of non-dominated solutions. We adapt a memory/search algorithm to make the proposed algorithm capable of storing new and distributed solutions. The stored solutions are reused in later stages to reinitialize part of the population when an environment change happens. The proposed algorithm, MNSGA-II algorithm, was implemented by enhancing the commonly-used NSGA-II algorithm. The MNSGA-II algorithm performance is validated using three benchmarks, SJY1, SJY2 and SJY3 test from a framework against five other DMOE algorithms, which are the dCOEA, the DNSGA-II-HM, the DNSGA-II-Rim, the DNSGA-II-A and the DNSGA-II-B algorithms. The results show that MNSGA-II is competitive with other state-of-the-art dynamic algorithms in terms of tracking the true Pareto front and maintaining the diversity.

Acknowledgements. The authors would like to thank Dr. Chi Keong Goh for providing the source code for the dCOEA algorithm. The authors also would like to thank Prof. Shengxiang Yang and Shouyong Jiang for the help regarding their Dynamic Test Problems for Dynamic Multi-objective Optimization.

References

1. Srinivas, N., Deb, K.: Multiobjective optimization using nondominated sorting in genetic algorithms. IEEE Trans. Evol. Comput. **2**(3), 221–248 (1994)
2. Deb, K., Pratap, A., Agarwal, S., Meyarivan, T.: A fast and elitist multiobjective genetic algorithm: NSGA-II. IEEE Trans. Evol. Comput. **6**(2), 182–197 (2002)
3. Zitzler, E., Thiele, L.: Multiobjective evolutionary algorithms: a comparative case study and the strength Pareto approach. IEEE Trans. Evol. Comput. **3**(4), 257–271 (1999)
4. Hu, X., Eberhart, R.C.: Multiobjective optimization using dynamic neighborhood particle swarm optimization. In: Proceedings of Congress on Evolutionary Computation, Honolulu, HI, pp. 1677–1681 (2002)
5. Zhang, L.B., Zhou, C.G., Liu, X.H., Ma, Z.Q., Ma, M., Liang Y.C.: Solving multi objective problems using particle swarm optimization. In: Proceedings of the 2003 Congress on Evolutionary Computation, Canberra, Australia, pp. 2400–2405 (2003)
6. Hu, X., Eberhart, R.C., Shi, Y.: Particle swarm with extended memory for multiobjective optimization. In: Proceedings of IEEE Swarm Intelligence Symposium, Indianapolis, IN, pp. 193–197 (2003)
7. Reddy, M.J., Kumar, D.N.: An efficient multi-objective optimization algorithm based on swarm intelligence for engineering design. Eng. Optim. **39**, 49–68 (2007)
8. Coello, C.C.A., Pulido, G.T., Lechuga, M.S.: Handling multiple objectives with particle swarm optimization. IEEE Trans. Evol. Comput. **8**(3), 256–279 (2004)
9. Rossi, C., Abderrahim, M., Daz, J.C.: Tracking moving optima using kalman-based predictions. Evol. Comput. **16**(1), 1–30 (2008)
10. Deb, K., Rao N., U.B., Karthik, S.: dynamic multi-objective optimization and decision-making using modified NSGA-II: a case study on hydro-thermal power scheduling. In: Obayashi, S., Deb, K., Poloni, C., Hiroyasu, T., Murata, T. (eds.) EMO 2007. LNCS, vol. 4403, pp. 803–817. Springer, Heidelberg (2007)
11. Farina, M., Deb, K., Amato, P.: Dynamic multiobjective optimization problems: test cases, approximations, and applications. IEEE Trans. Evol. Comput. **8**(5), 425–442 (2004)
12. Yang, S., Yao, X.: Evolutionary Computation for Dynamic Optimization Problems. Springer, Heidelberg (2013)
13. Grefenstette, J.: Genetic algorithms for changing environments. In: Proceedings of International Conference Parallel Problem Solving from Nature, pp. 137–144 (1992)

14. Cobb, H.: An Investigation into the use of hypermutation as an adaptive operator in genetic algorithms having continuous, time-dependent nonstationary environments. Technical Report, Naval Research Laboratory (1990)

15. Vavak, F., Jukes, K., Fogarty, T.: Adaptive combustion balancing in multiple burner boiler using a genetic algorithm with variable range of local search. In: Proceedings of 7th International Conference on Genetic Algorithms, pp. 719–726 (1997)

16. Bui, L.T., Nguyen, M.H., Branke, J., Abbass, H.A.: Tackling dynamic problems with multiobjective evolutionary algorithms. In: Knowles, J., Corne, D., Deb, K., Chair, D.R. (eds.) Multiobjective Problem Solving from Nature, pp. 77–91. Springer, Heidelberg (2008)

17. Lechuga, M.S.: Multi-objective optimisation using sharing in swarm optimisation algorithms, Ph.D. Dissertation, University of Birmingham, Birmingham, UK (2009)

18. Greeff, M., Engelbrecht, A.P.: Solving dynamic multi-objective problems with vector evaluated particle swarm optimisation. In: Proceedings of World Congress on Computational Intelligence (WCCI): Congress on Evolutionary Computation, Hong Kong, pp. 2917–2924 (2008)

19. Helbig, M., Engelbrecht, A.P.: Archive management for dynamic multi-objective optimisation problems using vector evaluated particle swarm optimisation. In: Proceedings of Congress on Evolutionary Computation, New Orleans, USA, pp. 2047–2054 (2011)

20. Helbig, M., Engelbrecht, A.P.: Dynamic multi-objective optimisation using PSO. In: Alba, E., Nakib, A., Siarry, P. (eds.) Metaheuristics for Dynamic Optimization. Springer, Heidelberg (2013)

21. Goh, C., Tan, K.: A competitive-cooperative coevolutionary paradigm for dynamic multiobjective optimization. IEEE Trans. Evol. Comput. 13(1), 103–127 (2009)

22. Jiang, S., Yang, S.: A framework of scalable dynamic test problems for dynamic multi-objective optimization. In: CIDUE, pp. 32–39 (2014)

23. Goldberg, D., Smith, R.: Nonstationary function optimization using genetic algorithm with dominance and diploidy. In: Proceedings of 2nd International Conference Genetic Algorithms and Their Applications, pp. 59–68 (1987)

24. Ramsey, C., Grefenstette, J.: Case-based initialization of genetic algorithms. In: Proceedings of 5th International Conference Genetic Algorithms, pp. 84–91 (1993)

25. Branke, J.: Memory enhanced evolutionary algorithms for changing optimization problems. In: Congress on Evolutionary Computation, CEC 1999, pp. 1875–1882 (1999)

26. Zhang, Z., Qian, S.: Artificial immune system in dynamic environments solving time varying non-linear constrained multi-objective problems. Soft. Comput. 15(7), 1333–1349 (2011)

27. Zhou, A., Jin, Y., Zhang, Q., Sendhoff, B., Tsang, E.P.: Prediction-based population re-initialization for evolutionary dynamic multi-objective optimization. In: Obayashi, S., Deb, K., Poloni, C., Hiroyasu, T., Murata, T. (eds.) EMO 2007. LNCS, vol. 4403, pp. 832–846. Springer, Heidelberg (2007)

28. Muruganantham, A., Zhao, Y., Gee, S.B., Qiu, X., Tan, K.: Dynamic multiobjective optimization using evolutionary algorithm with Kalman filter. In: 17th Asia Pacific Symposium on IES, pp. 66–75 (2013)

29. Li, H., Zhang, Q.: Multiobjective optimization problems with complicated pareto sets, MOEA/D and NSGA-II. IEEE Trans. Evol. Comput. 13(2), 284–302 (2009)

Hybrid Dynamic Resampling Algorithms for Evolutionary Multi-objective Optimization of Invariant-Noise Problems

Florian Siegmund[1($\boxtimes$)], Amos H.C. Ng[1], and Kalyanmoy Deb[2]

[1] School of Engineering Science, University of Skövde, Skövde, Sweden
{florian.siegmund,amos.ng}@his.se
[2] Department of Electrical and Computer Engineering,
Michigan State University, East Lansing, USA
kdeb@egr.msu.edu

Abstract. In Simulation-based Evolutionary Multi-objective Optimization (EMO) the available time for optimization usually is limited. Since many real-world optimization problems are stochastic models, the optimization algorithm has to employ a noise compensation technique for the objective values. This article analyzes Dynamic Resampling algorithms for handling the objective noise. Dynamic Resampling improves the objective value accuracy by spending more time to evaluate the solutions multiple times, which tightens the optimization time limit even more. This circumstance can be used to design Dynamic Resampling algorithms with a better sampling allocation strategy that uses the time limit. In our previous work, we investigated Time-based Hybrid Resampling algorithms for Preference-based EMO. In this article, we extend our studies to general EMO which aims to find a converged and diverse set of alternative solutions along the whole Pareto-front of the problem. We focus on problems with an invariant noise level, i.e. a flat noise landscape.

Keywords: Evolutionary multi-objective optimization · Simulation-based optimization · Noise · Dynamic resampling · Budget allocation · Hybrid

1   Introcution

Simulation-based optimization of real-world optimization problems is usually run with a limited time budget [13]. The goal of Evolutionary Multi-objective Optimization is to provide the decision maker with a well-converged and diverse set of alternative solutions close to the true Pareto-front. Covering the whole Pareto-front is usually hard to achieve within the available time, dependent on the optimization problem. Besides limited optimization time, Simulation-based Optimization entails the challenge of handling noisy objective functions [2,4,11,22]. To obtain an as realistic as possible simulation of the system behavior, stochastic system characteristics are often built into the simulation models. When running the

© Springer International Publishing Switzerland 2016
G. Squillero and P. Burelli (Eds.): EvoApplications 2016, Part II, LNCS 9598, pp. 311–326, 2016.
DOI: 10.1007/978-3-319-31153-1_21

stochastic simulation, this expresses itself in deviating result values. Therefore, if the simulation is run multiple times for a certain system configuration, the result value is slightly different for each simulation run.

If an evolutionary optimization algorithm is run without noise handling on a stochastic simulation optimization problem, the performance will degrade in comparison with the case if the true mean objective values would be known. The algorithm will have wrong knowledge about the solutions' quality. Two cases of misjudgment will occur. The algorithm will perceive bad solutions as good and select them into the next generation (Type II error). Good solutions might be assessed as inferior and might be discarded (Type I error). The performance can therefore be improved by increasing the knowledge of the algorithm about the solution quality.

Resampling is a way to reduce the uncertainty about objective values of solutions. Resampling algorithms evaluate solutions several times to obtain an approximation of the expected objective values. This allows EMO algorithms to make better selection decisions, but it comes with a cost. The additional samples used for Resampling, are not available for the optimization. Therefore, a resampling strategy which samples the solutions carefully according to their resampling need, can help an EMO algorithm to achieve better results than a static resampling allocation. Such a strategy is called Dynamic Resampling, DR. In this paper, we run Dynamic Resampling strategies for Evolutionary Multi-objective Optimization, which were proposed for Preference-based EMO in our previous work [17–19], and which can be used on any EMO algorithm. They are evaluated together with the EMO algorithms Non-domination Sorting Genetic Algorithm-II (NSGA-II) [7], and the Hypervolume Estimation algorithm (HypE) [1].

The resampling need varies between solutions and can be calculated in many different ways [19]. One approach is to assign more samples to solutions close to the Pareto-front. Since in real-world problems, the Pareto-front is not known, this can be achieved approximatively by assigning more samples to solutions as the optimization time progresses. If the Ideal Point can be estimated, fitness-based Dynamic Resampling algorithms can be used, which can assign higher samples to solutions with good objective values [8,16]. Another approximation strategy is to assign more samples to solutions that dominate more other solutions, or are dominated by fewer other solutions, respectively. This is done by, for example, Confidence-based Dynamic Resampling and the MOPSA-EA algorithm [21], or Rank-based Dynamic Resampling [17–19]. EMO algorithms have a selection operator which determines if solutions will be part of the next population or be discarded. This approach is used by the MOPSA-EA algorithm [21] which compares pairs of solutions using Confidence-based DR (CDR) or the MOCBA and EA approach [5] (Multi-objective Optimal Computing Budget Allocation), which compares sets of solutions. CDR and MOCBA show promising results [5,16], but they have the disadvantage of limited applicability to only special types of Evolutionary Algorithms. CDR is limited to steady state EAs and MOCBA requires an EA with high elitism. Other examples for DR algorithms which have limited applicability, due to direct integration with a certain

EA, were proposed [10] and [15]. The mentioned algorithms, except for MOPSA-EA/CDR, do not consider the limited sampling budget as in our situation. In this paper, we show how Dynamic Resampling and EMO performance can be improved by considering a fixed time limit for optimization.

The paper is structured as follows. Section 2 provides background information to Dynamic Resampling and an introduction to the NSGA-II and HypE Evolutionary Algorithms. Different Resampling Algorithms for time-constrained optimization are explained in Sect. 3. Section 4 describes a stochastic production line problem. In Sect. 5, numerical experiments of Dynamic Resampling algorithms on the two Evolutionary Algorithms are performed and evaluated with different performance metrics for multi-objective optimization. In Sect. 6, conclusions are drawn and possible future work is pointed out.

2   Background

In this section, background information is provided regarding Resampling as a noise handling method in Evolutionary Multi-objective Optimization. Also, an introduction to the multi-objective optimization algorithms NSGA-II [7] and HypE [1] is given, which are used to evaluate the Dynamic Resampling algorithms in this paper.

2.1   Noise Compensation by Sequential Dynamic Resampling

To be able to assess the quality of a solution according to a stochastic evaluation function, statistical measures like sample mean and sample standard deviation can be used. By executing the simulation model multiple times, a more accurate value of the solution quality can be obtained. This process is called resampling. We denote the sample mean value of objective function F_i and solution s as follows: $\mu_n(F_i(s)) = \frac{1}{n}\sum_{j=1}^{n} F_i^j(s)$, $i = 1 \ldots H$, where $F_i^j(s)$ is the j-th sample of s, and the sample variance of objective function i: $\sigma_n^2(F_i(s)) = \frac{1}{n-1}\sum_{j=1}^{n}(F_i^j(s) - \mu_n(F_i(s)))^2$. The performance degradation evolutionary algorithms experience caused by the stochastic evaluation functions can be compensated partly through resampling.

In many cases, resampling is implemented as a Static Resampling scheme. This means that all solutions are sampled an equal, fixed number of times. The need for resampling, however, is often not homogeneously distributed throughout the search space. Solutions with a smaller variance in their objective values will require less samples than solutions with a higher objective variance. Solutions closer to the Pareto-front or preferred areas in the objective space require more samples. In order to sample each solution the required number of times, Dynamic Resampling techniques are used, which are described in the following section.

Dynamic Resampling. Dynamic Resampling allocates a different sampling budget to each solution, based on the evaluation characteristics of the solution. The general goal of resampling a stochastic objective function is to reduce the standard deviation of the mean of an objective value $\sigma_n(\mu_n(F_i(s)))$, which increases the knowledge about the objective value. A required level of knowledge about the solutions can be specified. For each solution a different number of samples is needed to reach the required knowledge level. Resampling can be performed dynamically in the way that each solution is allocated exactly the required number of samples, up to an upper bound. In comparison with Static Resampling, Dynamic Resampling can save sampling budget that can be used to evaluate more solutions and to run more generations of an evolutionary algorithm instead [17].

With only a limited number of samples available, the standard deviation of the mean can be estimated by the sample standard deviation of the mean, which usually is called standard error of the mean. It is calculated as follows [8]:

$$\mathrm{se}_i^n(\mu_n(F_i(s))) = \frac{\sigma_n(F_i(s))}{\sqrt{n}}. \tag{1}$$

By increasing the number of samples n of $F_i(s)$ the standard deviation of the mean is reduced. However, for the standard error of the mean this is not guaranteed, since the standard deviation estimate $\sigma_n(F_i(s))$ can be corrected to a higher value by drawing new samples of it. Yet, the probability of reducing the sample mean by sampling s increases asymptotically as the number of samples is increased.

Sequential Dynamic Resampling. An intuitive dynamic resampling procedure would be to reduce the standard error until it drops below a certain user-defined threshold $\mathrm{se}_n(s) := \mathrm{se}_n(\mu_n(F(s))) < \mathrm{se}_{th}$. The required sampling budget for the reduction can be calculated as in Eq. 2.

$$n(s) > \left(\frac{\sigma_n(F_i(s))}{\mathrm{se}_{th}}\right)^2. \tag{2}$$

However, since the sample mean changes as new samples are added, this one-shot sampling allocation might not be optimal. The number of fitness samples drawn might be too small for reaching the error threshold, in case the sample mean has shown to be larger than the initial estimate. On the other hand, a one-shot strategy might add too many samples, if the initial estimate of the sample mean was too big. Therefore Dynamic Resampling is often done sequentially. For Sequential Dynamic Resampling often the shorter term Sequential Sampling is used. A Sequential Dynamic Resampling algorithm which uses the $\mathrm{se}_n(s) < \mathrm{se}_{th}$ termination criterion is Standard Error Dynamic Resampling, SEDR proposed in [8].

Sequential Sampling adds a fixed number of samples at a time. After an initial estimate of the sample mean and calculation of the required samples it is checked if the knowledge about the solution is sufficient. If needed, another fixed number of samples is drawn and the number of required samples is recalculated. This is repeated as long as no additional sample needs to be added. The basic pattern

Algorithm 1. Basic sequential sampling algorithm pattern

1: Draw b_{min} minimum initial samples of the fitness of solution s, $F(s)$.

2: Calculate mean of the available fitness samples for each of the H objectives:
$\mu_n(F_i(s)) = \frac{1}{n} \sum_{j=1}^{n} F_i^j(s)$, $i = 1, \ldots, H$.

3: Calculate objective sample standard deviation with available fitness samples:
$\sigma_n(F_i(s)) = \sqrt{\frac{1}{n-1} \sum_{j=1}^{n} (F_i^j(s) - \mu_n(F_i(s)))^2}$, $i = 1, \ldots, H$

4: Evaluate termination condition based on $\mu_n(F_i(s))$ and $\sigma_n(F_i(s))$, $i = 1, \ldots, H$.

5: Stop if termination condition is satisfied or if maximum number of samples b_{max} is reached; Otherwise sample the fitness of s another k times and go to step 2.

---

of a sequential sampling algorithm is described in Algorithm 1. Through this sequential approach, the number of required samples can be determined more accurately than with a one-shot approach. It guarantees to sample the solution sufficiently often, and can reduce the number of excess samples.

Final Samples. After the optimization is finished and the final non-dominated result set has been identified, a post-optimization resampling process is needed. For each solution that is presented to the decision maker, we want to be confident about the objective values. Therefore, we guarantee a high number of samples b_f for each solution ($b_f \geq b_{max}$) in the final population after the EA selection process. Thereby, we can also guarantee that the non-dominated solutions in the final population have been identified correctly. The total number of extra samples added to the last population is $B_F \leq (b_f - 1)|P|$, where $|P|$ is the population size.

2.2 Evolutionary Multi-objective Optimization

In this section, two popular Evolutionary Multi-objective Optimization algorithms are presented. In this article, those algorithms are used for evaluation with Multi-objective Dynamic Resampling algorithms.

NSGA-II Algorithm. The NSGA-II algorithm was proposed in [7]. NSGA-II is a population-based evolutionary algorithm for optimization problems with multiple objectives. In the selection step NSGA-II sorts the population and offspring into multiple fronts and selects front by front into the next population as long as the fronts can be selected as a whole. If the next front can only be selected partially NSGA-II uses a clustering method called Crowding Distance to determine which solutions of the front shall be selected, in a way that increases population diversity. The parental selection for offspring generation follows this fitness hierarchy. The NSGA-II selection step is depicted in Fig. 1. The same fitness hierarchy, i.e., dominance first, then reference point distance, is used for parental selection.

HypE Algorithm. The Hypervolume Estimation Algorithm HypE was proposed in [1] and is based on the NSGA-II algorithm [7]. It is designed to optimize the hypervolume of the final population and can achieve higher hypervolume values than NSGA-II [1]. It was chosen for evaluation in this paper, since it is a popular Evolutionary Multi-objective Optimization algorithm its goal is to find a distribution of solution alternatives which maximizes the hypervolume measure. This distribution is different that the NSGA-II distribution, and we hope to see a performance difference even in optimization with noisy objective values.

HypE works according to the same principal structure as NSGA-II. For both parental selection and environmental selection, first a non-domination sorting is performed. In the environmental selection step, HypE selects those fronts into the next population that fit as a whole. If the next front can only be selected partially, instead of using a clustering method like NSGA-II to maintain population diversity, HypE uses a measure called Shared Hypervolume to determine which solutions of the front shall be selected. This measure indicates how much a solution contributes to the hypervolume of the whole population. The HypE selection step is depicted in Fig. 1. The same fitness hierarchy, i.e., dominance first, then Shared Hypervolume, is used for parental selection.

The Shared Hypervolume is a better indicator of the hypervolume contribution of a solution marginal hypervolume used in SMS-EMOA [3]. In this paper, only bi-objective optimization problems are run. For a higher number of objectives, HypE uses a hypervolume estimation algorithm based on Importance Sampling for the Shared Hypervolume, hence the name. This makes HypE suitable for Many-objective Optimization.

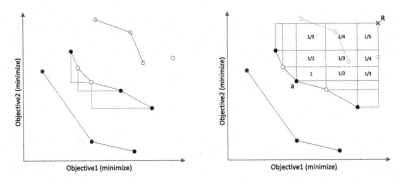

Fig. 1. Environmental selection step. Six out of twelve solutions are selected. The left figure shows the NSGA-II selection result. The complete first front and the extremal solutions from the second front are selected. The one remaining solution is selected due to its highest Crowding Distance value. — The right figure shows the HypE selection result. The complete first front and the solution set of three solutions with the highest overall hypervolume contribution are selected from the second front. The rectangles which are dominated by solution a are marked with a's hypervolume contribution.

3   Resampling Algorithms

This section presents the resampling algorithms that are evaluated in this paper.

3.1   Static Resampling

Static Resampling assigns the same fixed number of samples to all solutions involved in an optimization run. It is popular among optimization practitioners since it requires a relatively small implementation effort. The disadvantage is that accurate knowledge of the objective vectors is not needed for all solutions: Not all solutions have objective values with high variability, and in the beginning of an optimization the benefit of accurate values often does not justify their cost.

3.2   Time-based Dynamic Resampling

Time-based DR [17] is a generally applicable dynamic resampling algorithm which can be used on any optimization algorithm. It assumes that the need for accurate knowledge of objective values increases towards the end of the optimization run. Time-based DR makes only one sampling decision for each solution (one-shot allocation) according to the elapsed time, the current overall number of simulation runs B_t. However, if a solution survives into the next generation of an evolutionary algorithm the decision is made again. The optimization is run until the total number of solution evaluations B is reached, leavning only B_F samples for the final population. The acceleration parameter $a > 0$ allows to speed up or slow down the sampling allocation.

$$x_s^T = \min\left\{1, \frac{B_t}{B - B_F}\right\}^a. \tag{3}$$

3.3   Time-Step Dynamic Resampling

We propose another time-based DR algorithm which uses an allocation function different from Time-based DR. Instead of gradually increasing the allocation budget, Time-Step Dynamic Resampling allocates only b_{min} samples in the beginning of the optimization runtime, like Static Resampling. At a certain time threshold B_{thr}, the allocation is increased to b_{max} within a short period of time. This allocation function cannot be created by changing the acceleration parameter a in Time-based DR. The sudden raise of the allocation function can be done with the step function, as in Eq. 4.

$$x_s^{TS} = \begin{cases} 0 & \text{if } B_t < B_{thr}, \\ 1 & \text{otherwise.} \end{cases} \tag{4}$$

Another, smoother, way to rapidly increase the sampling allocation to b_{max} is the Sigmoid function or, more general, the Logistic Growth [17,23]. The allocation function of Time-Logistic Dynamic Resampling is given in Eq. 5, where $\gamma > 0$ is the growth rate, $B_{thr} \in [0,1]$ the point of highest growth, and $\nu > 0$ determines if the maximum growth occurs close to the lower or the upper asymptote.

$$x_s^{TS} = \frac{1}{(1 + e^{-\gamma(B_t - B_{thr})/(B - B_F)})^{\frac{1}{\nu}}}. \tag{5}$$

3.4   Rank-based Dynamic Resampling

Rank-based DR [17] is a dynamic resampling algorithm which can be used on any multi-objective optimization algorithm. It measures the level of non-dominance of a solution and assigns more samples to solutions with lower Pareto-rank. It can have a parameter allowing only the solutions with a Pareto-rank of n or less to be allocated additional samples (RankMaxN-based DR) [18]. Since after each added sample, the dominance relations in the population of an MOO algorithm changes Rank-based DR is performed sequentially.

The normalized rank-based resampling need of RankMaxN-based Dynamic Resampling, x_s^{Rn}, is calculated as in Eq. 6, where $a > 0$ is the acceleration parameter for the increase of the time-based and rank-based sampling need. n is the maximum considered Pareto-rank. Only solutions with Pareto-rank n or less are considered for allocating additional samples. Furthermore, S is the solution set of current population and offspring, R_s is the Pareto-rank of solution s in S, and $R_{max} = \max_{s \in S} R_s$.

$$x_s^{Rn} = 1 - \left(\frac{\min\{n, R_s\} - 1}{\min\{n, R_{max}\} - 1} \right)^a. \tag{6}$$

Rank-Time-based DR. Rank-based DR can be used in a hybrid version as Rank-Time-based DR [18]. Thereby, it can use the information given by the limited time budget B. A Rank-Time-based sampling allocation can be achieved by using the minimum or the product of the Rank-based x_s^R or Time-based x_s^T allocation functions (Eq. 7).

$$x_s^{RT} = \min\left\{x_s^T, x_s^R\right\}, \quad x_s^{RT} = x_s^T x_s^R. \tag{7}$$

3.5   Domination Strength Dynamic Resampling

We proposed Domination Strength Dynamic Resampling in [17], and present an improved version here. Like for Rank-based DR, we are interested to know the accurate objective values of Pareto-optimal solutions, or solutions which are only dominated by a few other solutions. In Domination Strength DR, a higher sampling budget is added to those solutions, but only if they dominate many other solutions. The idea behind this is that we need to know the accurate objective values for Pareto-solutions which are surrounded by dominated solutions, in order to make good selection decisions in the Evolutionary Algorithm. In order to reduce the sampling allocation for dominated solutions which themselves dominate many other solutions, we subtract the number of other solutions by which the solution is dominated. The allocation function is given in Eq. 8, where $\text{dom}(s, S)$ is the number of solutions in the population which are dominated by solution s, and $D_{max} = \max_{s \in P} \text{dom}(s, P)$. $\text{dom}(P, s)$ is the number of solutions in the population which dominate solution s, or in other words, to which solution s is inferior. The maximum count of dominating solutions is therefore called $Inf_{max} = \max_{s \in P} \text{dom}(P, s)$.

Similar to RankMaxN-based DR we introduce an upper limit n for counting the number of dominated or dominating solutions. Equation 8 defines the allocation function of DSMaxN-based DR.

$$x_s^{DSn} = \max \left\{ 0, \frac{\min\{n, \text{dom}(s, P)\}}{\min\{n, D_{max}\}} - \frac{\min\{n, \text{dom}(P, s)\}}{\min\{n, Inf_{max}\}} \right\}^a. \tag{8}$$

DS-Time-based DR. In the same way as for Rank-Time-based DR, for DS-based DR we propose to define a hybrid version which gradually increases the Domination-Strength-based allocation based on the elapsed optimization runtime. Using the information about the time limit, we are able to define hybrid Dynamic Resampling algorithms with better performance. A DS-Time-based sampling allocation can be achieved by using the minimum or the product of the Domination Strength-based x_s^{DS} or Time-based x_s^T allocation functions (Eq. 9).

$$x_s^{DST} = \min \left\{ x_s^T, x_s^{DS} \right\}, \quad x_s^{DST} = x_s^T x_s^{DS}. \tag{9}$$

4   Stochastic Production Line Model

The benchmark optimization problem used for evaluation is a stochastic simulation model described in [19], consisting of 6 machines (Cycle time 1 min, $\sigma = 1.5$ min, CV = 1.5, lognormal distribution) and 5 buffers with sizes $\in [1, 50]$. The source distribution is lognormal with $\mu = 1$ min and $\sigma = 1.5$ min, if not stated otherwise. The basic structure is depicted in Fig. 2. The simulated warm-up time for the production line model is 3 days and total simulated time of operation is 10 days. The conflicting objectives are to maximize the main production output, measured as Throughput (TH) (parts per hour) and to minimize the sum of the buffer sizes, TNB = Total number of buffers. This is a generalization of the lean buffering problem [9] (finding the minimal number of buffers required to obtain certain level of system throughput). In order to consider the maintenance aspect the machines are simulated with an Availability of 90 % and a MTTR of 5 min, leading to a MTBF of 45 min.

Fig. 2. A simplistic production line configuration.

In this stochastic production line model the cycle time of machine M4 has a higher standard deviation ($\sigma = 25$ min, CV = 25) than the other machines, causing a high standard deviation of the average throughput objective measured during the whole simulation time. The CV of TH becomes 0.2089, i.e. around 20 % noise. We call this model PL-NM-20 % in the following. Dynamic Resampling algorithms are required to compensate for the noisy Throughput values.

Automated machine processes are essentially deterministic so that an automated production line has in principle very low variability. But in practice, a single, stochastic, manual workstation is enough to add very high variability to an automated production line [14]. The high variability may be caused by a manual operator which is needed to start an automated process, e.g. after tool changes. Or more commonly, the variability may be cause by a manual quality inspection workstation in which the times for visually checking if there are any defeats in the work-pieces vary significantly.

5   Numerical Experiments

In this section, the described resampling algorithms are evaluated in combination with the NSGA-II and HypE algorithms on benchmark functions and a stochastic production line simulation model.

5.1   Problem Settings

Two function characteristics are important for experimentation. The function complexity and the noise level. Therefore, we test the DR algorithms on the ZDT1-5 % function, which has low complexity, and the ZDT4-5 % function with high complexity. In order to test a high-noise function, we use the stochastic production line model PL-NM-20 %. We configure it with CV=25 for machine 4, which gives an output noise of around 20 %.

The used benchmark functions ZDT1 and ZDT4 [25] are deterministic in their original versions. Therefore, zero-mean normal noise has been added to create noisy optimization problems. For example, the ZDT4 objective functions are defined as $f_1'(x) = x_1 + \mathcal{N}(0, \sigma_1)$ and $f_2'(x) = g(x)\left(1 - \sqrt{x_1/g(x)}\right) + \mathcal{N}(0, \sigma_2)$, where $g(x) = 1 + 10(n - 1) + \sum_{i=2}^{n}(x_i^2 - 10\cos(4\pi x_i))$.

The two objectives of the ZDT functions have different scales. Therefore, the question arises whether the added noise should be normalized according to the objective scales. We consider the case of noise strength relative to the objective scale as realistic which can occur in real-world problems, and therefore this type of noise is evaluated in this article. For the ZDT functions, the relative added noise 5 % is $(\mathcal{N}(0, 0.05), \mathcal{N}(0, 5))$ (considering the relevant objective ranges of $[0, 1] \times [0, 100]$). In the following, the ZDT benchmark optimization problems are called ZDT1-5 % and ZDT4-5 %.

5.2   Algorithm Parameters

The limited simulation budget is chosen as $B = 10,000$ replications for both the ZDT problems and the production line problem. NSGA-II and HypE are run with a crossover rate $p_c = 0.8$, SBX crossover operator with $\eta_c = 2$, Mutation probability $p_m = 0.07$ and Polynomial Mutation operator with $\eta_m = 5$.

We set the minimum budget to be allocated to $b_{min} = 1$ and the maximum budget to $b_{max} = 15$ for all Dynamic Resampling algorithms. Static Resampling is run in configurations with fixed number of samples per solution between 1 and 5. Time-based Resampling uses a linear allocation, $a = 1$. The allocation threshold of Time-Step DR is set to $B_{thr} = 0.8$. The allocation function of Time-Step Logis-tic DR is configured as $x_s^{TS} = \frac{1}{(1+e^{-40(B_t - 0.6)/(B - B_F)})^{0.5}}$. Rank-based Resampling and Rank-Time-based Resampling are run as RankMax5-based Dynamic Resam-pling and use linear allocation $(a = 1)$ for both the rank-based and time-based cri-teria. Rank-Time-based Resampling uses the minimum of the Pareto-rank-based allocation and the time-based allocation: $x_s^{RT} = \min\{x_s^T, x_s^R\}$. The same para-meters are chosen for Domination Strength Dynamic Resampling and DS-Time-based Dynamic Resampling.

5.3   Performance Measurement

In this paper, the Hypervolume measure (HV) [26] is used for measuring conver-gence and diversity of the optimization results. For the HV metric, a Hypervolume reference point HV-R and a Hypervolume base point HV-B for normalization must be specified. For ZDT1-5 %, we use $HV\text{-}R=(1,1)$ and $HV\text{-}B=(0,0)$. For ZDT4-5 %, we use $HV\text{-}R=(1,20)$ and $HV\text{-}B=(0,0)$. The HV performance of the PL-NM-20 % model is measured with $HV\text{-}R=(100,25)$ and $HV\text{-}B=(10,45)$.

In order to get a second measurement value, the Inverse Generational Dis-tance Metric (IGD) [6] is used to measure convergence and diversity for problems where a reference Pareto-front is available. In order to measure convergence sep-arately, the Generational Distance (GD) [24] is used, for optimization problems with known true Pareto-front. For measuring the diversity of the population separately, we use a new metric which is described in the following section.

Population Diversity (PD). In order to measure the diversity of a popu-lation, we use a metric based on the NSGA-II Crowding Distance [7], which we proposed in [19]. It measures the diversity of not only the first front, but also of all other fronts in the population. This is because the diversity of the fronts $2 \ldots n$ influences the future development of the diversity of the first front. Another reason for measuring the diversity of all fronts in the population are the noisy objective vectors. Some solutions that appear to be part of the first front of the population would be identified as dominated solutions, if the true objective values were known.

Our goal is to calculate a diversity measure allowing us to compare the diver-sity of populations, therefore called Population Diversity PD. In order to make the value independent of the population size, the diversity measure δ based on the Crowding Distance is calculated for each solution within the population; all values are summed up and divided by the population size as in Eq. 10.

$$PD(P) = \sum_{s \in P} \delta(s) / |P|. \tag{10}$$

The formal definition of $\delta(s_k)$ is given in Eq. 11, where N_j are neighboring solutions and we write $F_i(s_k) := \mu_n(F_i(s_k))$ to simplify the notation.

$$\delta(s_k) = \sum_{i=1}^{H} \begin{cases} |F_i(s_k) - F_i(s_N)| & \text{if } F_i(s_k) \text{ extremal,} \\ |F_i(s_{N1}) - F_i(s_{N2})| & \text{otherwise.} \end{cases} \tag{11}$$

5.4   Results

The measured optimization performance results are shown in Table 1 for the ZDT1-5 % benchmark problem, in Table 2 for the ZDT4-5 % benchmark problem, and in Table 3 for the stochastic production line problem PL-NM-20 %.

Table 1. Performance measurement results on the ZDT1-5 % benchmark function for NSGA-II and HypE with different Dynamic Resampling algorithms. The measurement is performed on the last population where $b_f = 25$ final samples have been executed.

	HV	IGD	GD	PD	HV	IGD	GD	PD
	NSGA-II				HypE			
Static1	0.7493	0.3416	**0.0558**	0.0438	0.4030	0.0780	0.0182	0.0570
Static2	0.7141	0.3365	0.1239	0.0233	0.4658	0.1282	0.0078	0.0693
Static3	0.6225	0.3687	0.1191	0.0183	0.5328	0.0489	0.0090	0.0641
Static4	0.6159	0.3294	0.0678	0.0206	0.5306	**0.0369**	0.0104	0.0723
Static5	0.5218	0.3719	0.2711	0.0463	0.5977	0.0560	0.0082	0.0446
Time 1–10	0.7178	0.2928	0.2678	0.0538	0.6278	0.0446	0.0076	0.0538
TS 1–10	0.6854	**0.2755**	0.2279	0.0573	0.4372	0.1108	0.0163	0.0492
TSL 1–10	0.6617	0.3085	0.2162	0.0364	0.5682	0.0513	0.0085	0.0458
Rank 1–10	0.4387	0.4042	0.3887	0.0507	0.5249	0.0644	0.0095	0.0376
RT 1–10	0.7496	0.2861	0.2848	0.0569	0.5927	0.0548	**0.0071**	0.0382
R5 1–10	0.6677	0.3307	0.2737	0.0553	0.5927	0.0834	0.0129	0.0603
R5T 1–10	**0.9100**	0.3135	0.1048	0.0228	0.6738	0.0502	0.0133	0.0610
DS 1–10	0.5108	0.3576	0.3225	0.0345	0.4840	0.1602	0.0109	0.0354
DST 1–10	0.7576	0.2937	0.2507	**0.0593**	0.5437	0.1070	0.0134	0.0416
DS5 1–10	0.7231	0.3305	0.1818	0.0532	0.4585	0.2718	0.0084	0.0247
DS5T 1–10	0.8618	0.3151	0.1973	0.0445	**0.6924**	0.0517	0.0152	**0.0866**

Table 2. Performance measurement results on the ZDT4-5 % benchmark function for NSGA-II and HypE with different Dynamic Resampling algorithms. The measurement is performed on the last population where $b_f = 25$ final samples have been executed.

	HV	IGD	GD	PD	HV	IGD	GD	PD
	NSGA-II				HypE			
Static1	0.5594	0.2825	0.1048	0.0321	0.4304	0.3635	0.0607	0.0177
Static2	0.3014	0.4077	0.1301	0.0398	0.4288	0.4160	0.1598	0.0468
Static3	0.5434	0.3535	0.1075	0.0223	0.2198	0.4255	0.1897	0.0367
Static4	0.0614	0.3596	0.1006	0.0169	0.2433	0.1592	0.0895	0.0485
Static5	0.0000	0.4360	0.1244	0.0105	0.0000	0.2451	0.2649	**0.0518**
Time 1–10	0.0720	**0.1573**	0.0903	0.0549	0.2016	0.2458	0.1245	0.0282
TS 1–10	0.6331	0.2406	0.0816	0.0422	0.6981	0.3978	**0.0391**	0.0087
TSL 1-10	0.2359	0.2248	0.0964	0.0373	0.2963	0.3956	0.1373	0.0436
Rank 1–10	0.0559	0.1879	0.1677	0.0467	0.0000	0.3628	0.1512	0.0270
RT 1–10	0.4020	0.3741	0.1158	0.0249	0.0000	0.4069	0.0958	0.0095
R5 1–10	0.1043	0.4240	0.1209	0.0187	0.0000	0.4259	0.1332	0.0430
R5T 1–10	0.5297	0.4057	0.0961	0.0196	0.3863	**0.1284**	0.0634	0.0275
DS 1–10	0.4754	0.3450	0.1495	0.0332	0.0567	0.3775	0.1512	0.0225
DST 1–10	0.0022	0.3929	0.1294	0.0236	0.4546	0.3876	0.1157	0.0221
DS5 1–10	0.2885	0.4135	**0.0625**	**0.0080**	0.0644	0.4140	0.1367	0.0263
DS5T 1–10	**0.6530**	0.3802	0.0648	0.0164	**0.6983**	0.4007	0.0498	0.0153

For the ZDT1-5 % function, it can be seen that considering the elapsed optimization time in sampling allocation increases the performance of both NSGA-II and HypE, compared with DR algorithms that do not consider time. The R5T and DS5T DR algorithms, combining dominance and time, show the best performance. The highest diversity could be achieved with the Domination Strength allocation and the HypE algorithm.

Similar observations can be made on more complex ZDT4-5 % problem. However, not as high diversity values could be achieved. This is due to the fact that the algorithms are not able to explore the whole Pareto-front within the $B = 10,000$ function evaluations. A final population close to the Ideal Point is found. Rank-based and Domination Strength-based DR both show good results, with an advantage for Domination Strength-based DR.

On the stochastic production line model PL-NM-20 %, NSGA-II and HypE show a similar performance. Time-based DR algorithms show the best results, whereof Time-based Logistic DR, with an increase of samples around $B_t = 0.6$ shows the best results. The reason for the superiority of time-based Dynamic Resampling is shown in the #Sol. column in Table 3. Time-based DR algorithms save function evaluations in the beginning of the optimization runtime, using less function evaluations for resampling. These evaluations can be used to explore the

Table 3. Performance measurement results on the production line model PL-NM-20%
for NSGA-II and HypE with different Dynamic Resampling algorithms. The measurement is performed on the last population where $b_f = 25$ final samples have been
executed. The #Sol. column shows how many unique solutions have been simulated.

	HV	PD	#Sol	HV	PD	#Sol
	NSGA-II			HypE		
Static1	0.4899	0.0501	8800	0.4728	0.0505	8800
Static2	0.4860	0.0512	4400	0.4888	0.0353	4400
Static3	0.4667	0.0581	2900	0.4638	0.0485	2900
Static4	0.4614	0.0606	2200	0.4661	**0.0599**	2200
Static5	0.4390	0.0650	1750	0.4735	0.0373	1750
Time 1–10	0.4910	0.0552	2400	0.4602	0.0423	2400
TS 1–10	0.4688	0.0554	6800	0.4891	0.0330	6800
TSL 1–10	**0.5079**	0.0563	3850	**0.5152**	0.0334	3850
Rank 1–10	0.4355	0.0530	1150	0.4062	0.0359	1100
RT 1–10	0.4739	**0.0718**	2500	0.4674	0.0337	2550
R5 1–10	0.4891	0.0480	1900	0.4490	0.0367	1950
R5T 1–10	0.4995	0.0368	3650	0.4851	0.0479	3600
DS 1–10	0.4149	0.0642	2650	0.5110	0.0451	2400
DST 1–10	0.4883	0.0613	4800	0.4744	0.0460	4950
DS5 1–10	0.4824	0.0557	2050	0.4860	0.0445	2100
DS5T 1–10	0.5067	0.0594	4450	0.4817	0.0417	4600

objective space and Pareto-front instead. It can be observed that Domination-
Strength-based DR algorithms allocate less samples and have more unique solution evaluations available for objective space exploration.

6   Conclusions and Future Work

This paper has provided a number of existing and possible methodologies of
handling noise in multi-objective optimization algorithms. Different possibilities from Static to several Dynamic Resampling strategies have been discussed.
Two stochastic, numerical two-objective problems and an uncertain production line optimization problem have been chosen to compare different sampling
methodologies. As a conclusion, we find that in EMO with limited time budget,
Hybrid Time-based Dynamic Resampling algorithms are superior to DR algorithms which do not consider the elapsed optimization runtime. The proposed
Domination Strength Dynamic Resampling variants show comparable results to
Pareto-Rank-based Dynamic Resampling algorithms.

Future Work. In this paper, we focused on invariant-noise problems. In a future study, we will evaluate Dynamic Resampling for general EMO with limited budget on optimization problems with variable output noise. This study will compare Multi-objective Standard Error Dynamic Resampling [20], which allocates samples until a certain objective accuracy is reached, with other Multi-objective Dynamic Resampling algorithms, like the MOCBA algorithm [5] or Confidence-based Dynamic Resampling [21], on optimization problems with different noise landscapes.

Acknowledgments. This study was partially funded by the Knowledge Foundation, Sweden, through the BlixtSim and IDSS projects. The authors gratefully acknowledge their provision of research funding.

References

1. Bader, J., Zitzler, E.: HypE: An algorithm for fast hypervolume-based many-objective optimization. In: Computer Engineering and Networks Laboratory (TIK), ETH Zurich (2008)
2. Bartz-Beielstein, T., Blum, D., Branke, J.: Particle swarm optimization and sequential sampling in noisy environments. In: Doerner, K.F., Gendreau, M., Greistorfer, P., Gutjahr, W., Hartl, R.F., Reimann, M. (eds.) Metaheuristics - Progress in Complex Systems Optimization. Operations Research/Computer Science Interfaces Series, vol. 39, pp. 261–273. Springer, Heidelberg (2007)
3. Beume, N., Naujoks, B., Emmerich, M.: SMS-EMOA: Multiobjective selection based on dominated hypervolume. Eur. J. Oper. Res. **181**(3), 1653–1669 (2007). ISSN 1049–3301
4. Branke, J., Schmidt, C.: Sequential sampling in noisy environments. In: Yao, X., Burke, E.K., Lozano, J.A., Smith, J., Merelo-Guervós, J.J., Bullinaria, J.A., Rowe, J.E., Tiño, P., Kabán, A., Schwefel, H.-P. (eds.) PPSN 2004. LNCS, vol. 3242, pp. 202–211. Springer, Heidelberg (2004)
5. Chen, C.-H., Lee, L.H.: Stochastic Simulation Optimization - An Optimal Computing Budget Allocation. World Scientific Publishing Company, Hackensack (2010). ISBN 978-981-4282-64-2
6. Coello Coello, C.A., Reyes Sierra, M.: A study of the parallelization of a coevolutionary multi-objective evolutionary algorithm. In: Monroy, R., Arroyo-Figueroa, G., Sucar, L.E., Sossa, H. (eds.) MICAI 2004. LNCS (LNAI), vol. 2972, pp. 688–697. Springer, Heidelberg (2004)
7. Deb, K., Agrawal, S., Pratap, A., Meyarivan, T.: A fast and elitist multi-objective genetic algorithm: NSGA-II. IEEE Trans. Evol. Comput. **6**(2), 182–197 (2002)
8. Di Pietro, A., While, L., Barone, L.: Applying evolutionary algorithms to problems with noisy, time-consuming fitness functions. In: Proceedings of the Congress on Evolutionary Computation 2004, vol. 2, pp. 1254–1261 (2004)
9. Enginarlar, E., Li, J., Meerkov, S.M.: How lean can lean buffers be? IIE Trans. **37**(5), 333–342 (2005)
10. Fieldsend, J.E., Everson, R.M.: The rolling tide evolutionary algorithm: a multi-objective optimizer for noisy optimization problems. IEEE Trans. Evol. Comput. **19**(1), 103–117 (2015). ISSN 1089–778X

11. Goh, C.K., Tan, K.C.: Evolutionary Multi-objective Optimization in Uncertain Environments: Issues and Algorithms. Springer, Heidelberg (2009). ISBN 978-3-540-95976-2

12. Jin, Y., Branke, J.: Evolutionary optimization in uncertain environments - a survey. IEEE Trans. Evol. Comput. **9**(3), 303–317 (2005). ISSN 1089–778X

13. Lee, L.H., Chew, E.P.: Design sampling and replication assignment under fixed computing budget. J. Syst. Sci. Syst. Eng. Syst. Eng. Soc. China **14**(3), 289–307 (2005). ISSN 1004–3756

14. Papadopoulos, C.T., O'Kelly, M.E.J., Vidalis, M.J., Spinellis, D.: Analysis and Design of Discrete Part Production Lines. Springer Optimization and its Application Series, vol. 31. Springer, New York (2009). ISBN 978-1-4419-2797-2

15. Park, T., Ryu, K.R.: Accumulative sampling for noisy evolutionary multi-objective optimization. In: Proceedings of the Conference on Genetic and Evolutionary Computation, Dublin, Ireland, pp. 793–800, ISBN 978-1-4503-0557-0 (2011)

16. Siegmund, F.: Sequential sampling in noisy multi-objective evolutionary optimization. Master thesis, University of Skövde, Sweden and Karlsruhe Institute of Technology, Germany (2009). http://urn.kb.se/resolve?urn=urn:nbn:se:his:diva-3390

17. Siegmund, F., Ng, A.H.C., Deb, K.: A comparative study of dynamic resampling strategies for guided evolutionary multi-objective optimization. In: Proceedings of the IEEE Congress on Evolutionary Computation 2013, Cancún, Mexico, pp. 1826–1835. ISBN 978-1-4799-0454-9 (2013)

18. Siegmund, F., Ng, A.H.C., Deb, K.: Hybrid dynamic resampling for guided evolutionary multi-objective optimization. In: Proceedings of the 8th International Conference on Evolutionary Multi-Criterion Optimization, Guimarães, Portugal, pp. 366–380. ISBN 978-3-319-15934-8 (2015)

19. Siegmund, F., Ng, A.H.C., Deb, K.: Dynamic resampling for preference-based evolutionary multi-objective optimization of stochastic systems. Submitted to European Journal of Operational Research (2016). http://www.egr.msu.edu/kdeb/papers/c2015020.pdf

20. Siegmund, F., Ng, A.H.C., Deb, K.: Standard Error Dynamic Resampling for Preference-based Evolutionary Multi-objective Optimization. Submitted to Computers & Operations Research (2016). http://www.egr.msu.edu/kdeb/papers/c2015021.pdf

21. Syberfeldt, A., Ng, A.H.C., John, R.I., Moore, P.: Evolutionary optimisation of noisy multi-objective problems using confidence-based dynamic resampling. Eur. J. Oper. Res. **204**(3), 533–544 (2010). ISSN 0377–2217

22. Tan, K.C., Goh, C.K.: Handling uncertainties in evolutionary multi-objective optimization. In: Zurada, J.M., Yen, G.G., Wang, J. (eds.) Computational Intelligence: Research Frontiers. LNCS, vol. 5050, pp. 262–292. Springer, Heidelberg (2008)

23. Tsoularis, A.: Analysis of logistic growth models. Res. Lett. Inf. Math. Sci. **2**, 23–46 (2001). ISSN 0377–2217

24. Van Veldhuizen, D.A.: Multiobjective Evolutionary Algorithms: Classifications, Analyses, and New Innovations. Ph.D. thesis, Department of Electrical and Computer Engineering, Graduate School of Engineering, Air Force Institute of Technology, Wright-Patterson AFB, Ohio, USA (1999)

25. Zitzler, E., Deb, K., Thiele, L.: Comparison of multiobjective evolutionary algorithms: Empirical results. Evol. Comput. **8**(2), 173–195 (2000)

26. Zitzler, E., Thiele, L.: Multiobjective optimization using evolutionary algorithms - a comparative case study. In: Schoenauer, M., Schwefel, H.-P., Eiben, A.E., Bäck, T. (eds.) PPSN 1998. LNCS, vol. 1498, p. 292. Springer, Heidelberg (1998)

Author Index

Printed in the United States
By Bookmasters